The Department of Education Battle, 1918–1932

Notre Dame Studies in American Catholicism

Sponsored by the
Charles and Margaret Hall Cushwa Center
for the Study of American Catholicism

The Department of Education Battle, 1918 – 1932

PUBLIC SCHOOLS, CATHOLIC SCHOOLS, AND THE SOCIAL ORDER

Douglas J. Slawson

University of Notre Dame Press
Notre Dame, Indiana

Notre Dame, Indiana 46556
www.undpress.nd.edu

Published in the United States of America

Library of Congress Cataloging-in-Publication Data

Slawson, Douglas J.

The Department of Education battle, 1918/1932 : public schools, Catholic schools, and the social order / Douglas J. Slawson.

p. cm. — (Cushwa studies in American Catholicism)

Includes bibliographical references and index.

ISBN 0-268-04110-5 (cloth : alk. paper)

1. Education and state—United States—History—20th century. 2. Catholic Church—Education—United States—History—20th century. I. Title. II. Series.

LC89.S52 2005

379.73′09′042—dc22

2005002502

To my wife, Linda Lepeirs,

and to my mother and father,

Sheila and James Slawson

Contents

Abbreviations

AAB	Archives of the Archdiocese of Baltimore
AAC	Archives of the Archdiocese of Cincinnati
AASF	Archives of the Archdiocese of San Francisco
AASL	Archives of the Archdiocese of Saint Louis
ACBCCU	Archives of the Central Bureau of the Catholic Central Union
ACE	American Council on Education
ACUA	Archives of The Catholic University of America
ADCh	Archives of the Diocese of Charleston
ADCl	Archives of the Diocese of Cleveland
ADA	Archives of the Diocese of Albany
ADP	Archives of the Diocese of Portland, Maine
AGU	Archives of Georgetown University
AKC	Archives of the Knights of Columbus
ANCEA	Archives of the National Catholic Educational Association
AUND	Archives of the University of Notre Dame
CEA	Catholic Educational Association

MCLC	Manuscript Collection of the Library of Congress
NA	National Archives of the United States
NACE	National Advisory Committee on Education
NEA	National Education Association
NCEA	National Catholic Educational Association
NCCM	National Council of Catholic Men
NCCW	National Council of Catholic Women
NCWC	National Catholic Welfare Council/Conference
PTA	Parent-Teachers' Association
SAB	Sulpician Archives, Baltimore

Preface

A cardinal tenet of the American educational tradition has been the local control of schooling, a right reserved to states by the Tenth Amendment. Thus, during the early national period (1789–1815), diversity and voluntarism characterized education. Private individuals, religious denominations, ethnic groups, and benevolent societies provided tuition-financed schools that frequently enjoyed additional funding from state and local governments.[1] Reformers, however, were soon on the horizon. During the first half of the nineteenth century, Horace Mann, Henry Barnard, and others led a movement for free, universal education through a system of tax-supported, state-controlled schools to assimilate America's burgeoning immigrant population and to eradicate social distinctions and poverty. Lower-class children were to attend these common schools where they might absorb the civic virtues of the republic and the economic virtues of capitalism. Because religion buttressed both sets of values and served as the foundation of good morals, it was imperative that it be taught. To circumvent denominationalism, reformers advocated a nonsectarian, in fact pan-Protestant, form of Christianity—the "Religion of Heaven," as Mann called it. The sole repository of this nondenominational Christianity was the Bible, to be studied daily without note or comment. Children were to learn sectarian doctrines at home or church.[2]

The common-school movement was imbued with anti-Catholicism. Protestant America had a deep, abiding fear of the Roman church, which was identified with the Antichrist and believed to be bent on papal domination of the country. Religious antagonism was not the only factor in anti-Catholicism. Class, ethnic, and cultural differences were also significant, especially because the American Catholic church was composed largely of immigrants.[3]

Initially, Catholics sought in varying ways to come to terms with the common schools, but successful accommodation was usually short-lived.[4] More typically, attempts at accommodation spawned political battles drawn along religious lines. In New York City, for instance, Catholic attempts to secure public funds for parochial schools led to the creation of an effective Catholic party, fueling Protestant fears of papal domination. In Philadelphia, the church's success in securing permission for Catholic children to use the Catholic version of the Bible for scripture reading in public schools led to wholesale religious rioting.[5]

As a result of the troubles in Philadelphia and New York, Protestants formed the American party, whose platform included a twenty-one-year residency for naturalization, the exclusion of the foreign-born from holding office, a proscription against the use of public funds for religious institutions, and the promotion of the Bible as the American way of life. Eclipsed by the Mexican-American War and the difficulties over slavery, the movement resurfaced as the Know-Nothing party in the mid-1850s.[6]

Given Protestant hostility, Catholics grew increasingly committed to the establishment of their own system of education. The First and Second Plenary Councils of Baltimore, held in 1852 and 1866, respectively, urged each parish to erect a parochial school and pay teachers out of parish funds.[7] Despite the resultant increase in school numbers, the proportion of parishes with schools stayed constant.[8] Thus, a great number of Catholic children remained in the public system. The Third Plenary Council of Baltimore in 1884 bound parents to procure a Catholic education for their children. Schools were to be erected near every church within two years.[9] Initial response to this legislation was gratifying. By 1890 fully 75 percent of the parishes had schools. Thereafter, the building program slowed dramatically, failing to remain on par with the rapid growth of a Catholic population swollen by immigration. The proportion of churches with schools dropped to 62 percent in 1900 and declined to 55 percent five years later, a level maintained through 1920.[10]

The church's inability to provide enough schools for all its children was not lost upon a group of churchmen known as Americanists, led by Archbishop John Ireland of Saint Paul. They sensed compatibility between the fundamental principles of Catholicism and the fundamental principles of America.

They believed that if America became Catholic, the American Catholic church would lead the world into a new age of Catholic democracy. America would become the redeemer nation. To facilitate the process of conversion, the Americanists believed that the church must adapt itself to American culture as far as possible.[11]

The issue that bitterly divided them from their coreligionists was education. According to traditional Catholic teaching, education belonged by right to parents and the church; the state could enter that field only when invited or when the other two neglected their responsibility. The Americanists took a different view, conceding educational rights to all three. The government had the right to exercise all power necessary to ensure an educated citizenry for the welfare of the state. In a controversial address to the National Education Association in 1890, Ireland proclaimed the right of the state to educate and declared his willingness to allow states to make public schools openly Protestant as long as states paid for the secular instruction delivered in Catholic schools. His words created a furor among conservatives. When the controversy grew to fever pitch, the Vatican intervened with a compromise, upholding the position of the Americanists in principle without prejudice to the educational decrees of the Third Plenary Council of Baltimore.[12] In 1895 Pope Leo XIII struck at this accommodation of the Americanists by urging American Catholics to remain apart from society and associate only with each other "unless forced by necessity to do otherwise." Four years later he condemned a French distortion of Americanism that made it seem that the movement distinguished erroneously between passive and active virtues, exalted individual guidance by the Holy Spirit over external church authority, and advocated diluting doctrine in order to win converts. The condemnation sent a shockwave through the American Catholic church.[13]

The condemnation set the American church on a trajectory that would mold Catholics into a self-conscious subculture. Several factors propelled this trajectory. First was the papal admonition that American Catholics associate only with each other. Second was a loss of professionals from the church, which the noted Catholic sociologist Father William Kerby and his former student, Father John J. Burke, C.S.P., sought to stem by promoting Catholic group consciousness.[14] The final factor was the papal condemnation of Modernism (1907), a religious movement that undermined confidence in the traditional understanding of the Bible and dogma. Modernism also eroded trust in hitherto unchallenged assumptions: a belief in a rational, predictable cosmos; the certainty and immutability of moral values; the inexorable upward course of humanity; and the conviction that education and art were to reinforce these assumptions. While historians have described these beliefs as the assumptions of innocence or Victorian mores, the people of the early twentieth century spoke of them as traditional values. The condemnation of Modernism, notes historian William Halsey, "locked [Catholics] into those nineteenth-century assumptions which Modernism (not simply the theological, but the philosophical and cultural dimensions as well) had proceeded to smash."[15] Catholics increasingly came to view themselves as the protectors and bearers of traditional values.

With America's entry into the First World War, the trajectory toward the creation of a Catholic subculture accelerated. Catholics established multiple organizations, most of them paralleling already existing ones in the host culture, and all of them aimed at preserving the distinctive features of Catholicism while injecting Catholic values into society.[16] An important element of this separate-but-equal subculture was the parochial school, the bearer and transmitter of the Catholic faith and world view.

While Catholics organized for the preservation and spread of their faith and traditional values, a new phase in public education took shape, a phase intimately linked to Progressivism, a pervasive reform movement. Though some historians contend that the movement subsided after the First World War, Arthur Link argued more than forty years ago that Progressivism perdured through the 1920s, albeit under serious handicaps, among them the defection of intellectuals and many middle-class urbanites, the former disillusioned by the war and the latter intent on enjoying prosperity. Prosperity, however, had bypassed farmers during the decade, and Progressivism remained alive among southern and western agrarians.[17]

While Progressivism promoted political and social reforms, it also had a coercive side best illustrated by attempts at social control, like the eugenics movement, Americanization, Prohibition, and immigration restriction. At its fringe were erstwhile reformers who lapsed into out-and-out bigotry, like Tom Watson of Georgia and William Frank Phelps of Missouri, who launched anti-Catholic crusades in their journals, *Watson's Magazine* and the *Menace*, respectively. Both warned of a papal plot to take over the country. Whereas earlier waves of anti-Catholicism were largely confined to urban areas, these magazines spread it to rural folk. As historian John Higham has pointed out, anti-Catholicism "finally found its most valiant champions among the hicks and hillbillies."[18] With America's entry into the First World War, the nation united and anti-Catholicism subsided, only to resurface with a vengeance in the postwar period in what some contemporaries described in the 1928 presidential election as a Catholic assault on Protestant civilization, leaving a legacy of suspicion and ill will that the church countered for decades through street preaching in farm towns.[19]

The purpose of the present study is to shed light on an issue that has received mention in recent historical studies, but as yet has failed to gain the full-scale examination it deserves, namely, the effort of a group of progressives within the educational community to secure a reform that aimed at the establishment of a department of education with a massive aid package to implement a national program of schooling. An important corollary to this movement was compulsory public education, a reform aimed at establishing a cohesive American identity and community through schooling. This drive began shortly before America's entry into the war and lasted into the first years of the Great Depression.

Educational progressives transformed the National Education Association (NEA) into a lobby to promote this program. More often than not, the NEA chose congressmen from rural states to sponsor its legislation. Throughout the decade, agrarians comprised the part of the nation that represented old-stock America and traditional values at a time when modern mores and pluralism were increasingly challenging the former. Reliance on sponsors from agricultural regions was natural for educational progressives because most of them hailed from rural America and shared the values of its people, values their educational program sought to promote. Given that educational progressives and Catholics each viewed themselves as the bearers of traditional values, the stage was set for a struggle over what it meant to be a real American, at least on the education question.

Outside of Congress, the NEA enjoyed the support of two organizations not usually associated with reform: Masonry, especially the Southern Jurisdiction of the Scottish Rite, and the Ku Klux Klan. Recent historical studies have indicated that both shared progressive values and concerns. With respect to education, each stood for a department of education, federal aid, and compulsory public education, reforms that placed them squarely in the camp of the NEA. Because both Masonry and the Klan found their strongest support among agrarian Americans, the issue was surely to be religiously drawn. In fact, the more the Catholic church opposed the federal advancement of public schools, the more it provoked religious animus against itself.

This study will show that the program of the NEA ultimately failed for several reasons. First, educational administrators were divided over the proper method of attaining federal participation in education. Second, by the time they found common ground, the movement faced an antistatist backlash to wartime centralization. Third, the drive for educational conformity ran headlong into the reality of pluralism, namely, that various ethnic and religious groups sought to define themselves as American in ways that did not entail the abandonment of ancestral cultures.

The present volume will focus on the pluralism espoused by the American Catholic community, for this religious body was the principal opponent of the NEA. Shortly after the war, the church established its own lobby, the National Catholic Welfare Conference, to protect Catholic interests. At the heart of its educational interest was the parochial school. Increasingly, Catholics came to view the program of the NEA as inimical to the parish system of education. The more this became apparent, the more each side used the rhetoric of combat. By 1924 the Welfare Conference declared that a state of war existed between the church and the forces supporting the NEA. At the height of the controversy, Catholics came to believe that religious freedom in America would survive or perish with the fate of parochial education.

I thank my wife Linda for her patience and support over the past six years while I worked on this manuscript and its revisions. She has endured evenings and weekends spent by herself in the absence of a husband who was in the study or in libraries. Her tolerance was borne of a belief in the importance of history for an understanding of and the advancement of the church.

Douglas J. Slawson
National University, April 2004

1

A National Program for Education

The twentieth century dawned with the rise of a widespread reform movement called Progressivism. A mixed and shifting combination of self-interest groups, Progressivism sought to correct the shortcomings and evils in modern industrial society. Imbued with the values of Victorian culture and small-town America, and spurred on by Evangelical Protestantism and a commitment to the scientific solution of problems, progressive reformers were old-stock American, middle-class businessmen, lawyers, editors, and professionals who clung to the genteel culture of the late nineteenth century. At the same time they sought to provide clean and efficient government at the state and municipal levels and to curtail the abuses of "predatory wealth" through government intervention. Progressives were divided over the nature of desirable federal intervention. Viewing the growth of huge corporations as both inevitable and an antidote to destructive competition that retarded the national economy, some progressives believed that the federal government had an obligation to supervise and regulate business in the national interest through a commission of administrative experts. This was the New Nationalism espoused by Theodore Roosevelt and his followers. Opposed were progressives who viewed individualism and competitiveness as the supreme economic virtues. They believed that the federal government ought to establish industrial freedom through trust-busting, forcibly reopening the marketplace to competition. This was the New Freedom espoused by Woodrow Wilson and his followers.[1]

While promoting economic and political reforms, progressives espoused social reforms aimed at reinforcing white, Anglo-Saxon, Protestant culture. Three such reforms were interrelated: immigration restriction, Americanization, and Prohibition. The first proposed to limit the number of newcomers to the nation, the second sought to promote the assimilation of those admitted, and the third intended, at least in part, to increase the wealth and work habits of the lower social orders, which included most immigrants, by outlawing the production and sale of alcoholic beverages. Given that much of the working class comprised immigrant Catholics with a culture of drinking, these reforms would eventually be perceived by the Catholic church as an Anglo-Saxon Protestant attack on members of the Roman faith. Historian Lynn Dumenil has observed, "There was a strong undercurrent throughout the [progressive] movement to use the law to control and assimilate immigrants to American Protestant morality and standards."[2]

While the mood of Progressivism remained confident prior to the First World War, these social reforms were relatively benign. Proponents of immigration restriction sought and secured the exclusion of only illiterate foreigners. Americanization programs were haphazard and consisted mainly of voluntary evening classes in civics. Constitutionally, prohibition belonged to the states. By 1912 only six had outlawed the manufacture and sale of alcoholic beverages, legislation rendered porous by a Supreme Court ruling that upheld the right of citizens in dry states to import alcohol for personal use, thus giving rise to a mail-order booze industry. In 1913 progressives in Congress passed the Webb-Kenyon Act, which took interstate commerce in liquor out of federal hands, thus permitting dry states to halt the mail-order liquor business.[3]

Among educators, Progressivism took two interrelated though distinct forms: one curricular and the other administrative. Curricular progressives believed that schooling was to focus not on the teacher and textbook but on children. The curriculum had to be transformed to prepare them for life in the larger world; it was to initiate them into a cooperative community interested in and capable of the ongoing reform of society.[4] More important to the present study were the administrative progressives, dubbed the "educational trust" by historians. These were superintendents and university professors interested in the municipal reform of education and in rooting the superintendency in the "scientific management" of schools. Hailing from small-town, rural America, they sought to promote Protestant, Victorian values through a modernized school system. They also applied the cost-cutting mechanisms of industry to education, a key element of which was the establishment of standards against which to measure performance. Eager to build on educational improvements of the late nineteenth century—gradation, standardized curricula throughout the city, promotion based on examination—administrative progressives sought to replace the decentralized decision-making of local school boards composed of lay people with centralized, bureaucratic, top-down direction by professional educators, chosen by a corporate board of directors selected in citywide elections. Policy was to be set by the superintendent with the board playing an advisory role. Thus, education was to be "taken out of politics" by removing it from the contentiousness of local boards and entrusting it to the hands of an objective professional.[5]

The new system of administration opened the way for the educational trust to exert informal control over schooling through "hidden hierarchies" consisting of professional educators like Ellwood Cubberley of Stanford University; Charles Judd of the University of Chicago; George D. Strayer of Teachers College, Columbia University; and Frank Spaulding of Yale University; all of whom maintained regular contact with each other while training a generation of school superintendents and finding placements for them. In 1915 Cubberley, Judd, Strayer, and Spaulding formed the Cleveland Conference, an informal association originally consisting of about twenty colleagues who met annually. Serving as the conference's guiding light in its early years, Judd believed that both the curriculum and the structure of schools needed a major overhaul. In his view, it was incumbent upon the Cleveland Conference to take up "the positive and aggressive task . . . of a detailed reorganization of the materials of instruction in schools of all grades." This reform must be "as broad and democratic as possible," a grassroots crusade with the members of the conference supplying the impetus.[6]

Most members of the conference were adherents of the New Nationalism with regard to schools. They considered it appropriate for the federal government to participate in and to regulate education. Yet the conference would split over the sort of agency the federal government should establish for education and over the terms on which federal aid ought to be granted. Democracy, moreover, meant one thing to Judd and something quite different to Strayer. Judd understood it to be a widespread, open discussion of the issue of federal participation by the educational community as a whole and the public at large, leading to congressional action. Strayer understood it as a top-down reform led by a cadre of educators who would determine the form of federal participation and then conduct a drive for it with members of the National Education Association (NEA) providing grassroots (that is, democratic) support. The NEA's Department of Superintendence became an important forum for this segment of the educational trust. In the years immediately following the First World War, Strayer and other members of the trust would struggle to restructure the NEA itself into a representative assembly that might nonetheless be controlled by the trust.

Educational reformers in the South had to deal with basic matters, such as the improvement of the region's school buildings. At the turn of the century, progressive educators spearheaded a crusade to secure local taxation for schools, compulsory attendance laws, lengthened school terms, and increased salaries for teachers. Industrial education and vocational training were important elements of the campaign because they would provide the skills necessary to convert the South from a provider of raw materials into a producer of finished goods. As in the North, administrative progressives attempted top-down, "democratic" reform, seeking to persuade rural folk of the need to improve their schools. As historian William Link notes: "Reformers believed in a democracy in which public opinion reacted to, rather than initiated, policy changes." The crusade experienced success as the illiteracy rate among whites was halved in the first fifteen years of the new century. Still, the South lagged behind the rest of the nation educationally.[7]

While southern progressives improved their region's schools, they also made important contributions to educational breakthroughs at the federal level. Senator Hoke Smith, a progressive Georgia Democrat, and Representative Asbury Lever, a South Carolina Democrat, cosponsored a bill that provided federal aid for the diffusion of useful knowledge regarding agriculture and home economics. Funds were to be appropriated on a matching basis, that is, if a state accepted the federal money, it had to devote an equal amount of its own for the purpose. For the first time controls were imposed on the aid. Any state that accepted federal money had to submit its program of implementation for approval by the Department of Agriculture and make annual reports to a federal director. In 1914 Congress passed the Smith-Lever Act.[8]

Smith also secured passage of another significant piece of federal education legislation. Under his aegis, President Woodrow Wilson appointed a commission to study federal participation in vocational education. Smith, Representative Dudley Hughes (a Georgia Democrat), and Commissioner of Education Philander Claxton crafted a bill that embodied the fruit of the commission's labor. In 1917 Congress passed the Smith-Hughes Vocational Education Act, establishing the Federal Board of Vocational Education and mandating the creation of local boards by those states that chose to receive the benefits. Funds were again appropriated on a matching basis to train vocational instructors and pay their salaries. Vocational programs were to be conducted in public high schools. Once again federal control followed the aid, with the requirement that each participating state submit its program for approval by the federal board.[9] The next attempt at federal education legislation would come at the hands of the educational trust during the crisis of the First World War.

In April 1917, after two and a half years of precarious neutrality, Wilson led the United States into the First World War. Spurred to action by Germany's resumption of unrestricted submarine warfare and by his own desire for the establishment of a new world order, Wilson articulated for the nation the purpose of its entry into battle. Careful to distinguish between the German government and the German people, the president denounced the former as the "natural foe to liberty," while sympathizing with the latter as "pawns and tools" of their ambitious leaders. "The world," said Wilson, "must be made safe for democracy." With "civilization itself seeming to be in the balance," the United States would fight "for democracy, for the right of those who submit to authority to have a voice in their own Governments."[10] These lofty words inspired the nation with a sense of mission and steeled its resolve.

Patriotism is a supple thing, capable of serving many ends. With the nation a belligerent, notes David Kennedy, "Special-interest groups of all kinds . . . sought to invest America's role in the war with their preferred meaning, and to turn the crisis to their particular advantage."[11] Of interest to this study are two such groups: the Catholic church and the progressives of the educational trust. Despite the fact that many Catholics had sympathized with the Central Powers during the period of neutrality,[12] once America went to war, they placed loyalty to the nation above ancestral feeling. Catholics were motivated by patriotism, but they were motivated by something more. Because their patriotism had long been questioned, they viewed the war as an opportunity to prove their citizenship in blood by rallying to the cause in numbers disproportionately large compared with their size in the population at large.[13] Nor were Catholics to be outdone on the home front. Just as the YMCA set up recreation centers for soldiers in military camps, so too did the Knights of Columbus.[14]

This was not all. Father John J. Burke, editor of *The Catholic World* and a promoter of the militant, separate-but-equal Catholicism referred to in the preface, believed that for the Catholic war effort to be effective, the church must be organized nationally. He invited every bishop and each

Catholic lay society to send a representative to a meeting at The Catholic University of America for the purpose of unifying the church's war work. Delegates from sixty-eight dioceses and twenty-seven societies participated in the establishment of the National Catholic War Council, which gave the church national structure under an administrative committee composed of four bishops. The War Council not only organized the church's war effort but also represented Catholic interests before the federal government. More significant, it served as the model for a postwar Catholic organization that would become a church lobby and the agent of a militant Catholicism. As recent historians have pointed out, the Catholic church and its fraternal organizations, like the Knights of Columbus, incorporated immigrants and their descendants into the political life of the nation in a way that allowed them to be American citizens while retaining their religious and cultural identity. Ironically, as Dumenil observes, the fact that Catholics now appeared to be "organized and powerful" fueled fear in their countrymen that the church was intent on overwhelming Protestant America.[15]

Like Catholics, the majority of progressives accepted Wilson's idealism and followed him to war. Viewing the fight as an ugly fact that might be turned to good ends, the educational trust hoped to use the conflict as a means of furthering reform. Mobilization involved the federal government in unprecedented intervention in the national economy through the wartime regulation of agriculture, fuel, industry, and the railroads.[16] Given the expanded federal presence in business, the moment seemed opportune for the establishment of a federal department of education and a national system of schooling.

The trial of mobilization revealed the shortcomings of the nation's educational system. The first shock was the alleged low mental ability of the citizenry. Using the rapidly developing science of IQ testing to identify men of officer caliber and to categorize the rest for social efficiency within the military, psychologists erred by assigning mental ages to the scores. The results were disturbing, indicating that the nearly 2 million men examined had an average mental age of fourteen. Subscribing to the hereditarian school of interpretation, which held that people were genetically unchangeable, the testers concluded that the average American was defective.[17] Psychologists holding an environmentalist view might well have challenged this conclusion, for the teaching profession and educational institutions were in a deplorable state.

The war accentuated and accelerated an already acute problem: the flight of teachers to other livelihoods. At the turn of the century, men comprised

30 percent of the profession, but by 1916, scarcely 20 percent. During the war, the number of male teachers dropped an alarming 41 percent.[18] While some entered military service, many others left the classroom for the factory or the office where pay was substantially higher. According to government statistics for select occupations, only agricultural laborers earned less than teachers. It is not surprising that many industrial workers drafted into the military were replaced by educators, especially those from rural areas.[19] Male teachers were joined in the exodus by their female counterparts, who, in the face of the rising cost of living, felt the lure of high wages in wartime business.[20]

As seasoned educators found new professions, their places were taken by the young and inexperienced. At the end of 1918 over a quarter million of the nation's teachers were under the age of twenty-five, and of these, tens of thousands were still in their teens. The typical rural educator had only three years of schooling beyond the primary grades; the town or city teacher had few more. Although the war exacerbated the problem of teacher flight, chronic low pay lay at the root. "As a natural result of this condition the teaching population is transient and unstable," concluded the educational trust. "For the overwhelming majority of teachers, what should be a serious and permanent profession becomes a temporary and casual occupation."[21]

Equal to the problems of the teaching profession were those of the schools themselves. Rural ones suffered most. Their teachers were generally younger than those in urban settings, and they labored under difficult classroom conditions. While most city schools were divided into grades, the majority in the country had but one room in which the instructor faced children at various stages of growth and learning. From the students' point of view, the one-room schoolhouse made education more difficult and less rewarding. Rural children were at a further disadvantage because school terms tended to be shorter than in cities—sometimes by months—and fewer years were required. To compound matters, attendance laws were poorly enforced, if enforced at all.[22]

The draft highlighted the need for physical education. There was nothing new about rejecting men from military service because of physical defects. On this ground the army had refused, without public outcry, a staggering 76 percent of the 400,000 men who had tried to enlist between 1914 and 1917. The nation was appalled, however, when it learned that of the more than 2.7 million men examined in the first draft, 47 percent were physically defective, and more than half of them were medically discharged. Of those rejected, about one-third could have been healed in childhood if their impairments had been detected through a physical education program. Yet in 1915 only three states required schools to offer one. Although eight more enacted such legislation during the war, thirty-seven remained without it.[23]

Conscription also disclosed an alarming amount of illiteracy—25 percent of those inducted—which caused public dismay. In reality, the war simply focused attention on a problem that the discerning observer could have noticed in census reports. Nor was it as pronounced as military figures indicated. According to 1910 statistics, 5.7 million Americans aged ten or over (7 percent) were unable to read or write. Of those, 1.6 million were immigrants; the rest were mainly southerners, both white and black, a largely rural people. Administrative progressives in the South had made significant strides in redressing the problem. Still, wartime scrutinizers of education considered the existing amount a "peril to free institutions."[24]

A corollary to the problem of illiteracy, especially as it related to immigrants, was Americanization. Between 1890 and 1914, 21 million people, mainly from eastern and southern Europe, entered the United States, though net immigration was less because of the return home of several million. Indeed, many of these migrants had come to America simply to make money for their families back home or for themselves so that they could buy land upon repatriation. The nation's liberal residency laws enabled them to incorporate easily into the economic life of the country without necessarily having to incorporate politically through citizenship. These immigrants settled in urban ethnic enclaves where they maintained their language and culture and retained strong ties to the communities in their homelands.[25]

The repercussion of this situation was felt in the war effort as about 8 percent of those drafted could scarcely understand commands given in English, and another 8 percent could understand only orders delivered in a foreign language, often that of the enemy: German. This raised concern about the education of immigrants who often attended foreign-language parochial schools. Because such students were beyond the pale of public education, the NEA believed that they would "grow to manhood and womanhood not only ignorant of our laws, our institutions, even of our language, but actually nurtured upon alien ideals brought to them through the medium of an alien tongue."[26]

The greatest wartime weakness of the educational system, at least as far as the government was concerned, was the absence of a federal agency to mobilize the schools. Limited by law to gathering and disseminating information, the Bureau of Education lacked authority to organize the nation's educational war effort. When the armed forces needed personnel expertly trained in medicine, chemistry, and engineering, the War Department had to fill the gap by appointing a committee to develop courses of study and establish educational programs for the universities.[27] The Committee on Public Information (CPI), established by Wilson to disseminate information about the Allied cause, quickly fell victim to the old axiom that truth is the first causality in war. Seeking to use the public schools to propagandize America's aims, the CPI set up a Division of Civic and Educational Cooperation to write patriotic pamphlets that set forth Wilson's purposes for schoolchildren while portraying the conflict in black and white. The propaganda campaign ran amok when the CPI tried to put the literature in the hands of teachers. So decentralized was the school system that the Bureau of Education was unable to supply a complete list of institutions even after a year of concerted effort. In the end, the CPI had to advertize its pamphlets in *History Teacher's Magazine.*[28]

In summary, the wartime condition of education was far from comforting. Many teachers were immature and untrained, most were underpaid, and a number were disinclined to remain in the profession. Rural schools were backward and responsible for most of the nation's native-born illiterates. A sizable proportion of immigrants were either illiterate or untutored in English. Nor was there a government agency to cope with these problems. The mobilization of schools fell to ad hoc federal offices established specifically for that purpose. This situation turned the thoughts of some educators and congressmen to comprehensive legislation that would include the creation of a department of education and federal aid to remedy educational woes.

The drive for a department unfolded along bifurcated paths: one followed by the Association of American Colleges (AAC), the other by the NEA. Because educational institutions of all levels were experiencing increasing and confusing pressures from all sides regarding the war effort, the AAC devoted its annual meeting in January 1918 to a discussion of higher education's contribution to the cause. The association passed a series of war-related resolutions calling for, among other things, the establishment of a federal department of education and for Wilson to take steps toward the immediate comprehensive mobilization of the nation's educational forces under a centralized agency to coordinate efforts and stimulate defensive activity.[29] From this and what will follow, it seems certain that the AAC endorsed the establishment of a department of education without having in mind a comprehensive overhaul of the educational system. Rather, it viewed a department as a means for coordinating schools for the war effort and as a voice for national educational concerns.

An upshot of the convention was an informal conference of representatives from various educational associations that took place in Chicago a week after the AAC meeting. Owing to an unprecedented blizzard, only members of the AAC, the Association of American Universities, the National Association of State Universities, and the Catholic Educational Association (CEA) were able to attend. Although none present had authority to endorse the resolutions recently passed by the AAC, all agreed that the federal government must coordinate the activities of elementary schools, high schools, and colleges for effective participation in the war effort. In their view, only the president of the United States could achieve this goal. Agreeing to reconvene at the end of the month in Washington, D.C., the conference sent delegates to advise Wilson of the nation's educational difficulties and to ascertain his intentions.[30]

When the Chicago conferees reconvened in the capital in late January, their numbers were swelled by representatives from the Association of Urban Universities, the American Association of University Professors, the Society for the Promotion of Engineering Education, and the NEA. In meetings presided over by Commissioner of Education Claxton, the attendees learned that the delegation from Chicago had met, not with Wilson, but with Secretary of the Interior Franklin K. Lane, whose department housed the Bureau of Education. Greatly interested in the effective coordination of the schools, Lane promised to give careful consideration to any plan proposed. On learning this, the Washington conferees established the Emergency Council on Education, an organization comprising their respective associations and dedicated to placing all educational resources at the disposal of the government during the war and through the period of postwar reconstruction.[31]

The council's attention quickly focused on the creation of a department of education. President John McCracken of Lafayette College, seconded by President J. H. T. Main of Grinnell College, both representing the AAC, moved that the council endorse a bill recently introduced by Senator Robert Owen, a progressive Democrat from Oklahoma, to elevate the Bureau of Education to department status. While the majority supported the idea, the two representatives of the CEA, Monsignor John B. Peterson and Father James Burns, C.S.C., both members of the CEA's Advisory Committee, moved to table the resolution. Although the minutes are silent as to their motive, McCracken's reaction suggests that they felt an endorsement was premature. When the council refused to table, McCracken withdrew his resolution and recommended that a committee be sent to Senator Smith "to ascertain more fully the nature of the bill and the chances of its passage." Instead, the council, supported by James W. Crabtree, secretary of the NEA, appointed McCracken, President Harry Pratt Judson of the University of Chicago, and President Paul Campbell of the University of Oregon to set forth reasons in favor of the Owen bill and to present them to Smith. At the same time, the council postponed the vote on an endorsement of the Owen bill until its member organizations could be polled on the issue.[32]

McCracken, Judson, and Campbell marshaled arguments that were limited in scope. They emphasized the necessity of a department of education for its power to coordinate schools to support the war effort, its ability to serve as a liaison on equal footing with national educational ministries in foreign countries, and its capacity to become the mouthpiece through which the states, while preserving "all the old measure of autonomy in their own educational systems," might express themselves regarding policies of national and international scope. Finally, they viewed a department as a means of conveying national ideals that were "to be crystallized into definite form, and to become the well defined directing motives in the national consciousness."[33] With the exception of this last argument, these motives were almost identical to those already embodied in the resolutions of the AAC—one of the bifurcated paths—in the drive for a department of education. Whatever else the last argument might have meant, it did not, as will be seen, imply a national system of education.

While the AAC and the Emergency Council on Education were seeking the establishment of a department for limited purposes, Strayer viewed the war as the opportunity to develop a national program for education. So, too, did Mary C. Bradford, president of the NEA, who in February 1918 appointed a committee to direct the association's campaign for the "rebuilding of civilization thru a war-modified education." In the following month, the NEA established the Commission on the Emergency in Education, which boasted four founding members of the Cleveland Conference: Strayer, as chairman; Lotus D. Coffman of the University of Minnesota, as secretary; Cubberley; and Spaulding, then superintendent of schools in Cleveland.[34] The commission was to study the problems of mobilization and postwar reconstruction. Under Strayer, however, its investigation concentrated on the educational weaknesses highlighted by the war. Because in the commission's view these debilities called for comprehensive legislation rather than a modest proposal like the Owen bill, the NEA began drafting its own measure, which embodied not only the establishment of a department of education but also a federal boost to the nation's educational system.[35]

The NEA's course of action alarmed the Emergency Council on Education. Concerned less that a member organization was drafting its own bill for a department of education, the council feared more that the NEA might propose its measure independently and thereby weaken a concerted effort for the establishment of such a department. In the council's view, the success of the drive depended on unity of purpose and action. To ensure this, it passed a resolution declaring that "the presentation of all matters approved by the Council to legislative committees or other Government agencies should be through the Emergency Council and not by separate action of component groups." Although Strayer affirmed the need to present a united front, neither he nor the NEA would honor the resolution.[36]

The movement for a department of education caught the Catholic church off guard. Pressure to confront the issue came almost immediately. In preparation for the Emergency Council's poll on the question of endorsing the Owen bill, the Association of American Universities solicited from member institutions their views on the legislation. Bishop Thomas Shahan, who was both rector of The Catholic University of America and president of the CEA, delayed responding because the purposes and work of the university were "so closely bound up with the interests of the entire Catholic system of schools that it would be unwise and perhaps unfair, for the University to express itself as in favor of the proposed Department without having secured the consent of all concerned." So he referred the matter to the CEA.[37]

Catholics had much to be concerned about. The nation's long period of neutrality had prompted considerable anti-German sentiment, much of it directed inward. Entry into the conflict had unleashed a drive for 100 percent Americanism. Equating patriotism with conformity, this movement increasingly focused on the public school as the instrument for shaping all Americans in the same mold. Two corollaries followed. The first was the eradication of German-language schools and even the teaching of the German language. The second was the end of parochial education, which allegedly divided allegiance between Rome and Washington. With respect to the first, both the Emergency Council on Education and Claxton felt compelled to defend the Teutonic tongue as a worthy academic subject. On the other hand, Claxton advised that teaching in a foreign language was "not in harmony with the interests of the country," a point bearing directly on German-Catholic parochial schools.[38] A few in the church agreed with him. The Jesuit weekly *America* carried an article recommending that instruction in all schools be in English, a position seconded by the German-born Joseph Schrembs, bishop of Toledo. As historian Philip Gleason notes, "[German-language] newspapers and societies were 'scoured away' in the storm of anti-German feeling that swept the country."[39] While Catholics, even many German ones, might support the first corollary, which demanded English as the language of instruction in all schools, their ranks would be unbroken with regard to the second, which sought to abolish parochial education. In the spring of 1918 patriots in Michigan circulated a petition to amend the state constitution to place children between the ages of five and sixteen in public schools, thereby outlawing parochial ones.[40] The Michigan amendment was the harbinger of a resurgent nativist movement that was to endure long after war's end.

Catholics in the upper echelon of the CEA were not quick to hew to the standard of educational patriotism that gripped the nation. Rather, they viewed the movement with caution because of the suspicion it cast on parochial schools. Bishop Louis Walsh of Portland, Maine, a founder of the CEA, expressed his concern to Father Francis Howard, a fellow founder and the secretary general of the association. Walsh had attended the February 1918 meeting of the NEA Department of Superintendence at which Claxton had sounded the message of "Federalization, Nationalization, and Americanization," a theme picked up by Bradford who set the wheels in motion that led to the appointment of the NEA's Commission on the Emergency in Education. Walsh thought that the advocates of centralization were using the war to further their own ends. The object was federal control of education through federal aid. Speakers had taken care, said Walsh, to affirm the rights of states and local communities over educational matters that would not be touched by federal aid, while subjecting them "to federal control to the extent of that aid." In his view, the nationalism propelling the drive for a department posed a threat to parochial schools, which would come under attack, "not in a direct way, but insidiously." He believed that Claxton was the "skillful, even foxy," mastermind behind the whole plan.[41] To be sure, Claxton was an educational nationalist, but his form of nationalism was not identical with that of the NEA. Opposed to federal control and the politicization of education through the appointment of a cabinet-level secretary, he believed that the nation's educational ills could be remedied through federal aid offered with no strings attached. If a department of education were to be established, it ought to be under an independent board of education rather than a secretary.[42]

At Shahan's urging, Howard convened the CEA Advisory Committee to discuss the centralization of education. Ecclesiastically and politically, Howard was a conservative representing the staunchest form of separate-but-equal Catholicism, the type that held aloof from all participation with non-Catholics. Despite repeated urging by Walsh for the CEA to send representatives to meetings of the NEA's Department of Superintendence so that they could inject Catholic principles into that body, Howard assiduously ignored the advice.[43]

Prior to the meeting of the Advisory Committee, Howard outlined his strategy for Shahan. He proposed a waiting game for several reasons. Given an alleged "strong current of opposition [to centralization] in influential labor circles," he suggested that this constituted a sufficient retardant on the movement, precluding the necessity of the CEA's taking a stand at the time. The labor groups Howard had in mind are unknown, but contrary to his information, two influential ones, the American Federation of Labor (AFL) and the American Federation of Teachers (AFT), favored the establishment of a department. If the CEA referred the matter to the hierarchy, Howard believed that some bishops would demand that it oppose the movement. If, on the other hand, the CEA decided to favor centralization, it would "incur the ill will" of many on whom it depended for support. "It seems to me best to keep things as they are for the time being," concluded Howard.[44] Endorsing this course, Shahan wanted the committee to formulate a position on the question of educational centralization so that the church would "be ready to face the issues it raises with unity of view and harmony of action."[45]

Shahan got from the committee far less than he had hoped. Instead of formulating a policy and proposing a plan of action that the church could follow, the Advisory Committee intended to pass the buck at some future moment to the archbishops of the country. In a memorial that was to be sent to them only when the situation warranted, the committee warned that "the establishment of Federal control of education offers no advantages, so far as can be perceived, for Catholic educational interests, while, in the light of the experience of the Church in other countries, it involves some very distinct dangers." The committee left to the archbishops any decision about a course of action.[46] In keeping with Howard's waiting game, this memorial would not be sent to the board for a year and a half.

While the CEA bided its time, Strayer's commission prepared a tentative draft of its bill. The measure proposed a department of education to encourage and help to develop public schools. It apportioned to the states on a matching basis $100 million annually for the promotion of literacy, immigrant Americanization, teacher training, rural schools, and physical education. None of the money was to further private education.[47] In May Strayer presented the draft to the Emergency Council on Education in the hope that the council would eventually submit the bill to Senator Smith. Claxton offered the most cogent criticism, predicting that the legislation had little hope of success because it was actually two measures in one: a bill to create a department of education and a bill for federal aid. In his view, "the chances of getting either of them through would be better apart." The council referred the draft for further criticism to McCracken, Judson, and Campbell. After having conferred with the NEA's commission, Judson reported that his committee supported the portion of the bill proposing a department of education, but "utterly disapprove[d] of . . . making large appropriations for education to the states." In the committee's view, the most important thing to accomplish was the establishment of a department. With that achieved, "a careful study should be given to the educational needs and possibilities, before asking appropriations for specific purposes."[48] Clearly, McCracken, Judson, and Campbell thought that the NEA's commission had rushed to judgment on the matter of federal aid, an issue that in their view required "careful study." Their position would enjoy the later support of Judd, dean of the Cleveland Conference.

Rejecting the advice of both Claxton and the Committee on a national department of education, the NEA decided to proceed on its own.[49] This was a fateful decision that in the short run made sense. To capitalize on wartime urgency, quick action was necessary. Time would be lost while the Emergency Council polled member associations on the establishment of a department (in fact, it took almost three years), and even more time would be lost thereafter in a careful study of educational conditions. By forging ahead, the NEA could count on the Emergency Council's support for at least the department feature of the bill.[50] In the long run, however, the financial aspect of the bill proved to be its Achilles' heel by dividing the majority of educators, who otherwise favored a department.

At the NEA's annual convention in July 1918, the Commission on the Emergency in Education launched its campaign for a "National Program for Education." Strayer's keynote address echoed the Wilsonian theme of a world struggle between autocracy and democracy. "Universal conscription in education," he declared, "is the only sensible method of perpetuating democracy, just as universal military conscription is the only democratic method of raising an army." Educational conscription was necessary because democracy rested on universal intelligence, which in turn depended upon literacy. It was also necessary because the nation had, by failing to Americanize immigrants, tolerated "the formation and continuation of racial and language groups with ideals and practices inimical to our free institutions." The problem lay in an absence of resolve. "We have been unwilling," said Strayer, "to accept the kinds of discipline and control necessary for a people which is to defend its own freedom and to fight for the establishment of world-democracy." The solution lay in the development of a "more adequate system of public education," one in which "every schoolhouse must be a community center of true democracy," where English was the common language, literacy was universal, and all "*unreservedly*" subscribed "to the principles and ideal of democratic government." Concluded Strayer: "May we realize now that democracy's greatest safeguard is the public school. May we recognize the necessity for the development of a more efficient public-school system costing vastly greater sums of money. The hope of humanity rests upon the education of the children of our democratic society."[51]

Strayer's fellow commissioner and colleague at Columbia University, William C. Bagley, developed the theme. Though schooling was traditionally a local affair, the crisis of war had awakened people to an awareness "that educational backwardness in any part of the country may handicap the progress and imperil the safety of the nation as a whole." The system of local support for and administration of schools had made for "glaring inequities" in education that the war revealed: 700,000 illiterates subject to the draft; "boys and girls . . . grown to manhood and womanhood in utter ignorance of American ideals and institutions, ignorant of the very language of our country, and even nurtured upon alien ideals brought to them thru the medium of an alien tongue"; and military personnel unable to understand commands given them in any language "save that of our principal enemy." In Bagley's view, the dependence of the schools on local revenue and local control acted as a retardant. "Of all our collective enterprises," he declared, "education alone remains hampered and constrained by the narrow confines of an obsolete conception." Bagley had no fear "that the nationalization of our schools will Prussianize our people." The NEA's Commission on the Emergency in Education had drafted its bill for a department of education on the conviction "that the nation may participate in the support of education without involving the dangers of bureaucracy and autocratic control."[52]

The convention tendered strong support to the commission's plan. Strayer was elected president of the NEA. The convention endorsed a resolution in favor of the establishment of a department of education and urged the federal government to adopt of a policy of "encouraging all the states to establish uniform minimum standards of health service, training for citizenship, and preparation and compensation for teachers, thru financial aid distributed to the states enforcing these standards." Yet nothing in the program was "to weaken the local responsibility or initiative or to subtract from the power of the state to . . . supervise the schools of that state." Finally, the convention authorized the hiring of a full-time field secretary to promote the program through correspondence, publication, and travel.[53] Not long thereafter, the Commission on the Emergency in Education appointed a subcommittee on enlistment to enroll new members in the NEA. The membership drive had three purposes: to enlarge grassroots support for the bill among teachers; to ensure that the teachers who would have to carry out the program once it was enacted would cooperate in doing so; and, through dues, to increase revenue in order to finance the structures and personnel necessary to secure enactment of the program.[54]

In effect, the NEA was gearing up to participate in what contemporaries referred to as the "new lobbying." In the previous century, lobbying had been conducted secretly and in the interest of individual corporations. In the early twentieth century, given the progressive reliance on governmental solutions to problems, the centralization brought about by the war, and the increasing organization of American life, lobbying underwent a transformation. Institutionalized, public, and conducted by voluntary organizations, it became an important means of wielding power. Lobbies, like the NEA, instituted legislation, sought friendly congressmen to introduce it, exerted pressure on congressional committees by supplying facts and expert testimony to support the legislation, and attempted to secure public backing through publicity campaigns.[55] As will be seen, the NEA became a masterful, if ultimately unsuccessful, participant in the new lobbying.

It would be easy to dismiss Strayer's and Bagley's words as wartime rhetoric. To be sure, they were that, but they were also instructive. Strayer had spoken of the nation's unwillingness to accept the discipline and control necessary to preserve its freedom, of the need for conscription in education, and of the necessity for every public school to become a community center for democracy. Implicit in this was the call for compulsory public schooling. Bagley argued that education was no longer a local matter, but a national concern. Like political and economic progressives of the New Nationalism, he looked to the federal government for the solution to educational problems. Both men had spoken of the need for cultural conformity through the eradication of alien ideals and customs. Finally, the NEA endorsed minimum standards in health education, citizenship, and teacher training. The establishment of standards was an important component of the administrative progressives' drive for efficiency in education.[56] States were to be induced to accept these federally imposed minimum standards through the carrot of federal funds. Thus, the program would be accomplished without diminishing the power of the states over their own schools. Recent precedents supported this approach: the Smith-Lever and Smith-Hughes Acts that offered federal money to states on a matching basis for specific educational purposes.

Accepting the money meant accepting federal regulation regarding use of the funds.

After the convention, Strayer took the case to the public in a half-page article published in the *New York Times.* Rehearsing the manifold problems revealed by the war, he asserted that "the burden of education should rest upon the nation" because the results to be secured were of national, rather than local, significance. The remedy to these problems lay in "a national system of education which will provide for the complete Americanization of millions of foreigners"; the establishment of "ideals of physical fitness"; compulsory education to the age of eighteen for the eradication of illiteracy and for the inculcation of "the meaning of liberty" and a willingness "to cooperate in the establishment of a democratic society"; the equalization of educational opportunity by uplifting rural schools to urban standards; and the payment of "a living wage" to teachers to attract the "choicest of our young men and our young women." All this was to be accomplished through the NEA's bill to establish a federal department of education and offer federal aid for these purposes on a matching basis. Admitting "that the administration and control of public education should be left in the hands of the several States," Strayer argued that the federal government alone had the financial resources "necessary for the maintenance of an efficient system of public schools."[57]

The case made by Strayer and Bagley had merit. Throughout the nation there were, in the latter's words, "glaring inequities" in schooling. The federal government had ample funds at its disposal to help equalize the system. Because the backwardness of schools in any part of the country did affect the nation as a whole, one could plausibly argue that the general welfare clause in the Constitution offered sufficient ground for a federal contribution to education. There was a certain brilliance in the plan to offer federal money on a matching basis to induce states and local communities to spend more on education. The sticking point in all of this was the issue of control. Constitutionally, the regulation of education was a state and local matter. A federal grant to education with no strings attached would leave control of education with the states and local communities. Though it is doubtful that Congress would ever have donated money without strings attached, Strayer and the NEA had other intentions. They wanted to establish minimum standards in education and use federal funds to induce states to accept them. In his argument to the public in the *New York Times,* Strayer had posed, but left unresolved, the thorny issue of control. He had admitted that education was essentially a state matter, yet asserted that the federal government alone had both the financial resources and the obligation to rectify the problems in the school system. A reader would have had to have been blind not to see where he and the NEA were headed.

Despite growing misgivings about the drive for a department of education, Howard kept a tight rein on the CEA's 1918 convention and held it to the policy of biding time on the issue. His recommendation to its Executive Board was that it authorize the Advisory Committee to meet more often. Accordingly, the committee was to meet frequently to keep an eye on the federal education question, but this merely continued the CEA's policy of vigilant inaction.[58]

While Howard held the CEA to a waiting game, Father Richard Tierney, S.J., a member of the CEA's Advisory Committee and editor of *America* magazine, felt no such compulsion. Having already advised Howard that the proposal for a department of education was "an obnoxious plan," Tierney set the magazine on a course that would make it the most intransigent foe of the scheme for a decade to come.[59] An article by L. F. Happel fired the opening salvo. He warned that a department, no matter how modest its scope, would push for an enlargement of its powers, ultimately leading to the state monopoly of education. Equally dangerous would be federal aid granted only to public schools. Because Catholic pocketbooks could not maintain a parochial system on par with a federally funded one, church schools would either be forced to attach themselves to the government or wither away.[60]

To sponsor its bill the NEA turned to Smith, chairman of the Senate Committee on Education and Labor. Southern progressives were committed to the consolidation of rural schools, the centralization of authority, and compulsory education laws, all features of the NEA bill. Moreover, Smith had a proven track record of progressive educational legislation.[61] On 10 October 1918 he introduced the NEA bill in Congress. It established a department of education with a cabinet-level secretary, three assistant secretaries, a solicitor in the Justice Department, and an unspecified number of educational attachés at U.S. embassies abroad. The new secretariat was to house the Bureau of Education and any other educational commissions or boards transferred to it by the president. Its purpose was "to cooperate with the States in the development of public educational facilities." To accomplish this end, the department had an annual appropriation of $100 million. Though seemingly small by modern standards, the sum amounted in present-day terms to $1.247 billion. The bill earmarked $7.5 million for the eradication of illiteracy through instruction that was to inculcate "the duties of citizenship"; an equal amount for the instruction of immigrants in "the English language and the duties of citizenship" and for development among them of "an appreciation of and respect for the civic and social institutions of the United States"; $50 million for the equalization of educational opportunity through the improvement of public elementary and secondary schools, the extension of the academic year, the gradation and consolidation of rural systems, and the centralized supervision of the latter; $20 million for physical and health education; and $15 million for the training of teachers. Each sum was to be divided among the states proportionately. For example, Iowa would receive the same proportion of the $7.5 million for the eradication of illiteracy as the number of its illiterates bore to the national total. The proportion for Americanization was to be based on the number of foreign-born, the proportions for equalization and teacher training on the number of public school teachers, and the proportion for physical education on a state's population.[62]

A progression of steepening conditions applied to the financial features of the bill. The measure authorized the secretary of education "to frame rules and regulations for carrying out the provisions. . . ." It further stipulated that to qualify for funds, a state had to secure from the secretary prior approval of all plans for implementing the various purposes of the bill. The plans must "specifically show courses of study and the standards for teacher-training preparation to be maintained." A state had to match the federal grant dollar for dollar. Furthermore, none of the money could go to private or parochial education. The foregoing constituted the minimum requirements a state had to meet to qualify for cash under the literacy and Americanization provisions of the measure. To qualify for funds under the remainder of the bill, a state must additionally establish an adequate program of teacher education within two years of the measure's passage. Finally, to qualify for the equalization fund, a state had to meet two further stipulations: it must legislate a public-school term of at least twenty-four weeks enforced by an adequate compulsory attendance law and require that English be the language of instruction in all common schools, both public and private.[63]

Several observations are in order. First, the secretary of education had wide discretionary power, not only to establish rules and regulations for implementing the aims but also to approve or disapprove of a state's plans for achieving the various ends. Those plans had to include the curriculum to be followed in classrooms as well as the standards to be maintained in teacher training. The NEA had officially endorsed having the federal government "encourage" states to adopt uniform, minimum standards in education, and the bill certainly offered that possibility through the discretionary power of the secretary. The level of that minimum would depend greatly on who sat in the secretary's chair. The bill obliquely recognized the existence of private and parochial schools through the clause that froze them out of the benefits, a policy pursued by state governments for more than a century. While not party to the financial rewards of the legislation, those schools would be subject to the compulsory attendance restriction and the requirement of English-only instruction.

With the NEA bill in Congress, the time had come for Catholic educators to formulate a policy. Howard summoned the CEA Executive Board into a special session at Catholic University in mid-November. The members engaged in a free discussion of federal control that failed to reach agreement about the issue. There seems to have been considerable willingness to entertain the possibility of some type of centralization. Although the minutes record only the general sentiment of the group, documents dating from either side of the meeting reveal the thinking of key members like Burns, Shahan, and Edward Pace. Mentally gifted, all three had contributed to the blossoming Catholic intellectual life that occurred in America during the decade and a half prior to the condemnation of Modernism (1907). Burns, a founder of the CEA and president of Holy Cross College in Washington, D.C., considered the American public school system "scandalously inefficient and superficial" and confided to Howard that "there ought to be a great change." He preferred a radical one. "I would gladly see complete reorganization, but I do not believe that this will be even attempted seriously." Rather, he foresaw a gradual reform that would begin with the elevation of the Bureau of Education to departmental status and then "the federalization of the common-school system."[64] A supporter of the Americanist movement of the 1890s, Pace could, as will be seen shortly, discover significant benefits in federalization for the Catholic educational system, though he was not blind to potential dangers.[65] Shahan, too, had supported the Americanist movement. An octogenarian, he remained remarkably open-minded, sharing Pace's view about federalization, even conditionally favoring the centralization of education.[66] Given such thinking, it is hardly surprising that while the Executive Board agreed that the Smith bill, as the NEA measure became known, posed potential dangers to education in general and to parochial schools in particular, it did not rule out nationalization altogether. Sending copies of the bill to the fourteen archbishops of the country, the board alerted them to the potential threat to education contained therein and asked them to "consider whether any modified plan of Federalization would be satisfactory."[67]

Not all Catholic educators were as restrained in their appraisal of the situation. When Brother John Waldron, who missed the meeting, learned of the board's action, he fired off a letter to Howard in which he questioned whether his colleagues understood what federalization meant for Catholic institutions. The wide, discretionary power of the proposed secretary was apparently not lost on Waldron. Appealing to the example of France, where a department of education had been used to decimate church schools, he urged Howard, "Think of the $100,000,000 at the disposal of one man evilly inclined to works of Christian and Catholic education!" Waldron protested against the bill to two prominent Democratic congressmen from his state of Missouri: Senator James Reed and Speaker of the House Champ Clark.[68]

Initial Catholic editorial opinion on the Smith bill was thoroughly negative, ranging from conspiracy theories to denunciations of state absolutism. In the conspiratorial vein, the Dubuque *Catholic Tribune* printed a letter written by a person claiming to be "in a position to know what is going on in Congressional circles." It warned of a movement afoot to legislate Catholic schools out of existence as a "'war measure of safety'" because they were "un-American, being under the control of a foreign potentate, the Pope of Rome." The anonymous informant advised that the proposed bill would not do this "in so many words." In fact it would "not mention the subject at all, but will reach its object just the same." Arthur Preuss, editor of the *Fortnightly Review* (Saint Louis), reprinted the letter with the recommendation that either the hierarchy or the American Federation of Catholic Societies (1901), a national organization of laymen, establish a "vigilance committee" in Washington to scrutinize legislation for threats to parochial schools. His article was in turn reprinted by papers in Texas and Michigan.[69]

German-American Catholics, whose schools were already the object of ill will, were alert to the danger. The Central Bureau of the Central Verein, a national federation of German Catholic societies promoting mutual aid while engaging its members with the social questions of the day, warned of a two-pronged attack led by the NEA and supported by the U.S. Bureau of Education. On the state level the assault aimed at "menacing the rights of parents and freedom of education"; on the national level it sought "federal control of the entire school system." The Central Bureau considered the scheme "an unwise plan fraught with grave perils to the spirit of American liberty." If states were permitted to drive private activity from the field of education, the same could happen with other endeavors, ending in "despotic Socialism."[70]

On a less conspiratorial note, the Cincinnati *Catholic Telegraph* believed that exclusion of parochial education from the benefits of the Smith bill "would put the stamp of at least quasi Federal disapproval upon religious schools," thereby magnifying the difficulties under which they labored.[71] The Saint Paul *Catholic Bulletin* took the idea a step further, predicting that the exclusion would stigmatize parochial schools as alien institutions: "An atmosphere of suspicion would grow up around private schools, they would come to be looked upon as un-American, and their death would not be far distant."[72] In a practical vein, several papers believed that the bill would starve parochial education out of existence, for Catholics could not maintain their institutions on the same level as federally supported public ones, especially when parents would have to pay increased taxes for the Smith bill subsidy.[73]

The threat to parochial schools was only one aspect of a larger danger perceived by Catholic editors. Another was the "Prussianization" of education. According to the Boston *Pilot,* the Smith bill was based on the theory that the child was the creature of the state, to be schooled as the latter saw fit, the theory by which Germany had achieved such an effective autocracy.[74] No Catholic journal developed this theme more forcefully than *America* magazine. It hammered at the section of the bill which required that standards for teacher training and classroom curricula be approved by the secretary of education. Said the editor, "A Government that can prescribe the training of teachers, as well as the studies to be followed by American children, is a Government at whose feet ancient Prussia might have humbly sat to beg further instruction in the now completely discredited art of making all children wards of the State, and all citizens mere puppets in the hateful game of selfish statecraft."[75] *America* warned that the autocratic tendency of the bill was linked to resurgent anti-Catholicism and argued that no matter how patriotic Smith's motives, "his plan embodies the extremist policies of foreign secularism."[76]

While Catholics warned of a plot against their schools, Freemasons came alive to the need for 100 percent Americanism. Masonry had traditionally avoided politics in favor of maintaining the lodge as an asylum of retreat from the conflicts and difficulties of society; the First World War, however, propelled the order into the secular arena. Like other Americans, Masons had answered Wilson's call for mobilization by raising funds, encouraging thrift among the citizenry, establishing clubs for servicemen belonging to the order, and promoting 100 percent Americanism. The last-mentioned activity continued after the war. An important key to understanding how Masons viewed the crisis facing the nation was their assumption about America's past. "They pictured a homogeneous society in which equality was real and democracy unchallenged," writes Dumenil. "Native, Protestant, middle-class Americans dominated this world—politically, culturally, and economically." These were the same assumptions underlying the progressive movement. While conservative Masons sought to foster social uniformity and harmony by influencing public opinion, radicals in the order advocated the use of the political process, and schools became the focus of their drive for homogeneity. No Masonic body was more active in this regard than the Southern Jurisdiction of the Scottish Rite, comprising the thirty-two states below the Ohio River and west of the Mississippi, the same territory in which the farm bloc, a coalition of agrarian progressives, would emerge.[77]

In October 1918 the Southern Jurisdiction alerted members to a plot against public education. The editor of the jurisdiction's *New Age* magazine observed that although the Catholic hierarchy considered public schools "godless . . . sinks of iniquity," bishops worked tirelessly to place "their instruments" in teaching positions in them and on boards of education. The editor accused prelates of trying to make the "schools as worthless as they say they are." Given the danger, the magazine considered it "the unquestionable and inexpugnable right of the state" to supervise the education of its people. Moreover, it was the duty of Americans to keep the schools "free from all sectarian enemies and influence." Finally, no one ought to be elected to any office who was not "a product of the *public schools*!"[78] By the turn of the year, the *Masonic Observer* (Minneapolis) proclaimed that private and parochial schools should not exist and would not long be tolerated.[79]

The reform branch of Masonry, particularly the Southern Scottish Rite, viewed public schools as the bedrock of American democracy. As will be seen in Masonic editorials throughout this study, the order considered them the quintessential levelers of society and promoters of Anglo-Saxon social order. In this regard the educational trust and Masonry were in complete accord.

In fact, the Masons were voicing an opinion that went beyond their ranks. As seen above, 100-percent Americans in Michigan had unsuccessfully attempted to amend their constitution to mandate compulsory public education. An initial failure would not deter them. Scarcely had the movement received a setback in Michigan than it sprang up in Nebraska. In January 1919, 100-percenters and anti-Catholics joined hands in that state in support of a bill that made attendance at public school compulsory for rural children between the ages of seven and sixteen. As in Michigan, the legislature defeated the measure, but proponents remained undeterred, carrying the standard to the state constitutional convention at the end of the year.[80]

Given Masonic sentiment and the movements against parochial schools in states like Michigan and Nebraska, it is little wonder that some Catholic editors saw a conspiracy against parochial schools behind the Smith bill, especially in view of the fact that Smith himself was a Mason. To be sure, Strayer and other administrative progressives viewed public schools as cradles of democracy and nurseries of citizenship. Just as surely, they also considered parochial schools, especially foreign-language ones, un-American seedbeds of alien values. If Smith, Strayer, and others of the educational trust actually intended the NEA bill to abolish parochial education—and it seems certain that at least some did—the attainment of that goal would be achieved only as a byproduct of promoting public schools. If privately funded Catholic schools could not keep pace with federally funded public ones, so be it. The exclusion of Catholic schools from the benefits of the Smith bill hardly constituted a conspiracy, for such exclusion was an accepted, though for Catholics a galling, practice.

November 1918 brought a significant change to the American political scene. On the first Tuesday of the month, voters sent a Republican majority to Congress for the first time in six years. A month earlier, Wilson had urged the nation to return Democratic majorities to both houses of Congress as an endorsement of his leadership. The return of a Republican majority to either house, he said, would be interpreted abroad as a repudiation of him and his ideals. This partisan call did little to endear him to wavering Republican supporters. Wilson had won reelection in 1916 largely because he had run on a peace platform and because the transformation of his policies from the New Freedom to the New Nationalism during his first term appealed to a coalition of southern and western progressives in both parties. In this midterm election, however, western Republicans began to return to the fold, partly because the president had failed to keep the nation out of war, but especially because of his preferential treatment of southern Democratic farmers, reviving sectional rivalry. This realignment gave the Republicans control of both houses of Congress beginning in 1919, something that did not bode well for the Democrats in the 1920 presidential election.[81] Close on the heels of this political triumph came the abrupt end of the war.

In December 1918 the Senate Committee on Education and Labor held a hearing on the Smith bill. The NEA Commission on the Emergency in Education organized the witnesses, most of whom were commission members themselves. As educational experts, they took the opportunity to present evidence in support of the need for various aspects of the bill (covered above in the discussion of educational problems revealed by the war). Although Strayer could not be present, his *New York Times* article was printed into the record. Dr. John Keith, president of the Pennsylvania State Normal School, explained that the appropriations for literacy and Americanization were aimed at eradicating those problems within a ten-year period. The requirement that states enact a compulsory education law to qualify for literacy funds would ensure the success of the campaign within that time. Reflecting the temper of 100 percent Americanism, he considered illiterate foreigners "not only a national menace but a state menace and a community menace." The bill would offer them a reasonable opportunity to become Americanized. If they failed to do so, then he favored further legislation to require their deportation, a sentiment that struck a chord with Senator William Kenyon, a Republican Iowa progressive soon to found the farm bloc. Fully $85 million of the $100 million annually appropriated under the bill, Keith explained, was intended for "the stimulation of the whole public school education of this entire country, which is a permanent thing and which will continue." Although he assured the committee that there was "only enough Federal control provided for in this bill so that the central government may be assured the States are . . . doing the things for which they are receiving the money from the Federal Government," neither he nor his colleagues addressed the wide discretionary powers of the proposed secretary of education, nor did any of the senators raise the issue.[82]

The only substantive criticism of the bill came from Republican Senator Carroll Page of Vermont. Though not opposed to federal aid to education, he believed that precedents like the Smith-Hughes Act must be built upon gradually. While Page supported the portions of the Smith bill that focused on particular issues, like literacy and Americanization, he had a serious problem with the equalization fund because of its general nature and his belief that the American people would be unwilling to support something so vague. He encouraged the NEA to develop specifics.[83] In the end, the Senate committee made no recommendation regarding the bill.

If the measure were to have any chance of success, it needed a cosponsor in the House. Whether in a bid for bipartisan support of the bill or in recognition of the new political reality of a soon-to-be, Republican-dominated Congress, the NEA's Commission on the Emergency in Education worked with the symbolically named Judge Horace Mann Towner, a progressive Republican congressman from Iowa, to revise the Smith bill so as to eliminate the "bad features" of the original. Basically, this meant minimizing bureaucracy and toning down the language of centralization. The latter change was an obvious attempt to avoid the constitutional issue of states' rights by making it clear that the federal government had no coresponsibility for schooling. The revised bill also included a forthright statement that money appropriated for the equalization of educational opportunity ($50 million) was to be used "for the partial payment of teachers' salaries." Although this had always been the NEA's intention, the association had left the matter mute in the original measure in order to avoid the appearance of self-interest. Now, however, both the AFT and AFL demanded a clear expression in return for their support. The two organizations even threatened to draft their own legislation unless the NEA made specific mention of wages for teachers.[84]

Towner introduced the bill into the House at the end of January 1919, and within a month Smith entered the companion measure in the Senate with only slight variations.[85] To boom the Smith-Towner bill and the national program for education, the NEA hired as field secretary Hugh S. Magill, an Illinois educator and Mason. He sent out 20,000 copies of a pamphlet describing the plan for national education, 30,000 press statements, and 80,000 summaries of the main features of the Smith bill.[86] In a speech to a conference on postwar reconstruction hosted by the National League of Popular Government, Magill argued that education was no longer a local matter but one of national concern. Outlining the features of the NEA's plan, he assured listeners that "our public schools should . . . remain under direct local control, and the autonomy of the state should be very carefully preserved."[87] An NEA pamphlet describing the revised bill made the same point. "The administration and control of education is left entirely to the states, and to local authorities, the Federal Government exercising supervision only to the extent necessary to see that the several amounts appropriated are used by the States for the purposes of the Act."[88] Though the NEA gave assurances that states would maintain control over education, the guarantees rang hollow in view of the power of the secretary of education to establish regulations for carrying out the bill's provisions and to approve state plans for its implementation, including school curricula and standards for teacher training.

Catholic archbishops awakened slowly to the issue of federalization. Having received a copy of the original Smith bill from the CEA, Archbishop Henry Moeller of Cincinnati leaned toward opposition but wanted the advice of Cardinal James Gibbons of Baltimore, the aged de facto primate of the American church. The cardinal in turn asked Pace for his view. At the end of January 1919 the latter submitted a memorandum. His cover letter noted that the NEA campaign had occasioned suspicions, surmises, and even exaggerations on the part of Catholics. Pace himself suspected that "those who are back of the movement have rather definite plans which they have not completely published, but which they will carry into execution as rapidly as circumstances permit." Catholics, therefore, must be alert to anything that might indicate the proponents' real intent.[89]

Suspicions aside, he wrote a dispassionate, objective analysis. He remarked that the NEA bill did not "explicitly or directly" deprive a state of control over its schools. With regard to the proffered federal aid, each state remained free to accept or reject any or all funds. If the money were accepted, a state must comply with the stipulations specified in the bill. This posed no direct danger to parochial schools because they were excluded from the benefits. Only public institutions, the beneficiaries of federal aid, would be subjected to federal control. Like Catholic editors, Pace admitted that church schools might suffer from an inability to compete with federally supported public ones, but little could be done about that. "Once we admit that the State has a right to maintain schools," he noted, "we cannot consistently object to larger expenditures for such schools, whether the funds be appropriated by each State or by the Federal Government."[90]

With regard to federal control of education, Pace observed that many Catholics feared the bills were simply the first step toward that end. Agreeing that such an interpretation was "plausible enough" in light of the wartime centralization of other matters, he predicted that with the return of peace and of Republicans to power in Congress, a backlash against centralization would set in. In the event that either bill should pass and should federal control eventuate, the situation might prove a boon to Catholics. For instance, the federal control of education might facilitate the prevention of discriminatory hiring practices by public schools against Catholic teachers because "the whole pressure of Catholic opinion could be brought to bear upon a single point, i.e., a small group of men in Washington" more easily than on a number of state agents. Good might also come to federally supervised parochial schools, provided of course they were allowed to maintain their religious character. For example, if the federal government established standards and rated all institutions, Catholics might show the superiority of their schools and demand federal funds on the basis of merit. On the negative side, federal control might be used to force all children into public institutions. Given the possibilities and uncertainties, Pace recommended that the hierarchy establish a committee in Washington to keep an eye on developments and to speak for the bishops. The war had occasioned the establishment of the National Catholic War Council to represent the church's interests in war work before the federal government. Some similar organization was now "still more necessary for the protection of our rights in the matter of education."[91]

Within days, an alarmed Bishop Walsh fired off a letter to Shahan. Wrongly informed that Pace and Shahan had conditionally approved the Smith bill and had persuaded Gibbons to do the same, Walsh pleaded against giving any endorsement of the measure to the government until bishops whose interest in education entitled them to an opinion on the matter had the opportunity to express it. Given "the present temper of the people in several parts of the country" (no doubt Michigan and Nebraska), he saw nothing but danger in federalization.[92] Shahan referred the letter to Pace for a response. Increasingly viewing the Smith bill as an antidote against the temper noted by Walsh, Pace directed his attention to the just-adopted Eighteenth Amendment. The success of the Prohibition movement had taken the Catholic church by surprise, as it did many congressmen who had voted for the amendment in the belief that the time limit they had set for ratification could not be achieved. They were stunned when in little more than a year the deed was done. The church now found itself having to negotiate for the uninhibited use of sacramental wine, an issue not altogether satisfactorily resolved until 1926.[93] "The success of the Prohibition movement shows," Pace told Walsh, "what can be done when each State makes its own regulations on a given subject, which is then taken up by Congress and worked into the Constitution." If 100-percenters in Michigan and Nebraska were successful, and if the movement were to spread from state to state, proponents would clamor for a constitutional amendment outlawing private and parochial education. The centralization of education could prove a bulwark against the drive for compulsory public education because the church would be able to bring pressure to bear on the federal government to prevent states from taking adverse action against parochial schools. If, on the other hand, Catholics could not forestall the drive for compulsory public education at the state level, and if that drive resulted in a constitutional amendment requiring compulsory attendance at public schools, concluded Pace, "We, as Catholics, will not be much concerned about Federal control for the simple reason that we will not have any Catholic schools to be controlled. I think that the alternative becomes clearer each day."[94]

On receiving the letter, Walsh telegraphed Shahan for a copy of the Smith bill. After reviewing it, Walsh responded that the measure ought to be entitled "for the strangulation of all privately endowed and religious schools and colleges" because that would be the effect of their exclusion from federal aid. He was not, however, averse to federal or state supervision of parochial education. In his opinion, if Catholics had "courted it years ago," the church could have won the confidence of authorities in several eastern states and could have had the "Catholic school system . . . incorporated more or less completely into the State system." Warming to the point of view of Pace and Shahan, Walsh thought that the church should labor to eradicate the section of the Smith bill that excluded parochial schools from sharing in the appropriations and gave the secretary of education authority to approve state plans for implementation, including school curricula. Like Pace, he wanted the hierarchy to establish a committee in Washington to monitor legislation.[95] Neither man had long to wait before that occurred.

After almost a year of playing a waiting game, Howard sent the CEA Advisory Committee's April 1918 warning about the dangers of federalization to the board of archbishops.[96] Having accepted Pace's advice, Gibbons had asked Bishop William Russell of Charleston to prepare a paper on the need for a Catholic agency in Washington to monitor legislation. Russell required no convincing. While a pastor in the national capital, he had served as Gibbons's representative to the federal government. As a bishop on the Administrative Committee of the National Catholic War Council, he had experienced first-hand the good that such an organization could perform.[97]

On 20 February 1919 the American hierarchy celebrated Gibbons's fiftieth episcopal anniversary in a gala affair in Washington, boasting about seventy-five bishops or three-quarters of the hierarchy. Pope Benedict XV sent Archbishop Bonaventura Cerretti, secretary of the Congregation of Extraordinary Ecclesiastical Affairs, who had served in the apostolic delegation in Washington (1906–14). After a luncheon banquet at Catholic University, he addressed the bishops about the postwar task facing the church. The pontiff, he said, was particularly concerned about adjustments in the areas of education and labor. "Rome," declared Cerretti, "now looks to America to be the leader in all things Catholic and to set the example to other nations."[98] The assembled bishops began a discussion about how they might do so. Taking this occasion to deliver his paper on the need for a national Catholic agency in Washington, Russell noted that until the crisis of world war, Gibbons had handled the church's interests in Washington. Age now prevented him from doing so. Protestants had already formed a lobby in Washington, the Federal Council of Churches, which not only monitored legislation, but often initiated it. Experience had convinced Russell that the federal government wanted to respect Catholic opinion. The only way to offset the influence of the Federal Council of Churches was to establish a permanent Catholic secretariat in Washington, that is, a lobby that would be "constructive rather than destructive or obstructive." "We must watch the storm in its growing," he warned. "The Prohibition movement shows what we may expect, if we disregard the distant murmurings of a storm that is brewing against our schools, hospitals, and asylums."[99]

The bishops appointed a committee to suggest a plan of action to be presented the next day when the board of archbishops held its annual meeting. The plan called for an annual convention of the hierarchy to coincide with the yearly meeting of the archbishops; it also proposed the establishment of the General Committee on Catholic Interests and Affairs to represent the hierarchy before the government. Having approved the plan, the archbishops left it to Gibbons to appoint five bishops for service on the new committee. They entrusted to it the letter from the CEA Advisory Committee warning of the dangers of federalization in education. More than a month passed, however, before Gibbons acted on the membership of the Committee on Catholic Interests. Finally, at Russell's urging, the cardinal placed himself in the chair and selected the four bishops of the War Council's Administrative Committee: Russell, Schrembs, Peter Muldoon of Rockford, and Joseph Glass of Salt Lake City.[100] As will be seen in the next chapter, this was the first step toward the creation of a permanent, national, Catholic agency in Washington, which, among its other duties, lobbied for church interests.

Within a fortnight of Gibbons's jubilee, the Sixty-fifth Congress expired, bringing a quiet death to the first Smith-Towner bill. Despite its demise, Catholic editors kept up a drumroll of criticism because the measure was surely to be resurrected. Their objections fell into three categories: centralization, the threat to parochial schools, and finances. With respect to the first, the Boston *Pilot* had come to view the drive for a department of education as a self-serving attempt by the NEA to control schooling nationwide. If the Smith-Towner bill carried, warned the editor, it would "make the National Educational Association, a private society of educators, a Department of American Government, and give them the power that will put every school, public and private, in the land under their thumbs."[101] There was also the ever-present fear for Catholic education. The Los Angeles *Tidings* warned that a concerted attempt would be made to wipe out parochial schools. The *Denver Catholic Register* linked this concern with the movement for a department of education. If the latter were created, predicted the editor, "We can rest assured that enemies of Catholic education are going to use the machinery created for the purpose of driving parish schools out of business."[102] Finally, other papers mined a new editorial lode: economics. During the war, the national debt had expanded nearly twenty-fold. With the House Committee on Appropriations predicting that government spending would outstrip revenue in the coming years, the Smith-Towner bill with its $100 million appropriation would simply cost too much. Moreover, it was a pork-barrel measure designed "to take money from those parts of the country which educate their children and spend it on those states which do not." To put it in sectional terms, northerners were being asked to pay for the schooling of southern children.[103]

The economic reality of the bill was far-reaching. With respect to education, the measure proposed a significant redistribution of wealth. Thirteen states stood to lose by the legislation, while the remaining thirty-five would come away big winners. To be specific, New York, Michigan, Rhode Island, Massachusetts, Illinois, Pennsylvania, California, Maryland, Connecticut, Delaware, Ohio, New Jersey, and North Carolina, in that order, contributed a greater proportion to federal revenue than they would reap in benefits from the bill. The unequal tax burden borne by these states was to accrue to the educational good of the remaining ones (see Appendix). With the exception of North Carolina and California, these fourteen represented the industrialized Northeast and North Central. Under the terms of the bill, thirty-five states would receive back more than 100 percent of their educational contribution to federal revenue, twenty-four of those would reap more than 200 percent, and seventeen of them would reap more than 300 percent. Although in several cases, the actual sums in question might be quite small, overall the proposed redistribution was massive and most beneficial to the South, Midwest, and West.[104] The NEA chose not to call attention to the economic impact of the measure, stressing instead the need for the equalization of educational opportunity, an equalization that would in fact be paid for by prosperous industrial and commercial states. Only at the end of the 1920s, after ten years of a depressed farm economy, would advocates of a department of education and federal aid openly argue for a redistribution of national wealth to promote equal educational opportunity.

Catholic editorial opinion stirred up considerable anxiety over the Smith-Towner bill. On 29 April 1919 Pace informed Russell about the many requests he had received for an article or statement on federalization and the proposed department of education. If he knew where the hierarchy stood, he would write "as strongly and as plainly . . . as anyone could desire," but he did not care to express himself if his ideas would not "be endorsed by the proper authority."[105] Russell was desperate. For weeks he and his confidant, Father Edward Dyer, vicar-general of the Sulpicians, had been trying to get Gibbons to endorse a comprehensive program entrusting major church affairs to the General Committee on Catholic Interests. The plan called for granting it authority to manage, among other things, postwar work and to maintain an agency in Washington to safeguard Catholic interests. Grown hesitant with old age, the cardinal refused to endorse the scheme.[106]

The impasse with Gibbons was finally overcome in May 1919, when Bishop Muldoon, vice chairman of the committee, took the initiative to call a meeting. The relieved cardinal turned matters over to him and gave tentative support to the program devised by Russell and Dyer. In a letter ghostwritten by them, Gibbons offered the program to the committee and declared that the issue of centralization was "the most pressing of problems, the one on which we can least afford to delay."[107] Though others on the committee wanted to temporize, Russell pressured them into asking Gibbons to invite all the bishops to attend the first annual meeting of the hierarchy in September 1919 in Washington, D.C., to consider the Russell-Dyer program. There, the prelates were to determine a position on the federalization of education. In the meantime, the committee would poll the hierarchy on the issue by means of a questionnaire. So, if an emergency arose demanding action, the Committee on Catholic Interests would "have the guidance and authority of a majority of the Hierarchy."[108] Thus, the Catholic church in America took a step closer to united episcopal action and the formation of a lobby.

This move toward the creation of a Catholic lobby came not a moment too soon. The NEA had launched a campaign to establish a federal department of education and promote a national program for schooling. While upholding the right of states to control education, it had gone on record as being in favor of federally encouraged minimum standards in schooling. Both the Smith bill and the Smith-Towner measure contained the machinery to produce such federally induced standards. The legislation offered a huge aid package to the states for specific purposes, while it gave the proposed secretary of education power to regulate how the law would be enforced, including approval rights over school curricula and standards for teacher training. Moreover, the NEA had all but openly stated that it favored mandatory public schooling. Finally, it was well along in the process of refashioning itself into a formidable lobby.

In the face of this drive, Catholic educators had yet to find common ground. Twice the CEA had confronted the issue federalization, and twice it had come up empty-handed regarding a policy. The Advisory Committee grappled with the subject, but could do little more than register a warning that the NEA plan held no benefits for Catholic education and might prove dangerous to parochial schools. When the Executive Board later took up the Smith bill, it agreed that the measure in its present form seemed dangerous to education in general and to Catholic schools in particular, but the committee remained open to the idea that some form of federalization might be appropriate, a position toward which progressive members of the Catholic educational community, like Shahan and Pace, were increasingly drawn. With the CEA having referred the matter to the board of archbishops, which passed the issue to the Committee on Catholic Interests, it remained to be seen what stand the hierarchy would take. As will be seen, the same indecision that plagued the CEA would plague the bishops for a year.

2

The Making of a Religious Issue

The sudden end of the First World War dramatically altered the political and social landscape. With the advent of peace, postwar conservatism gripped the nation with unexpected swiftness. The restoration of Republican rule in Congress placed a number of Old Guarders in key positions. For example, Pennsylvania Boss Boies Penrose took the chair of the potent Senate Finance Committee. Men of his stripe opposed Progressivism of either stripe: the New Freedom or the New Nationalism. They advocated tariff protection, tax cuts, and economy in government. They found allies among non-progressive Democrats of the South and Northeast, who wanted an end to wartime federal regulation of the economy and a restoration of balance between Congress and the presidency. In general, congressional Republicans were unenthusiastic about any legislation except that with party backing. Moreover, President Woodrow Wilson's lack of leadership during his prolonged illness left the government rudderless in the face of mounting social and economic problems. Realpolitik suggested that Republicans do little and let Wilson take the blame.[1]

To be sure, progressives of the New Nationalism remained alive as legislation like the Smith-Towner bill, the Sheppard-Towner Maternity bill to grant federal aid to needy mothers and their newborn, the Plumb Plan to nationalize railroads, the Industrial Rehabilitation bill to offer federal funds for the vocational retraining of workers injured on the job, and the Child Labor Amendment to outlaw the employment of minors bear testimony. But the movement ran headlong into an opposing force: a strong reaction against wartime centralization that expressed itself in a fear of encroaching federal power.[2] The educational trust and its progressive congressional supporters had to negotiate this new political terrain.

In some respects educational administrative progressives benefited from postwar social trends, which caused many to view public schools as the means of enforcing cultural conformity. During the late nineteenth and early twentieth centuries, modern intellectual currents called into question Victorian values and traditional understanding of humanity and the cosmos. Inextricably linked to this temper was the city because the modernist impulse found its readiest welcome among urban dwellers, rather than people of the soil.[3] And agrarian folk seemed to be a declining breed. The 1920 census revealed that for the first time more citizens lived in urban areas than in rural ones. By defining "urban" as a place with a population of at least 2,500, the census bureau had erased the considerable differences between urban Chicago, "urban" Iowa City, and "urban" Sac Prairie, Wisconsin. Even so, the real urbanization of America cannot be denied. In 1920 one-third of the nation dwelt in 144 cities of more than 50,000 people, double the number of such places since the turn of the century. This combination of reality and statistical sleight of hand left postwar folk with the impression that the nation was changing.[4]

While cities proved amenable to the modern temper, they also provided a home to newcomers to the shores. Since 1890 urban ranks had been swelled by an unprecedented number of immigrants. Unlike their counterparts earlier in the nineteenth century—northern Europeans and Britons, people who fit more or less comfortably in the Anglo-Saxon host culture—this new wave hailed from eastern and southern Europe. A number were Jews, but most were Catholics. As mentioned in the previous chapter, many had come to America for economic advancement and maintained strong ties with their homelands, a number of them intending to return. They incorporated into the nation through machine politics, labor unions, settlement houses, ethnic clubs, foreign-language newspapers, and especially the Catholic church. Given their retention of cultural ways, they seemed resistant to the melting pot. The eugenics movement lent a veneer of scientific respectability to progressive and popular prejudice that made it acceptable for old-stock Americans to view these newcomers as racially inferior, threatening dilution of the superior Nordic race. Fueled by concerns about communist unrest in eastern Europe, the formation of the Communist party in America, a rash of postwar strikes at home, and a series of bombings aimed at political and business figures, the Red Scare of 1919 fastened attention on the enemy within, the new type of immigrant who lived in the city. Though the scare and its worst by-products subsided relatively quickly, its effects lingered through the decade. Increasingly, many Americans viewed the otherness of these newcomers as a threat to be eradicated by severely restricting future immigration and by Americanizing in the public schools those immigrants already in the land.[5]

One need not view the looming cultural battle as simply an urban–rural conflict because the reality was more complex. During the 1920s the number of farms with automobiles doubled, resulting in an increasing dependence on towns for gasoline and service, bringing farmers in touch with the forces that were changing society. The radio, too, brought urban America into farm homes. During the decade the percentage of farms with electricity also rose from 1.6 to 10.4 percent, buttressing the fact that modernity was coming to the countryside, if slowly. On the other hand, immigrants were not the only ones to swell the ranks of urban dwellers. During the 1920s an estimated 6.3 million Americans abandoned the plow for the factory. There was class tension as urban prosperity touched city folk unevenly.[6]

Postwar political and social alterations cut two ways. Local communities increasingly viewed schools as agents of socialization for reinforcing Anglo-Saxon, Victorian values. Because the educational trust aimed at bolstering public education in order to Americanize immigrants and to imbue children with the duties of citizenship and traditional culture, it could count on a measure of popular support for its program of schooling.[7] Conversely, given that the trust comprised educational nationalists, its program would alienate those who reacted against wartime centralization and sought to restore the balance of power to the states and local communities. Similarly, because most of the new immigrants were Catholics, and because the church maintained its own system of education, Catholics could expect that their schools would increasingly be viewed with distrust by those who sought social conformity through education. On the other hand, given that the church stood for parental rights in education, which accorded with local autonomy, it could count on the support of conservative politicians. Since no less a value than the school hung in the balance for each side, the stakes were high and the matter would prove volatile, ultimately becoming a religious issue.

On 19 May 1919, the opening day of the Sixty-sixth Congress, Judge Horace Mann Towner introduced a revised bill for the creation of a department of education, and scarcely a week later Hoke Smith entered its companion in the Senate. The measure was a streamlined version of the previous Smith-Towner bill, rewritten to meet the objections of conservative critics who opposed centralization. In addition to Catholics, the NEA admitted that a number of state superintendents of education considered the original legislation a threat to the local control of schooling. So too did many teachers, whose representative body, the AFT, negotiated for changes in the measure, not the least of which was a forthright statement that money in the equalization fund be used to increase their salaries. The Smith-Towner bill evenly divided the $50 million equalization fund, half for the partial payment of teachers' salaries and half "for providing better instruction and extended school terms especially in rural schools." The measure now disavowed any intention of seeking uniformity in plans of implementation and forbade the secretary of education to supervise state systems of education. The only string attached to federal funds was the requirement that each state file a year-end report on expenditures, which needed the secretary's approval before the next year's funds could be released. Missing from the new bill was the disqualification of parochial schools from the benefits, a cosmetic gesture to appease Catholics because nearly every state had a constitutional restriction against granting public money to private education.[8]

These changes signaled an apparent retreat from George Strayer's remarks about the need for a more disciplined and controlled system of education as well as from the NEA's 1918 resolution calling for federally induced, uniform, minimum standards in education. In all likelihood, the revisions were a tactical deferral of these goals in order to achieve the establishment of a department and win federal aid.[9] That is certainly how a number of educators viewed the matter. When Representative Caleb Layton of Delaware solicited the views of a wide range of schoolmen, a number favored the revised bill precisely because they believed that it would lead to the nationalization of education.[10]

Adjustments to the legislation failed to appease the Catholic press. Editors still believed that it aimed at a federal monopoly of education. Paul Blakely of *America* asserted that the revisions made the bill even more dangerous by cloaking its real purpose, thus enabling propagandists to claim that it safeguarded states' rights. The measure, however, still required an annual report of spending, which needed the secretary's endorsement in order for a state to qualify for future funds. "If the programs and general conduct of the local schools do not win the approval of the Federal dictator in Washington," argued Blakely, "they must be changed under penalty of loss of all Federal aid."[11] The Boston *Pilot* referred to the legislation as the "Georgia Python." Like the snake, which concealed itself from unsuspecting victims until it could coil about them, the Georgia senator's new bill hid its real purpose but would just as surely coil around the schools and squeeze the freedom out of education.[12]

In June the Central Bureau of the Central Verein published a pamphlet critical of the bill. While acknowledging the NEA's attempt to ensure local autonomy in the rewrite, the bureau was convinced the measure would still restrict independence. The secretary's power to review and approve expenditures meant that the money would have to be used "exactly as stipulated." Based on the theory that public schools were the only legitimate ones, the bill also aimed at proscribing private and parochial education, to be achieved subtly through the financial provisions. Catholic parents would have to pay higher taxes to finance public schools, while at the same time supporting their own parochial ones. Revenue for the latter would necessarily decrease, stunting their growth or causing their deterioration. The bill, argued the Central Bureau, had the support of two types of people. The first were those who believed that for "business efficiency" a monopoly was far superior to individual enterprise. The second were those "who desire to eliminate all Christian ideas from education entirely." The end result would be secular socialism. The Central Bureau sent the pamphlet to every congressman.[13]

The revised bill also drew criticism from congressional opponents of federalization. Two advocates of states' rights fired the opening salvo of the postwar antistatist backlash by attacking the Smith-Towner bill as a bureaucratic boondoggle. During a debate on legislation for vocational rehabilitation of people injured on the job, Senator William King, a Utah Democrat, claimed that the matter under question was simply one part of a program to relieve states of their duty to provide education. The total scheme called for a department of education. The theory behind the movement, said King, was that "States are supposed to be mere appendages to the Federal Government, not sovereignties possessing sovereign powers and charged with sovereign responsibilities." He favored letting the people close to a problem solve it for themselves. Senator William Kenyon, the progressive Iowa Republican who now chaired the Committee on Education and Labor and a supported the Smith-Towner bill, objected: "We are not trying to take over the work of the states. It is entirely Federal stimulation, while the work is really carried on by the States." The precedent for such stimulation had been set by the Smith-Hughes Act. Waxing biblical, King responded: "We are being seduced by Federal aid and the States are clamoring for crumbs that fall from the table of this rich and powerful Dives, the National Government, which sits enthroned in this Capital City of the Nation." Senator Charles Thomas, King's fellow Democrat from Colorado, agreed. The proposed department was an attempt to have states relieve themselves of the burden of education that rightly belonged to them.[14] This exchange was important because it showed that the position of conservative politicians and that of Catholic editors had considerable overlap. Both favored local control of education.

In the late spring of 1919 the Committee on Catholic Interests learned that the House and Senate Committees on Education planned to hold a joint hearing on the Smith-Towner bill in July. If the hierarchy wished to send a representative, the bishops had to be polled on their position, so Cardinal James Gibbons commissioned Edward Pace to develop a questionnaire. Pace divided it into three sections, covering federal control, federal aid, and national standardization. With regard to the first, he asked about the degree of federal control tolerable over public schools and whether that control should be extended to include parochial education. In his view, there was no way to predict whether federal authority over Catholic education would prove helpful or harmful. On one hand, it worked well in England and Canada where church institutions received government funds and yet maintained their denominational character. On the other, it had disastrous consequences in France where government authority was used to secularize education and the state. Did the bishops wish to oppose control of any kind; if so, should it be done actively? Financial concerns linked standardization to federal aid. Noting that the aid package of the Smith-Towner bill would increase taxes, would be distributed unequally among the states, and would make it harder for parochial schools to compete with public ones, Pace put the question of approval or disapproval of federal funding. He advised, however, that if the church accepted federal standardization of its own institutions, it might be possible to secure government money for them. Would the bishops therefore be willing to accept federal control in return for a clause in the Smith-Towner bill stating that "a standard shall be established for all schools public and private; every school attaining said standard shall receive an appropriation from the Federal Government"?[15] The questionnaire was objective, displaying the same openness that had characterized Pace's previous memorandum to Gibbons. It met the approval of both the cardinal and Bishop Thomas Shahan.

The decision about sending it to the hierarchy was left to Bishop Peter Muldoon, vice chairman of the Committee on Catholic Interests. Both Pace and John Fenlon, secretary of the Committee on Catholic Interests, informed Muldoon that the annual convention of the CEA was scheduled for the end of June and that federalization was surely to be discussed. While the association's Executive Board ought to respond to the questionnaire, the two cautioned that the CEA should be advised that the bishops had the matter in hand and the association "ought not to be allowed to mess it [*sic*] or to speak out for the Church."[16] Muldoon took the advice and sent word to Francis Howard. He also ordered that the questionnaire go out to the hierarchy with a cover letter asking whether a delegate of the bishops should appear at the joint hearing to deliver a statement based on their responses.[17] Thus did the episcopacy begin to assert its control over the education issue.

When the CEA met in Saint Louis, Howard made it clear to the Executive Board that the hierarchy had charge of the federal education question. Noting that the movement for a department posed great danger to Catholic schools, he reported that the Committee on Catholic Interests had been formed "to outline a policy for Catholics." The CEA's duty was "to await the directions that may be given and to cooperate loyally." The board held a general discussion of the bill and voted uniformly negatively on all but two queries in Pace's questionnaire: one member broke ranks to vote against active opposition to the Smith-Towner bill, and two voted in favor of federal appropriations for public schools. The board expressed the simple opinion that "the Smith-Towner Educational bills now in Congress should be defeated."[18]

The topic of federalization received public attention. Archbishop John Glennon of Saint Louis addressed the issue in his sermon at the opening mass. Noting that the hierarchy was to set policy on the issue, he was concerned about the meaning of centralization for Catholic education. In his view, parochial schools offered society an educational variety that served the best interest of civilization as well as of Catholics. The ability to maintain such variety rested on the twin pillars of parental rights and academic freedom, now being sapped by two movements. The first attempted "to draw into the State system not alone all taxes for its support, but all law for its expansion"; the second, to merge state monopolies of education "into one grand monopoly under federal control."[19]

The keynote address, written by Cardinal William O'Connell of Boston and delivered by John Peterson, was a philosophical disquisition on the nature and limit of state authority. Without making specific reference to it, he clearly had in mind the Smith-Towner bill and legislation like the Michigan amendment for compulsory public schooling. Honoring God and religion, America had always respected the rights of individuals and checked the power of the state. Of late, the nation's democratic institutions were "endangered by the present tendency of the State to increase its powers and to absorb the individual in its paternalistic legislation." To counter what O'Connell viewed as creeping socialism, he urged a return to a proper theory of the state, which recognized the individual and the family as the basic units of society. The state existed solely to protect the rights of individuals and families and to promote their welfare. It had authority to act only when private initiative had proven inadequate and the common good therefore demanded it. The state could intervene in education only if parents failed to provide adequately for the same. O'Connell concluded by praising parochial schools as "the only schools in the land that harmonize with our national traditions" because they alone taught religion and consequently upheld the dignity and rights of human beings.[20]

This last remark was expressive of an attitude growing among Catholics, namely, that they were the few who had remained faithful to traditional American values and institutions in an age of shifting mores. As Catholics viewed matters, their religion was the one that harmonized best with republicanism and was the only one capable of sustaining it. They saw themselves as the true Americans. This attitude, described by some historians as Catholicism unbound, was a hallmark of the church's separate-but-equal subculture described in the preface, a position diametrically opposed to that of Protestant cultural conformists.[21]

After muzzling the CEA, the Committee on Catholic Interests bridled the Knights of Columbus. In July Supreme Knight James Flaherty informed Muldoon that the order might pass a resolution on the bill at its convention in August. He inquired if this action would be acceptable and what attitude should be taken. Muldoon referred the matter to Gibbons.[22] In a letter that was likely ghostwritten by Fenlon, the cardinal advised "that there be no denunciation and no misrepresentation of the bill." If the Knights wished to make a statement, they should declare themselves in favor of states' rights in education and against federal control. Positively, he recommended "carefully worded" resolutions in appreciation of parochial schools and the liberty that Americans have always enjoyed in education, which served as the fundamental safeguard of parental rights therein. The Knights should recognize the necessity of public education while praising the zeal of the American people in providing for the schooling of their children.[23] This letter marked the beginning of what might be called official Catholic restraint regarding the bill. Gibbons—Fenlon—recommended a fundamentally positive stand, namely, that Catholics favored states' rights, supported both parochial and public education, and opposed federal control of either. They were to avoid objecting to the measure as an assault on church schools, while affirming principles that their countrymen could endorse.

The hearing on the Smith-Towner bill began on 10 July. The Joint Committee had two Catholic members: Senator David Walsh of Massachusetts and Representative Jerome Donovan of New York, both Democrats. The first day of testimony was virtually a repeat of the 1918 Senate hearing. The same members of the NEA Commission on the Emergency in Education appeared, joined this time by Strayer. In their view, education was a national, not merely a local or state, concern. They observed that each of America's allies considered it so and had dignified it with a national department. Not content with European precedents, NEA spokesmen pointed to an American one: the Department of Agriculture. Hugh Magill, the NEA's field secretary, argued that although the constitution had left farming under state jurisdiction, the federal government had established the Department of Agriculture to help states promote farming without asserting federal control over it. If the federal government could establish that department, it was more imperative that it establish a department of education. Because weakness in education in any part of the country weakened the nation as a whole, wealthier states had an obligation to aid poorer ones in equalizing educational opportunity. Only the combination of federal taxes and federal aid could accomplish this end.[24]

Issues of Catholic concern were addressed directly. Towner asserted that the revised bill gave the federal government no power over curricula and forbade the secretary of education to interfere with a state's administration of the appropriations. The section ensuring local control was "as strong as it can be drawn," noted Towner, "or if it can be stronger I am in favor of making it stronger, in order to save to the States absolute control of the plans, means or methods by which these funds are to be expended in the respective States." Democratic Senator Josiah Wolcott of Delaware asked if the appropriations could be used to induce a state to adopt a specified curriculum; both Towner and Magill reiterated that the measure specifically forbade the secretary of education to exercise any authority in that regard. Having received "a number of letters from bishops of the Catholic church opposing this bill," Kenyon asked Strayer if there was anything in it "to injure in any way the parochial schools of any church?" Strayer responded that Catholics had misunderstood the legislation. It intended no centralization and no uniformity; it left everything to state control and touched parochial schools only in the provision requiring English as the basic language of instruction. When Hoyt Chamberlain, chairman of the National Civic Betterment League, testified that his organization favored the national standardization of both textbooks and grade levels in schools, Towner interrupted to say that if Chamberlain wished such standardization, "the bill would have to be changed."[25] Clearly, both congressional and NEA proponents of the measure believed that the revised legislation had met the objections against federalization. Not only their assurances but also a careful reading of the measure might lead to the same conclusion. What remained uncertain, however, was the depth and sincerity of the NEA's commitment to the state and local control of education.

With regard to the compulsory attendance clause, Magill declared that it in no way aimed at placing all children in public schools. Noting that the present measure had dropped the disqualification of private schools from the benefits, Kenyon asked if this meant that they could receive federal money. Magill replied that the bill left that decision to the local level. "If a State should make a private school a legally constituted educational authority, that would end it." Towner, however, considered that idea "utterly impracticable."[26] As will be seen, he opposed private schools and believed that they should be suppressed. Kenyon adjourned the hearing for ten days to give opponents time to organize a case.

When the hearings had begun, Catholic bishops scarcely had time to respond to Pace's questionnaire. Even with Kenyon's grace period, the response was thin. Less than half answered—forty-five out of 108. Thirty-seven rejected the legislation, and twenty-four of them favored active opposition to it. Nine bishops approved the bill outright or favored it conditionally, including Shahan and five archbishops: Edward Hanna of San Francisco, Henry Moeller of Cincinnati, Austin Dowling of Saint Paul, Arthur Drossaerts of San Antonio, and John Shaw of New Orleans. One-third of the respondents said that they would approve of federal control if a clause were inserted granting federal aid to any private school that attained a federally established standard.[27] Negativity aside, a greater proportion of respondents showed more openness toward the legislation than had the members of the Executive Board of the CEA, and more than half of the positive responses came from archbishops.

Even though the majority of respondents favored active opposition to the bill, friends in Congress counseled otherwise. No doubt they feared that an open fight would turn the matter into a religious issue, something that was already happening. Furthermore, the situation seemed safe enough for the time being. Although most members of the Senate and House Education Committees supported the bill, no action was anticipated during the current session or at least not before September when the hierarchy was to meet, offering all bishops the opportunity to discuss the educational situation. Therefore, the Committee on Catholic Interests decided against sending a spokesman to the second session of the joint hearing.[28]

Although the hierarchy sent no one to the hearing, Donovan had secured statements of opposition to the bill. Before presenting them, he cornered professor William Bagley of Columbia University into admitting that Nicholas Murray Butler, president of that university, opposed the measure. Donovan then produced evidence that Presidents Arthur Hadley of Yale, John Grier Hibben of Princeton, and John H. Finley of the University of the State of New York also objected to it. Here was the educational old guard lining up against the new. Administrative progressives had challenged the power of these pioneers of the NEA and sought to transform the organization into something more pliable to their own ends. Donovan had exploited the rift between the two groups to the detriment of the Smith-Towner bill, summed up in Hibben's curt dismissal: "There is no need for the proposed legislation submitted in the Towner educational bill. A secretary of education seems to me wholly unnecessary and undesirable."[29]

The educational old guard was not alone in objecting to the measure. No less an administrative progressive than Commissioner of Education Philander Claxton opposed it by indirection. Vainly had the NEA sought his endorsement of the legislation, but he rejected the idea of a department of education, especially one of the Smith-Towner variety with a secretary at its head. If one were established, the only type that could save schooling from politicization and the overlordship of one man was a department administered by an independent federal board of education.[30] Claxton believed that the federal government had the responsibility of offering "support, kindly and wise guidance" to education. A warm advocate of federal aid, he viewed it as the key to improving education, but in a way that really maintained local control.[31]

Rather than announce his opposition to the Smith-Towner bill, Claxton made his point by proposing an alternate plan, calling for an initial annual appropriation of $125 million, rising to $300 million by the decade's end. Thereafter, the amount was to be increased 4 percent annually to keep pace with population growth and to ensure the increased efficiency of schools. States would be required to match the appropriation by two dollars to one. The appropriation was not to be used for capital improvements in schools, but solely "for the pay of teachers and for the direct and necessary means of teaching." The purpose was to equalize educational opportunity, including the areas of health education, vocational training, and citizenship. To qualify for the funds, a state must require a minimum of 160 days (32 weeks) of schooling per year for children in the primary grades and 480 hours per year for students in high school. Beyond granting aid, the federal government should study educational problems and disseminate the results, refraining "severely and consistently from all meddlesome interference with State and local administration of schools." Such a plan, concluded Claxton, would result in "an effective national system of democratic education, with all the strength of national support and guidance and all the vigor and freshness and perennial youth of local support, control, and initiative."[32]

Claxton's plan broadly resembled the financial features in the Smith-Towner bill, though differing significantly in particulars. All funds were to be invested in the heart of education: in teacher training, teacher salaries, and the means and materials necessary for teaching. His program called for higher appropriations both at the federal and state level, and the amount was to increase sharply throughout the decade. State control of education was guaranteed in two ways. First, there was no department of education and no review of state spending. Second, if Congress or some other governmental agency later attempted to assert federal control, states could walk away from future federal money and still continue to fund education at what would have been the Smith-Towner level because they would already have been spending two dollars for every one contributed by the federal government.

There was also an important cultural difference in the two schemes. Claxton's proposal made no mention of the Americanization of immigrants and contained no requirement that English be the language of instruction. The latter was a remarkable concession, given his wartime belief that foreign-language schools were out of harmony with the national interest. While firmly committed to Americanization, Claxton held that patriotic love of country was not something that could be drilled into people. He also denied the existence of a prototypical Americanism rooted in people of Anglo-Saxon stock. Rather, Americanism was the sum of the best characteristics of all the peoples making up the nation's cosmopolitan population.[33] His attitude did not mean that he was a wholehearted supporter of parochial schools. As will be seen, he believed that such institutions should be brought under a measure of state control. Yet such thinking placed him beyond the pale of 100-percenters who sought to impose cultural conformity. Needless to say, his plan got little press other than the column or two given it by the Bureau of Education's official publication.

Although Catholics were not the only ones opposed to the Smith-Towner bill, some members of Congress had the impression that the case was otherwise. According to *America*, an unidentified representative wrote to a constituent to ask "why the opposition to the Smith bill seemed confined to members of the Catholic Church."[34] Not long after the joint hearing, Smith himself broached the matter in an address to the Senate. He had several inflammatory Catholic denunciations of the bill printed in the *Congressional Record*. One was a baccalaureate address recently delivered at Georgetown University by Eugene De L. McDonnell, S.J. It charged that the bill would place the entire educational system "under the control of one autocratic overseer" and deprive parents of their "sacrosanct right and duty" to educate their children. Claiming that the measure's ultimate purpose was the complete secularization of American life, McDonnell labeled it "the most dangerous and viciously audacious bill ever introduced . . . , having lurking within it a most damnable plot to drive Jesus Christ out of the land."[35] These were intemperate and unfortunate words.

Smith countered that a careful reading of the legislation would show that it deprived no parent of the right to send a child to a school of choice nor did it grant any federal overseer control of education. "The charge that this bill would banish God from the school is without the slightest foundation," he declared. "The bill can only be considered an assault upon religion by those who oppose public schools, . . . The charge is really an attack on public education." Denying that Catholic leaders opposed such schools, Smith hoped that they would stop the unwise antagonism of their coreligionists.[36] The *New York Times* printed the final paragraphs of his speech under the heading "Meets Catholic Criticism."[37]

As mentioned in the previous chapter, the Southern Jurisdiction of Scottish Rite Masonry had detected a Catholic plot against public schools. Smith, himself a Mason, held a tempered version of this view and wove it skillfully through his words. Citing the denunciations of extreme Catholics against the bill as evidence of an assault on public education, he denied that Church leaders agreed with them. Smith urged bishops to help their errant subjects see the error of their ways. He seems to have been offering the church the chance to back away from opposition to the bill under threat of invoking religious prejudice. In his mind, objection to the bill was clearly opposition to public schooling, something his legislative career had proven was dear to his heart. Despite his contention that the majority of Catholics and their leaders disagreed with extremists, the fact remains that many of them harbored the same thoughts.

Sending each bishop copies of the record of the joint hearing and Smith's speech, Cardinal Gibbons remarked that the church was the only large element opposed to the bill. "Unfortunately . . . much of the criticism by Catholics has been manifestly unwarranted," he remarked, "while the solid objections . . . have not been made known in a sufficiently forceful manner." He noted that Smith, Towner, and the Joint Committee maintained that the bill in no way infringed on parental rights or the freedom of parochial schools. They had agreed, moreover, to eliminate any such provision that could be pointed out. Although Smith and Towner believed that the bill's language amply protected states' rights, they were willing to go even further if a better formula could be suggested. The bishops were to study the bill, formulate amendments, and send them to Muldoon for presentation to Congress in case the measure seemed likely to pass.[38]

By this time, Pace had concluded that the hierarchy should drop the matter altogether and attempted to convince Muldoon of the wisdom of this course. He argued that the bill touched parochial education only in the requirement that English be the language of instruction in all schools. Moreover, two purposes of the bill—Americanization and teacher training—were laudable and were being pushed by the church for its own schools. Because the church had already admitted the right of the state to control public schools, the issue of federal control was "a political question which as citizens we may discuss and about which Catholics may differ among themselves." The same held true of federal aid. Pace suggested that political issues of this sort were not matters of concern to the hierarchy "as such." Given that Catholic opposition to the Smith-Towner bill had already been interpreted by some as evidence of the "anti-American spirit" of the church, it went without saying what the public reaction to the hierarchy's opposition to the bill might be. Pace reiterated his earlier contention that the more dangerous movement was the one for compulsory public schooling taking shape at the state level, a movement that the church might more easily block if there were a department of education because the hierarchy could then focus pressure on a single point rather than on multiple state legislatures. Noting the pressure that the NEA had been able to put on Congress, he suggested that the church might do well to establish an educational office in Washington, too.[39] The hierarchy already had plans to do so when it would convene again for the first time since 1884.

This historic meeting occurred in Washington in mid-September 1919 and proved an event of great moment for the American Catholic church. Ninety-four bishops gathered for the first of what would become annual meetings of the hierarchy. The chief item on the agenda was the replacement of the Committee on Catholic Interests with a multi-leveled organization to be known as the National Catholic Welfare Council (NCWC). The NCWC was to be the hierarchy in annual assembly acting as a deliberative body, implementing its decisions through an Administrative Committee of seven bishops who operated a secretariat with an Education Department, a Lay Activities Department, a Social Action Department, a Legal Department, and a Press Department. The idea met initial resistance from a vocal few, in part over a concern that the NCWC Department of Education might eclipse the CEA. In fact, the bishops interlocked the CEA and the NCWC Department of Education by making the Executive Board of the former the Executive Committee of the latter.[40]

In effect, the NCWC was a national Catholic lobby headquartered in Washington where it could assert influence on the government. Through its National Councils of Catholic Men and Women (NCCM and NCCW), it mobilized the laity for civic engagement in social and political questions. Although Bishop William Russell had argued that such an organization was necessary to offset the influence of the Federal Council of Churches, the NCWC soon discovered that its true rival was the NEA. There was irony in the establishment of the Welfare Council that would not be lost on some conservative Catholics (see chapter 5). While the NCWC fought the centralization of federal power in Washington, the organization itself centralized Catholic power there, much to the chagrin of those who believed that authority should remain dispersed among the diocesan bishops in the ecclesiastical equivalent of a states' rights church.

Once the hierarchy had established the NCWC, the floor was opened for debate on the Smith-Towner bill. Both Russell and Archbishop James Keane of Dubuque wanted the hierarchy to say nothing publicly against the measure. Friendly senators had urged the church not to attack it because the vast appropriation would ensure its failure without Catholic assistance. Convinced that the legislation would kill parochial schools, Bishop Louis Walsh urged the hierarchy to take some action, at the very least explaining the Catholic position. Because passage of alternative legislation might defuse the drive for a department of education, his preferred course was to opt for the lesser of two evils, namely, the Smith-Bankhead bill, which proposed to eradicate illiteracy among native-born people and to teach English, citizenship, and homemaking skills to the foreign-born, two programs in the Smith-Towner measure. The hierarchy agreed and appointed Archbishop Hanna, Archbishop George Mundelein of Chicago, and Bishop Thomas Hickey of Rochester, New York, to visit the House and Senate Education Committees to inform them that if objectionable features were removed from the Smith-Bankhead bill, the church would not oppose it. Still, Walsh wanted the hierarchy to express its mind specifically on the Smith-Towner measure. The bishops went on record that "the passage of such a Bill is inexpedient and unwise."[41]

After the convention, Hanna, Hickey, Mundelein, and Walsh visited Senator David Walsh to inform him that the hierarchy had decided to seek the defeat of the Smith-Towner bill by supporting the Smith-Bankhead measure. Walsh thought that the Senate committee would be interested in hearing the proposition and agreed to arrange for the bishops to appear before it at an opportune time.[42] That moment never came because the fortunes of the Smith-Towner bill had soured after the joint hearings, thus making a trade-off of support unnecessary. Nonetheless, the idea of supporting alternative measures became an important element in Catholic strategy for sidestepping the creation of a department of education.

By fall of 1919 the Smith-Towner bill had run into effective opposition in both houses on the conservative grounds of states' rights, anti-bureaucratization, and economy. As already mentioned, the two most vocal critics in the Senate were King and Thomas. In the lower chamber, two former speakers of the House objected to it. Democratic Representative Champ Clark of Missouri, then House minority leader, decried the trend toward federalization. "Everyone comes to Uncle Sam to get some of his money," said Clark; "He seems to be regarded as a general Santa Claus." Greater federalization meant higher taxation on an already burdened public. He vowed opposition. From the other side of the aisle he was joined by "Uncle Joe" Cannon, a redoubtable Republican Old Guarder from Illinois. "Under our Constitution local matters are controlled by the States, and national matters are controlled by the Nation," he declared; ". . . The States are caring for education." Cannon believed that a careful reading of the Smith-Towner bill would convince anyone that "the whole thing is to be controlled and managed from Washington."[43] Telling was the applause won by both men, indicating that their theme had fallen upon sympathetic ears.

Growing congressional opposition was sufficient to stall the legislation. Two senators whose identities were kept anonymous—one of them certainly Walsh—informed *America* that the bill would remain in the Committee on Education and Labor because Congress would not pass it. Smith himself had come to see that the country was unready to endorse the massive appropriations in the measure. Consequently, the committee deferred action on both the Smith-Towner and the Smith-Bankhead bills in favor of the Kenyon Americanization bill, which proposed to eradicate illiteracy among the native-born and to teach English to the foreign-born through a modest 200 hours of instruction per year. According to one senator, Smith believed that the Kenyon bill could serve as stop-gap to accomplish a few aims of his bill. After witnessing several years of that bill in operation, the public might be ready to endorse the ambitious Smith-Towner plan. In January 1920 the Senate passed the Kenyon bill, but the House took no action.[44]

Given the dimming prospects of the Smith-Towner bill, three Catholic Democratic senators, Joseph Ransdell of Louisiana and both Walshes—David of Massachusetts and Thomas of Montana—thought the time had come for Catholic propaganda against the bill to cease. Their Protestant colleagues advised that it made the fight look like a religious issue and would cause many who opposed the legislation to favor it. The three believed that the church ought to develop indirect methods of opposition. Ransdell suggested finding some way to get the Lutherans and Mormons involved, for both denominations conducted church schools.[45] As a matter of fact, Lutherans were becoming increasingly aware of the educational question; not yet, perhaps, of the federal issue of the Smith-Towner bill, but certainly the drive for compulsory public education.

While Congress dallied with Americanization, the Catholic church took vigorous steps in this regard, especially because so many recent immigrants were of that religion and of so much concern to old-stock Americans. The NCWC ran more than a dozen community houses, which, like settlement houses, sought to incorporate Catholic immigrants into American life and politics. The Social Action Department helped eastern-European American Catholics to launch the "Loyal Americans" federation that was aimed at Americanizing their newly arrived compatriots. By the end of the year, the department had published a *Civics Catechism,* which ran through its first printing in a week. Praised by the secular press, it was used in Americanization programs conducted by the Chicago Commons, Hull House, and various women's groups, both Catholic and non-Catholic.[46]

In addition to congressional opposition, the Smith-Towner bill ran into antagonism within the educational trust. In February 1920, at the meeting of the NEA Department of Superintendence, William P. Burris, dean of Teachers College at the University of Cincinnati, denounced the measure. Despite the earlier help he had given to the AFT in negotiating with the NEA for the inclusion of teachers' salaries in the measure, he had come to oppose the bill on multiple grounds. Burris had deep differences with most administrative progressives, whom he believed lacked a solid foundation in philosophy and were driven by public opinion and a desire for efficiency.[47] While favoring the establishment of a department to coordinate the offices of education scattered throughout the government, he opposed the appointment of a secretary because the position would politicize education. Burris believed that a department should be under an independent federal board of education. He considered it preposterous that the federal government would grant the liberal aid proposed in the Smith-Towner bill without controlling the expenditures in some way. In his view, the aid package was simply the fulcrum to be used in putting over a national program of education. Burris believed that such a program and the federal control necessary to effectuate it were unconstitutional and undesirable.[48] Though not a Catholic himself, he would soon aid the NCWC in its fight against the bill, especially by providing the organization with insider information otherwise unavailable to it.

If Burris opposed the bill because it would lead to the federal standardization of education, Charles Judd, dean of the Cleveland Conference, opposed it because it failed to do so. His position was the more significant because it revealed a rift in the conference between himself and Strayer. At the annual meeting of the Harvard Teachers Association, the two squared off in debate, with President Emeritus Charles W. Eliot of Harvard serving as moderator. Contending that centralization was the natural course of educational history, Judd considered the establishment of a national system of education perfectly normal. Strayer's commission had bungled the job, going from one extreme to the other. Although the original Smith bill would have standardized education, the standards would have been established arbitrarily by the secretary. The new Smith-Towner measure, revised to meet objections against federalization, abandoned standardization altogether. Judd accused the NEA of suppressing public debate on the bill in order to ensure its chance of speedy passage. The only way to secure beneficial educational legislation was through the widest possible discussion of the issues. In Judd's view the government should first create a department of education to study education and propose standards. The standards it proposed should then be submitted to the people for democratic debate and adoption, all of which would take place through the regular congressional process. The standards adopted by Congress would become the basis for grants of federal money. Such a plan avoided the arbitrariness of the original Smith bill and the retreat from standardization of its successor. Judd urged the proponents of the Smith-Towner bill to revise it accordingly, or those who believed in the necessity of such revisions would come forward with an independent bill.[49]

In defense of the new measure, Strayer argued that the issue was to give education the national recognition it deserved through the establishment of a department. The mobility of Americans made the massive appropriation for the equalization of educational opportunity a necessity. For Strayer, the point was national leadership in education: "If we believe in the cause in which we are engaged, and if we are satisfied that the welfare of the nation and of the world is dependent upon our action, we may well forget our differences."[50] Strayer avoided the issue of standardization, preferring instead to speak of leadership. Given that both Judd and Strayer were founding members of the Cleveland Conference, it is unlikely that their essential positions were far apart, namely, the standardization of education. They differed, however, over the means of attainment.

Like Butler, Hadley, and Hibben, Eliot represented the educational old guard. He found himself out of step with recent educational movements. While wartime experience had shown that education was a national concern, this had not convinced him of the necessity of a department. Education was a local matter. Traditionally, Congress had responded to new national problems by establishing bureaus to deal with them; there was already a Bureau of Education. Eliot proposed enlarging its funding and power. "That is the old-fashioned way of dealing with great problems," he concluded, "without creating new organizations which nobody has thought through yet."[51]

At the annual convention of the NEA in July 1920, Strayer made the final report of the Commission on the Emergency in Education in which he dismissed all criticisms of the Smith-Towner bill with the assertion that "none of these objections have, in the minds of those responsible for the administration of public education in the United States, been considered valid." Against Burris's idea of a National Board of Education rather than a secretary, Strayer responded that "if national leadership and a national sanction for the development of education" were to be provided, a department must be organized in the same way as others in the executive branch. Against the insistence of those who held that the Smith-Towner bill would lead to the federal control of education, Strayer noted that the "answer to this objection is found in the bill itself." Against Judd's contention that the bill did not go far enough in the standardization of education, he argued that the bill set the minimum requirements that all states could meet to qualify for the funds: a twenty-four-week school year, compulsory attendance at some school for that period, and mandatory use of English for instruction. Perhaps that was as much standardization as the NEA felt it could get at the time. It had sought more in the original Smith bill, but retreated to gain the support of teachers and politicians. To secure passage of the new bill, Strayer recommended, and the convention approved, the establishment of a Legislative Commission whose sole purpose was to promote the NEA's educational program.[52]

During the convention, members of the educational trust finally accomplished the top-down reorganization of the association begun two years earlier. From its foundation, the NEA had been a democratic organization in which every active member had the right to attend the annual meeting and vote. In 1918 the association had begun a massive membership drive to increase grassroots support for its bill and to fill its coffers with money to put the measure across. The administrative progressives who headed the NEA—"the gang," as Upton Sinclair unflatteringly called them—realized that as long as the annual convention was held in an eastern or midwestern city, the association was vulnerable to control by regional blocs of teachers who attended it. The leaders decided that the best way to avoid control by self-interest groups like teachers was to disperse representation across the widest possible area. So, the educational trust devised new bylaws that transformed the NEA from a democratic organization into a representative assembly. The new rules permitted state, regional, and local educational associations to send one representative for each 100 members or major fraction thereof up to 500. The officers of the NEA (numbering nearly fifty in 1918, including most members of the educational trust) and all state superintendents of education (many of them allied to the educational trust through "hidden hierarchies") became ex officio voting members of the annual convention. Thus, the new bylaws limited the number of teachers' delegates who might attend, while granting the educational trust a disproportionate voice in NEA affairs. When proponents brought the new bylaws forward at the 1919 convention in Milwaukee, a bloc of Illinois teachers scotched the plan by noting that changes required prior congressional approval. Not to be frustrated, the educational trust decided to hold the 1920 convention in Salt Lake City, where few midwestern or eastern teachers had the financial resources to go and where a bloc of local Mormon teachers pliable to church elders might be induced into voting for the new plan. In May 1920 Congress approved the proposed changes to the bylaws, and in July the Utah teachers adopted them.[53] Thus, the administrative progressives of the NEA were in a position to carry out a top-down reform of education that was seemingly democratic.

The political setbacks of the Smith-Towner measure proved a godsend to the Catholic hierarchy, which needed time to erect the machinery of the NCWC. An understanding of its formation and function is important because from this point, the NCWC managed the church's affairs regarding a federal department of education. In December 1919 the Administrative Committee chose John Burke as general secretary of the NCWC, a post that gave him oversight of the administrative arm of the organization in Washington.[54] His selection was ideal because he possessed both wisdom and common sense. Deeply loyal to the church, he believed that Catholicism held the answer to national problems and sought to impart that understanding to his fellow citizens. In this regard, he was an exponent of separate-but-equal Catholicism. Burke blended love of church with love of country, both of which he drew together in the NCWC.[55]

The task of organizing the NCWC's Department of Education fell to Archbishop Dowling, its chairman. Unlike departments of government, those of the NCWC were headed by chairmen, not secretaries, who were members of the Administrative Committee. Beneath the chairmen were executive committees that deliberated on matters of interest and recommended policies. The policies were implemented by an executive secretary through a Washington bureau run by a director. As general secretary, Burke supervised all the executive secretaries and directors in the name of Archbishop Hanna, chairman of the Administrative Committee.[56]

In February 1920 the Executive Committee of the NCWC's Department of Education established a bureau at the Washington headquarters to assemble information on proposed education legislation, both federal and state, to discover what provoked it, how it was promoted, and how to defeat it. The bureau was also to tutor the public on the nature and aims of Catholic education and to foster parochial schools by urging the establishment of state associations of Catholic education.[57] It should be noted that the lobbying function of the office was framed in purely negative terms, namely, the defeat educational legislation, the presumption being that such bills would be inimical to parochial schools. Despite this defensive posture, members of the bureau routinely assumed a proactive stance and regularly supported alternative legislation to bills of the Smith-Towner variety.

Dowling had difficulty finding an executive secretary to manage his department. In January 1920 Father James H. Ryan, professor at Saint Mary of the Woods College in Terre Haute, Indiana, had volunteered his services but withdrew the offer upon being named president of the school. When Dowling asked the Executive Committee for suggestions, the members recommended Howard, adding the stipulation that he must reside in Washington. The nomination impressed neither Howard nor Dowling, the former because he feared the NCWC Department of Education would eclipse the CEA, the latter because he felt Howard had "a crochet rather than a 'vision.'" Setting personal opinion aside, Dowling offered him the job but anticipated a refusal. As expected, Howard declined the post. With Howard out of the running, Dowling invited Pace to fill the position on a temporary basis.[58]

While the NCWC was developing, it had to wrestle with a problem that bedeviled the church throughout the 1920s, namely, the threat to parochial schools. As mentioned above, many Americans viewed public education as the means for instilling in hyphenates and immigrants conformity to Victorian values and Anglo-Saxon culture. A number believed there should be only public schools, and all children should be compelled to attend them so that they might drink deeply of 100 percent Americanism. In late 1919, 100-percenters in Nebraska made another bid to outlaw private education. Undeterred by the defeat of their bill earlier in the year, advocates carried the fight to the constitutional convention. There the cause was championed by Wilbur Bryant, who wanted compulsory public education written into the fundamental law of the state. In his view, the cry of parental rights was a camouflage because behind every parent was the ecclesiastic. Public schools were the melting pot of society and essential to national security. While the convention gave him a hearing, it refused to act on his proposal.[59] Michigan patriots made a similar move in the spring of 1920. They were concerned about foreign-language private schools and worried about Catholic schools fostering a divided allegiance between Rome and Washington. To counter these threats, James Hamilton, leader of the Public School Defense League, resurrected the 1918 initiative to amend the state constitution to compel all children to attend public schools for primary education. The state's Catholic bishops joined hands with Lutherans to defeat the measure.[60]

The drive for compulsory public schooling found national expression in two organizations: the Ku Klux Klan and the Southern Jurisdiction of Scottish Rite Masonry. Defunct since the 1870s, the Klan was resurrected in 1915 in Georgia by William Simmons to serve as a patriotic, benevolent association committed to 100 percent Americanism and the supremacy of the Caucasian race, which meant little more than supporting the system of segregation prevalent in the South. The organization struggled for members until two advertising agents convinced Simmons of the possibility of playing on the fear and prejudice of society, with the chance of making money in initiation fees. The Klan succeeded by pandering to racism, nativism, anti-Semitism, and anti-Catholicism. While various "isms" gave it a cohesive national ideology, recent studies have demonstrated that local issues were often its drawing card, offering a vehicle for redress and reform. Klansmen saw themselves as the guardians of law and order (especially Prohibition), family life, Protestantism, and Victorian values, issues that one historian has summed up under the rubric of "white, Protestant nationalism." Estimates of Klan strength have varied from 2 to 5 million, with probably half the Klansmen in cities of more than 50,000 people. Membership in the Klan appealed to a representative cross section of urban population, with initial recruitment occurring among middle-class professionals and white-collar workers, usually members of a masonic lodge, a prime hunting ground of Klan recruiters. As the decade wore on, recruits increasingly came from the laboring class. Insofar as the Klan had a political platform, it called for the restriction of immigration, Prohibition, compulsory public schooling, and a federal department of education to dispense federal aid—progressive issues all. Like the educational trust, the Klan was concerned about illiteracy, Americanization, and poorly trained, underpaid teachers. The Smith-Towner bill and compulsory public schooling were viewed as critical means of correcting these situations.[61]

While some features of the Klan's program were progressive, others were quite the opposite. Night riding, flogging, tarring and feathering to enforce a fraying Victorian moral code, and cross-burnings to instill terror soon turned progressives and non-progressives alike against the Klan.[62] Even the Southern Jurisdiction of the Scottish Rite, as will be seen, eventually attempted to distance itself from it, though the Rite's Klan wing remained strong. Blending recent and traditional scholarship, the emerging picture shows the Klan to be a curious brew of reform, moralism, hypocrisy, greed, and brutality. Of course these characteristics seldom coalesced in the individual Klansman. Rather, they seem to have appeared singly or in combination in various locales and at various levels of leadership. Suffice it to say, while some of the Klan's programs were progressive, many of its deeds were not.

The second organization behind the movement for compulsory public education was the Southern Scottish Rite. The drive for such schooling stood as the centerpiece of its program for achieving national cohesiveness. It was also expressive of the jurisdiction's traditional anti-Catholicism. "If we are to educate *all;* to bring *all* together in mutual bonds of sympathy and love; to bury sectarian and racial dislikes and differences; then the state must undertake the task of educating its citizens," pronounced the editor of the *New Age.* Religion had no place in the public school; it belonged in the home or Sunday school. The contention that ethics demanded a religious foundation—a standard Catholic argument for parochial education—was "a great error"; a person could be quite moral without any faith at all. "*The idea of the parochial school is all wrong,*" concluded the *New Age.* "It separates children into religious groups, fosters dislikes and enmity and destroys the very taproots of democracy."[63] The Rite stood for a progressive humanism that promoted interracial and inter-ethnic harmony. The means of achieving this harmonious cohesiveness—democracy—was the public school.

The jurisdiction went beyond editorialization. In May 1920 the Supreme Council declared itself in favor of "free and compulsory education . . . in public primary schools . . . in the English language only . . . as the only sure agency for the perpetuation and preservation of the free institutions guaranteed by the constitution of the United States. . . ." While pledging support for the creation of a department of education and the foundation of a national university, the council promised to oppose any who sought "to limit, curtail, hinder, or destroy the public school system of our land."[64] Expressed positively, the Rite viewed public schools as cradles of patriotism and democracy. Its opposition to parochial schools sprang not from religious prejudice, but from love of country. As the Rite saw it, parochial schools fostered religious prejudice and divisions within society. At heart, they were un-American. So much did the jurisdiction believe this that it advocated barring from political office anyone who had not attended public schools.[65]

By the summer of 1920 Catholics suspected that the campaign for compulsory public schooling and the drive for a department of education were linked. The Klan and the Southern Jurisdiction of the Scottish Rite endorsed both movements. Moreover, prominent figures had voiced opinions about Catholic education dangerously similar to those of 100-percenters. Claxton, for example, widely—and erroneously—believed by Catholics to be a supporter of the Smith-Towner bill, questioned the Americanism of parochial schools and believed the national interest demanded placing them under public control. In a debate among educators on the Smith-Towner bill, Professor Dallas Lore Sharp of Boston University argued that private schools "tended to perpetuate class distinction in American life . . . and interfered with the necessary development of public education."[66] Towner told a Des Moines audience "that private educational systems were a menace to our communities and should be suppressed."[67] Given expressions like these and the platforms of the Klan and Masons, Catholics found it increasingly difficult to believe that federalization would be as harmless to their schools as the Smith-Towner backers would have them think.

The first real test of the NCWC's effectiveness as a lobby came at the party conventions of 1920. Most political observers agreed that the Republicans had the upper hand. During the previous two years, the Wilsonian coalition of Democratic and Republican progressives completely disintegrated as the government withdrew from wartime support of collective bargaining for labor and price supports for agriculture. Moreover, the president and his administration had managed to alienate large segments of the population in and outside of the Democratic party through the European peace process, the handling of the Red Scare, and his petulance in the wake of a debilitating illness. Despite their bright prospects, Republicans were not without concern. The return of the progressives to the party renewed the tension between them and the anti-progressive Old Guard: senators like Penrose, James Wadsworth of New York, Harry New and James Watson of Indiana, Reed Smoot of Utah, and Medill McCormick of Illinois. Their aim was to secure the nomination of someone attuned to conservative interests. Furthermore, the question of America's participation in the League of Nations was an issue that threatened to sunder the party, just as the progressive revolt had done in 1912.[68]

In the autumn of 1919 Will Hays, chairman of the Republican National Committee, had established an advisory committee under Ogden Mills, a young New York congressman, to craft a platform that would be acceptable to both the progressive and conservative wings of the party. On 2 June 1920 the committee issued a report that tried to straddle the fence on the education question. Acknowledging public "irritation" over the wartime extension of federal powers and the diminishment of local autonomy, the committee went on to recommend federal aid for education. "It is intended," read the report, "that the Federal aid . . . should be without compulsion or control of . . . local educational administration, but conditioned upon the States which accept such aid, coming up to certain minimum standards in the use of Federal funds."[69] In essence, the committee was recommending Claxton's plan of aid without federal supervision, but conditioned upon meeting certain minimum standards.

Catholics viewed this as a serious step toward endorsement of the Smith-Towner bill. Shortly after the committee made its recommendation, an alarmed Dowling sent identical telegrams to Hanna and to Burke, warning that the NEA, certain labor groups, and women's organizations would bring pressure to bear on both parties "to frame an educational plank in line with the Smith-Towner bill with the implication of a constitutional amendment establishing some sort of positive federal control of education." He urged that the NCWC lobby both national conventions about the seriousness of handing education over to a political party, of abdicating states' rights, and of menacing parental authority in this matter.[70]

The NCWC mobilized quickly. Pace drafted a protest to be submitted to the platform committee of the Republican convention. In it, he warned of an attempt to insert a plank "advocating, in some form or degree, the federal control of education." While admitting the need of money for programs like those in the Smith-Towner bill, he asserted that the funds should come from local governments because a federal subsidy would eventually cause states to abdicate responsibility. In any case, Pace urged that the education plank include the following disclaimer: "That nothing therein proposed is to be construed as against the rights of private educational institutions. We believe these rights to be fundamental to American civic life."[71] The NCWC had Cardinal Gibbons, a respected figure in national political circles, who was to deliver an invocation at the convention, offer the protest as a personal letter to the platform committee.[72]

The real lobbying was done by NCWC agents at the convention in Chicago. The Welfare Council already had a man in place there, John A. Lapp, codirector of its Social Action Department, whose office was in the Windy City. In addition, Burke sent both Pace and Michael Slattery, a Philadelphia layman, who was executive secretary of the National Council of Catholic Men (NCCM). The three were later joined by Ryan, soon to succeed Pace as executive secretary of the NCWC Education Department. Fearful lest their efforts be discovered and exposed by their use of Western Union for communication with the secretariat in Washington, Burke and his colleagues assigned code names to each of the principal politicians.[73]

Slattery's lobbying began on the train where he met three national committeemen whom he convinced to support the Catholic position on the education. They promised to intercede with Hays to assert influence on Mills. Arriving two days before the convention, Slattery and Pace met with Lapp to lay strategy. Their plan was to bring pressure on two key individuals: Watson, temporary chairman of the platform committee and the Old Guard's choice for the permanent position; and Mills, Hays's candidate for that post. In a meeting with Watson, Lapp found him firmly opposed to legislation like the Smith-Towner bill and secured a promise that if elected permanent chairman, he would see that any such plank was killed. Lapp also saw California Senator Hiram Johnson, progressive candidate for the presidential nomination, who also pledged support.[74]

Meanwhile, unable to find Mills, Slattery visited John King, national committeeman from Connecticut, who held the proxy of party boss Penrose. Described by historian Francis Russell as a "gutter Nietzschean," Penrose was a Philadelphia blue-blood who chose the low life and power politics over high society. Although failing in health and forbidden by doctors to attend the convention, he remained abreast of developments through direct telephone and telegraph lines with the Pennsylvania delegation, led by Governor

William Sproul. Slattery told King that he did not like the outlook and wanted action. King told him to be in Sproul's room at six P.M. Arriving at the appointed hour, Slattery received another pledge, prompting him to state "straight from the shoulder" that he did not want any more promises; he wanted someone to lead the fight against federalized education. Sproul said that he would make the issue his own and offered to give his word personally to Cardinal Gibbons. Slattery assured him that this gesture was unnecessary. Sproul in turn urged him not to worry. Slattery confided to Burke that while having Sproul's assurance was fine, "yet I am too old in the game to stop at this."[75] So he was.

Slattery finally found Mills. Earlier in the day, Burke had attempted to prepare the way by having a letter of introduction and Gibbons's protest hand-delivered to Mills.[76] Yet Mills denied having received anything and "acted as if he was not with us," though he did say that he had been visited by numerous people sent from Hays to talk with him about the education question—the pressure promised by Slattery's traveling companions. Mills agreed to meet again with Slattery in the morning.[77] At 9:30 A.M. Slattery, Pace, and Lapp had a lengthy discussion with him. Mills assured them that while he opposed his committee's plank, his hands were tied. Slattery thought him "too young and inexperienced to realize the danger" and wired Burke to have the NCCM and NCCW telegraph their opposition to Mills.[78]

The rest of the day went more to the NCWC's liking. Its lobbyists contacted Old Guard Senators New, McCormick, and Smoot, all of whom pledged support. Ryan made a second contact with Watson, whom he "nailed . . . to his promise previously given." The picture brightened considerably when Watson defeated Mills for the permanent chairmanship of the platform committee by a vote of 41 to 3, thus showing "that the Old Guard is in the saddle." To draft the platform, Watson appointed a subcommittee of thirteen with himself in the chair. The subcommittee included Old Guarders Smoot and McCormick, anti-statist Senator William Borah, and California Catholic John Neylan. Leaving nothing to chance, Slattery had Congressman Joseph McLaughlin of Pennsylvania, a man with access to Penrose's private line in the Pennsylvania delegation, telephone the boss to say that Catholics would fight the Republican ticket at the polls if the party failed to protect parochial education. To make sure the message reached home, Slattery telegraphed a friend in Philadelphia to deliver it personally to Penrose.[79]

That night, Catholic prelates on the scene met to discuss the situation at the residence of the Archbishop of Chicago. Present were Mundelein, Dowling,

Moeller, Archbishop Patrick Hayes of New York, and Archbishop Giovanni Bonzano, apostolic delegate to the United States. All agreed that federal involvement in education was the first step toward Prussianizing the schools and that the movement for aid and central control was "an organized drive against private education and especially against Catholic Elementary Schools." All agreed that Dowling should see Watson in the morning.[80] These deliberations marked an about face for both Dowling and Moeller, who in varying degrees had favored federalization scarcely a year earlier when the hierarchy had been polled on the Smith-Towner bill. No doubt the mounting campaign for compulsory public education had chastened their views. Indeed, they had come to link the drive for such schooling with the movement for a federal department with liberal doses of aid to sponsor a national program for education.

On the opening day of the convention, the NCWC lobbyists kept up the pressure. Dowling saw Watson and came away greatly encouraged. Slattery and Joseph Scott, a prominent Catholic attorney from Los Angeles, argued the case to Neylan, their coreligionist on the platform subcommittee, as did local pastor Edward Kelly to his fellow Chicagoan McCormick. While the subcommittee met, the remainder of the platform committee held an open hearing. After conferring, Slattery, Pace, and Lapp decided against an appearance because they had made their case to party leaders, "and there was no use to let the public know we were here." That night, Sproul sent for Slattery to tell him that Penrose had ordered that nothing be done to offend Catholics. "I see you are taking no chances with this thing," was the governor's only comment.[81]

The next day the platform committee gave no offense to the church. Certainly it had greater concerns than Catholics and education to consider, especially the plank on the League of Nations. Yet the party went out of its way to be inoffensive to the church. The education plank retreated from the advisory committee's blanket sanction of federal aid and endorsed it only for agricultural and vocational education, the two areas in which the federal government was already involved. A plank that escaped Catholic attention at the time, but one that would soon greatly aid the cause, was a call for the reorganization of federal departments "with a view to securing consolidation, a more businesslike distribution of functions, [and] the elimination of duplication." Another plank that would prove a boon called for the reduction of public debt. With regard to the party's education plank, newspaper editor William Allen White, a member of the subcommittee, explained that the panel adopted it because of the protest submitted by Gibbons.[82] When the Catholic press carried White's story, Dowling feared that the news might exacerbate the religious dimension of the Smith-Towner fight. Knowing that the real work had been done by lobbyists on the scene, he confided to Pace, "You & I can smile at this but the story hurts us & does no good."[83]

The frontrunners for the presidential nomination were General Leonard Wood, a progressive conservative who inherited the mantle of Theodore Roosevelt; Senator Hiram Johnson, a California progressive; and Governor Frank Lowden, a moderate from Illinois. The convention was awash with business lobbyists and conservatives seeking a restoration of "normalcy." When voting for the three leading candidates deadlocked early, Senator Warren G. Harding, a party regular acceptable to the Old Guard, emerged as the choice of the convention.[84]

The fight against a Smith-Towner endorsement at the Democratic convention in San Francisco promised to be fiercer. Writing to John Wynne, S.J., founder of *America,* and to Bourke Cockran, a former congressman and a Catholic Democrat of national stature, Pace urged them to line up delegates against the Smith-Towner lobby. In mid-June he spoke with Senator David Walsh and contacted Garret McEnerney, a prominent West Coast attorney. A rumor—correct, it would turn out—convinced Pace that Senator Carter Glass of Virginia, an ardent prohibitionist unsympathetic to Catholics, would chair the platform committee, which meant that the church would have to apply pressure on members other than the chairman.[85] Once again, Burke sent Slattery to oversee Catholic interests, and again Slattery's lobbying began on the train, where Secretary of State Bainbridge Colby was a traveling companion. The two conversed often, and Colby promised to uphold the church's educational position.[86]

The Smith-Towner lobby appeared at the convention in force. According to one source, it brought "an innocent looking plank nicely camouflaged," which the National League of Women Voters presented to the platform committee.[87] The plank was camouflaged only in that it couched the call for a department of education and federal aid within the context of illiteracy:

> We recognize that the appalling percentage of illiteracy among both native and foreign-born in the United States is a blot upon our civilization; the lack of understanding of the essentials of good government, a menace to our future. We, therefore, advocate a federal department of education to be headed up by a member of the president's cabinet; federal aid where necessary for the removal of illiteracy and for the increase of teachers' salaries; instruction in the duties and ideals of citizenship for the youth of our land and the newcomer to our shores.

Having met defeat in Chicago, proponents of the Smith-Towner bill came to San Francisco to argue that illiteracy was a national, rather than a local, problem. They supported their case for federal participation in education by citing the government's wartime use of colleges for military training and its use of schools to build morale and patriotism.[88]

Slattery's first evening in San Francisco convinced him that his work would be difficult. He learned from agents of the League of Women Voters that they had canvassed most of the principal delegates, many of whom supported the plank. He called on Al Smith, Catholic governor of New York and favorite son of a delegation that was anti-Prohibition and anti-Wilson at a convention firmly in the hands of the administration. Though opposed to the proposed plank, Smith lamented that his delegation had as much chance of obtaining a favorable hearing "as it would have had at the Republican Convention." Finally, Slattery saw David Walsh, who informed him that there was tremendous pressure in favor of the plank. Walsh promised to do all in his power to defeat it.[89]

As discouraging as these discoveries were, they could not have prepared Slattery for those of the next day. He met Senator Thomas Walsh, a progressive Montana Catholic, who informed him that he favored federal aid to education, though he opposed the establishment of a department to control it. Arguing with him for some time, Slattery found the senator "anything but friendly." Walsh all but told him that "the federalization of education was coming and that Catholics . . . might as well face the situation."[90] The second blow came in Slattery's interview with Patrick Henry Callahan, a prominent Knight of Columbus and a southern progressive who marched to the beat of a different drummer, one that increasingly put him at odds with his coreligionists. Taking an unfriendly attitude, Callahan told Slattery that he did not believe all the "tommy-rot" published by *America* about the Smith-Towner bill and said that Slattery would have to "sell" him on opposition. Although it seemed a waste of time, the lobbyist drafted a list of arguments so that Callahan could never claim that he did not know the hierarchy's position.[91] Slattery discovered that many other "so-called Catholic leaders" differed from the NCWC. Some considered the organization " 'un-American' . . . 'behind the times' . . . 'selfish about our institutions.' " Others were apathetic, and still others thought that the NCWC was unduly alarmed.[92] Clearly, progressive Catholic Democrats did not see eye to eye with the hierarchy on the educational question. Slattery would have to look elsewhere for support.

He found help in several Catholic laymen: James Burke, brother of the NCWC's general secretary, and Edward Corcoran, both of whom had come to

San Francisco to argue the church's position to delegates, especially those on the platform committee. Slattery composed a flyer contending that a department of education with a cabinet secretary would politicize education, lead to its centralization, abridge states' rights, permit the federal government to determine how state tax dollars were spent, and handicap, if not eradicate, private education. He knew he was facing an uphill battle when someone as sympathetic to Catholics as Governor Albert Ritchie of Maryland admitted that he failed to see how the plank would hurt Catholic interests. After Slattery argued the case to him, however, Ritchie agreed that the church was justified in opposing it and promised his help.[93]

After winning Ritchie's support, Slattery managed to head off an endorsement that might have proven disastrous. He had learned that the AFL, which supported the Smith-Towner bill, had inserted the plank of the League of Women Voters in its own demands to the platform committee. Through the intervention of Peter Grady, a New York representative of the AFL, Slattery got the plank removed from labor's demands.[94]

Just prior to the opening of the convention, an event occurred that would be important in the coming fight, though its immediate significance escaped the NCWC lobbyists. Although handpicked by President Wilson for the chairmanship of the convention, Colby withdrew from the race for the post so that he could represent Wilson's views, especially regarding the League of Nations, on both the floor and the platform committee. The next day when the convention opened, Glass was elected permanent chairman of the platform committee. He then appointed a subcommittee of nine to draft the platform. The only Catholic on the subcommittee was Thomas Walsh, described derisively by Slattery as "a stick." Fortunately, Colby, who had promised to look after Catholic interests in education, was also on the subcommittee. Even so, Slattery considered the outlook for success doubtful and decided to bring the strength of numbers to bear. He wired the secretariat in Washington to have national, state, and local branches of the NCCM and NCCW telegraph him authorization to speak in their names. Within twenty-four hours, Slattery had 122 telegrams, which he put to good use.[95]

Slattery and his colleagues concentrated their attention not on Walsh but on Colby, William Pattengall of Maine, George Hodges of Kansas, and Vance McCormick of Pennsylvania, the last named a close political friend of Wilson, all of them members of the subcommittee. The lobbyists received help at a distance from Gertrude Gavin, president of the NCCW and daughter of railroad baron James J. Hill of the Great Northern. She supplied advice and applied pressure on committee members. Slattery then approached "the real bosses of the Convention" to state that for Catholics the education question "was one of principle." He showed them the telegrams authorizing him to speak in the names of more than a hundred Catholic organizations. He added a touch that was a brilliant political stroke. While the subcommittee was meeting, the remainder of the platform committee held public hearings. Two agents of the League of Women Voters appeared to argue for their plank. Upon checking with their home states, Slattery discovered that both were Republicans, information he put to good use in his conversations with Democratic delegates. In the end, the subcommittee rejected the plank supported by the League and the NEA and brought in one that harmonized with the views of Senator Thomas Walsh. "Cooperative federal assistance to the states," said the resolution, "is immediately required for the removal of illiteracy, for the increase of teachers' salaries, and instruction in citizenship for both native and foreign-born."[96] Clearly, the NCWC's lobbyists had won only a partial victory, but a partial victory was better than none.

With a member of the subcommittee threatening to bring in a minority report on education, even the partial victory hung in the balance for two more days as the platform committee remained locked in a struggle over both Prohibition and the League of Nations. During that time, proponents of the Smith-Towner bill brought pressure to bear. However, the courage of their champion on the subcommittee collapsed in the face of a strong representation of Catholic interests on the committee of the whole, especially in the person of David Walsh, the church's "16″ gun." The Smith-Towner proponent failed to bring in the minority report.[97]

Prior to the convention, the frontrunner for the nomination was William Gibbs McAdoo, Wilson's son-in-law. When the ailing president withheld his support in hope of winning a third term, McAdoo withdrew his candidacy. Still, in early balloting, his backers supported him in a three-way battle with Attorney General A. Mitchell Palmer and Ohio Governor James Cox. Palmer finally withdrew, and the convention turned to Cox, who was disassociated from the administration and backed by urban political bosses. The selection of a candidate distanced from Wilson did little to improve the party's chances because the Democrats were demoralized, disorganized, and indolent, virtually ensuring a Republican victory.[98]

Ironically, the Catholic church enjoyed less success in lobbying at the Democratic convention, wherein it possessed considerable political capital in the way of highly placed church members, than it did at the Republican convention, wherein it possessed virtually none. Clearly, the church's view on states' rights and parental rights in education harmonized well with the movement toward "normalcy," championed by the resurgent Old Guard. The Democratic party, on the other hand, possessed a more assertive group of progressives, some influential Catholics among them, like Thomas Walsh and Callahan, who differed from the hierarchy on the education question. It is interesting to note that at both conventions, Catholic lobbyists had to influence and depend on non-Catholics to protect the church's interest in education. In both instances, they found and won a receptive hearing.

That neither party endorsed the Smith-Towner bill greatly pleased Dowling, who was happy with the performance of the NCWC. The bill's defeat by both conventions, he reported to the hierarchy, "keeps out of the campaign what would have amounted to a religious issue."[99] His assessment, while correct, seems too sanguine. Although religion did not figure in the presidential race, the bill itself was fast becoming a religious matter, something that concerned the NCWC Department of Education. When its Executive Committee met in July, members were aware of the reputation that the church was gaining on the school issue. The committee recommended that the church narrowly circumscribe its opposition to the Smith-Towner bill by discriminating between what it could and could not accept in the control of education, and then limit its objection to the latter. In so doing it should employ arguments likely to gain the widest audience, while at the same time avoiding reasons that might fuel religious controversy.[100] This was sound advice, especially in view of the mounting antagonism toward the church.

The editor of the Scottish Rite's *New Age* voiced this antagonism, arguing that public schools were the proper place for children to learn to live in harmony with one another. "It is a pity," he declared, "that the Catholic citizens of the United States cannot see the truth of this proposition. The parochial school idea is undemocratic and makes for separation, 'wheels within wheels.' The principal attacks on the American public school system emanate from the Roman Catholic Church."[101] In the view of Scottish Rite Masonry, opposition to the Smith-Towner bill was opposition to public education. It was un-American. The NEA too vented against the church, if indirectly. It reaffirmed its conviction that federal aid alone could solve the problems facing education, urged immediate passage of the Smith-Towner bill, and condemned "the efforts of the enemies of the public schools to defeat the measure." In a thinly veiled reference to Catholics, the NEA identified the enemy as "a minority of people whose leaders are traditionally opposed to public education."[102]

The NEA resolutely continued its drive for a department. Though the association had failed to get either party to endorse its bill, perhaps the candidates might. So the NEA sent delegates to both presidential hopefuls, Democrat Cox and Republican Harding, to secure endorsements. Cox gave an innocuous statement that went no further than the party's platform. More encouraging, Harding espoused federal aid to education for each of the purposes outlined in the Smith-Towner bill. Instead of a department of education, however, he wanted one for public welfare to promote social justice and human benefit.[103] His break from the party's plank on aid is puzzling, the first of his reversals on the issue; the second would come a year after his election. Perhaps as on the League of Nations, he waffled in an attempt to appease all sides. A concerned Burke inquired of Hays if the party had authorized speakers to endorse the Smith-Towner bill, only to be informed that it had not.[104] Despite Harding's endorsement of aid, his commitment to a department of public welfare, combined with the party's commitment to governmental reorganization and debt reduction, meant that if he were elected, the NEA would have great difficulty in securing passage of its legislation. Putting the best face on things, the NEA hailed the statements of both candidates as a step forward for the measure and counseled supporters not to lose heart because legislation of such moment progressed but slowly.[105]

In mid-September 1920 the NEA decided to force its bill out of committee and insist that Congress vote on it, if possible, during the final session. In October, Fred Hunter, president of the NEA, made the measure's enactment the highest priority for the next several months. No efforts were to be spared. Within weeks, six national women's organizations lined up behind the bill, including the League of Women Voters, the Daughters of the American Revolution, the Congress of Mothers' and Parent-Teachers' Associations, and the Council of Jewish Women. All urged their memberships to write every congressman in support of the measure.[106] Moreover, the ACE had finally begun polling its member organizations on the measure. Incomplete returns (twenty-two of an eventual seventy) showed that the majority of those thus far responding favored the legislation.[107]

Faced with this drive, Catholics took countermeasures. Prior to the national election, Howard had a long conversation with Harding and found his views on the Smith-Towner bill quite acceptable. Watson, who had helped to defeat an endorsement of the measure at the Republican convention, was indebted to Ryan for supporting his bid for reelection to the Senate. "If we can hold some of the leaders like Harding and Watson on the right side," Pace told Dowling, "we may be able to carry our point without making a big fuss."[108] In further pursuit of this course, Ryan worked on Old Guard Senators Smoot and New.[109] To counter the NEA's work with women's organizations, the NCWC rallied the NCCW, which sent each local chapter and affiliated society a copy of O'Connell's keynote address to the 1919 CEA convention.[110] Boston College organized a corps of speakers to undertake a campaign tour against the legislation.[111]

The NCWC was not alone in lining up against the bill. Responding to the ACE poll in November 1920, the Senate of the University of Chicago came out publicly against the Smith-Towner measure, though not against federal participation in education. Following the thinking of Judd, who directed the university's School of Education, the Senate contended that the bill ambiguously (if not contradictorily) acknowledged the need for federal supervision of education, while upholding complete state control of the same, and it appropriated money without careful study on a scientific basis. Urging Congress to make a thorough investigation of the issue, the Senate called for "a plan of national participation in education" that would "equalize educational opportunities and improve standards in all sections of the country." A federal agency (a department of education) would be a desirable instrument for conducting scientific investigations of education and making recommendations to Congress, yet the latter itself should determine what aid should be granted and determine the standards to be attained.[112] Here was a bold endorsement of Judd's plan for national, standardized education under the aegis of Congress rather than a secretary.

Indirect opposition to the bill came from the Republican Old Guard. During the final session of Congress, McCormick made good on the party's plank calling for the reorganization of the executive branch with a bill to restructure it and create two new secretariats, a department of public works and a department of public welfare. President-elect Harding, who had won in a landslide, favored the measure, especially its latter department, which was to house the Bureau of Education without extending any of its functions.[113] Congress established a Joint Committee on Reorganization to study the issue and report by mid-December 1922.[114] The appointment of this committee was significant because it rendered unlikely any action on the Smith-Towner bill until that date. Confiding to Pace that he did "not like the eternal role of dog in the manger," Ryan advised that the NCWC could improve the image of its opposition to the NEA bill if it were to support the creation of some new department like one of those proposed in the reorganization measure.[115] As will be seen, he was not alone in such thinking.

By mid-December 1920 the NCWC had received so many inquiries about the Catholic position on the Smith-Towner bill that Burke and Pace decided to draft an official statement. The task would be difficult because the Welfare Council had recently endorsed the Sheppard-Towner Maternity bill, a piece of progressive legislation which, like the Smith-Towner measure, granted federal aid on a matching basis for a matter beyond Congress's constitutional authority. In other words, the NCWC had already conceded a principle at stake in the Smith-Towner bill because it felt compelled to aid indigent mothers, particularly in light of the church's stand on birth control and the family. Dowling and Ryan considered the endorsement a grave error. Father George Johnson, superintendent of parochial schools in the diocese of Toledo, complained that the NCWC was "cutting the ground from under our feet," rendering the church unable to argue with any consistency against the Smith-Towner bill. Agreeing that the endorsement meant that the church would find it difficult to oppose the measure on the basis of states' rights, Pace still believed it possible to object to the measure because of the nature and extent of federal participation that it contained.[116]

Over the span of a month he and Burke crafted a statement reflecting the advice given by the Executive Committee of the NCWC Education Department to offer arguments likely to win the widest hearing while giving the minimum of opposition. The statement praised the aims of the Smith-Towner bill, but took issue with the means of attainment. Having already conceded the issue of federal aid in the Sheppard-Towner endorsement, the NCWC quibbled about its application. "It is not . . . a question of approving or condemning such aid as a matter of principle," declared the pronouncement, "but rather of discriminating between cases in which it would be wise to extend Federal aid and other cases in which it would be unwise and contrary to the best interests of the nation." The Smith-Towner bill would increase taxation, the bulk of which would fall on prosperous states, with the result that they were to pay the educational costs of their weak counterparts. This was unjust. Federal grants to the schools were also the surest "way of destroying State autonomy . . . [and] responsibility in the matter of education." First, federal aid would reduce local spending for schools; second, it always came with strings attached, leading ultimately to federal control. Admitting that the bill's purposes were laudable, the NCWC made a positive recommendation, suggesting the establishment of a federal agency or commission—not a department—to investigate educational problems and disseminate the findings to the states, which were to remedy the situations on their own.[117]

The NCWC sent 5,500 copies of the statement to local chapters of the NCCW and 11,000 to affiliates of the NCCM. At the NCWC's signal, and using the statement as a guideline, all were to protest to Congress against the Smith-Towner bill. Slattery sent the statement to the nation's 21,000 parish priests with instructions to bring it to the attention of prominent persons, particularly friendly non-Catholics who were to be urged to protest the bill when the NCWC gave the word. In doing so, the clergy were to avoid any religious argument and any reference to parochial schools. Objections had to be limited to those mentioned in the statement, especially the financial ones.[118] Thus by mid-month, Catholics stood poised to launch a nationwide protest against the Smith-Towner bill.

The NEA's drive to have the measure reported proved successful. On 11 January 1921 the House Committee on Education decided, without an official vote, to bring it to the floor. Three members reserved the right to oppose it there, one of them Donovan, the only Catholic on the committee.[119] On 17 January, the bill came out of the Education Committee with a favorable recommendation, insisting that the measure safeguarded states' rights in the strongest possible way. Against the objection that the federal government had no constitutional power to grant funds for education, the committee rehearsed the history of national aid beginning with the Land Ordinance of 1785 down to the Smith-Hughes Act of 1917. Finally, the report expressed doubt that the legislation would significantly increase taxes because not every state would qualify or wish to qualify for the funds of the bill's five aid packages.[120] This last comment is difficult to believe. Congress would have to find the means to fund the huge appropriation in the bill whether or not all the states availed themselves of the benefits.

The report failed to convince the NCWC. Slattery issued a press release declaring that the bill still aimed at federal control. Similarly, *America* and the Boston *Pilot* argued that federal control followed federal money.[121] In a practical vein, Russell suggested to Burke a way to kill the measure without implicating the church. The NCWC might persuade a non-Catholic northern congressman to offer an amendment stipulating that in the distribution of funds, southern blacks must have equal rights with whites. To ensure this, southern states must have two supervisors of education, one for each race. "This will appeal to northerners," Russell told Edward Dyer, "and the southerners will not be so anxious for the bill."[122] Rather than engage in racial politics, Burke wisely let the idea drop.

On the twentieth of January, the Senate Committee on Education and Labor met in executive session to consider its course. Smith was eager for action because his recent failure in a bid for reelection necessitated his retirement from the Senate. Considering the bill the capstone of his legislative career, at least with respect to education, he wanted an opportunity to address the Senate on it before his departure. Kenyon, too, wanted to report the measure favorably, though he was less enthusiastic about it than Smith. According to David Walsh, a committee member, Republican Senators Thomas Sterling of South Dakota and Lawrence Phipps of Colorado, Democratic Senator Andreius Jones of New Mexico, and himself either opposed the bill outright or opposed bringing it to the floor at present. Consequently, the committee did nothing. Still, he cautioned Pace against slackening opposition because the proponents were extremely active and no one could tell when sentiment might shift in their favor.[123]

In early February 1921 the NCWC's General National Committee met without Burke or Pace. Consisting of departmental directors, the committee coordinated the activities and pronouncements of the several departments to ensure that the various agencies were not working at cross purposes. Arthur Monahan, director of the bureau of education, announced that the ACE had finally completed its poll. Although member institutions favored the establishment of a department of education, the vote was evenly split on federal aid, therefore the ACE would not endorse the Smith-Towner bill. Monahan proposed a plan for joining forces with the ACE. He noted that its director, Dr. Samuel P. Capen, favored the creation of a department of education and science without federal aid. If the NCWC were to support such a bill as a substitute for the Smith-Towner measure, both the NCWC and the ACE could block the latter while doing something constructive for education. The committee approved the idea. When Burke read the minutes, however, he ordered the committee to reverse its decision.[124] Only the hierarchy had authority to establish NCWC policy. This incident is important because it shows that agents of the NCWC were not obstructionists. Like Ryan, they believed the NCWC could improve the image of its opposition to the NEA bill by favoring a suitable alternative.

Several times in the final week of January 1921 Smith agitated the issue on the Senate floor. On one occasion, he noted there was "a small organized opposition" that did "not represent any real majority of those from whom it comes."[125] The opponents were later identified when he had an editorial from the Scottish Rite Bulletin printed in the *Congressional Record.* The Masonic editor averred that the only newspapers opposing the bill were those that "receive their spiritual and mental (?) pabulum from a foreign country." It was beyond comprehension "that any real American can object to such a measure." The editor hoped for a purge that would drive from public office anyone "whose Americanism can be called in question, or whose allegiance . . . permits of division with any other overlord, actual or pretended, spiritual or secular."[126] Obviously, Catholics were the culprits, and not even all of them. In Smith's view, the sole opponent of his bill was a small, organized portion of the Catholic population.

Smith's tactic was described by Pace as a "policy of gradual injection." "His plan, evidently," commented the priest, "is to impress the Senate with the idea that the whole country is clamoring for the passage of the Bill."[127] The method disturbed Burke enough that he ordered the clergy and laity to protest, using the reasons outlined in the NCWC's statement.[128] Within days, Smith noted the influx of telegrams to Congress. Now claiming that Catholics misunderstood the bill, he received permission to address the Senate to show the irrelevance of their criticisms. Burke had William Cochran secure Senator King to respond for the church.[129]

Smith argued that Catholic opposition had derived from McDonnell's baccalaureate address, which claimed that the bill would subject all education to an autocrat in Washington, negate parental rights, and drive religion from the schools. Section by section, Smith explained the bill's provisions, showed its respect for local control, and upheld its revenue aspects. Nothing in it abridged a parent's right to send his child to the school of his choice. King responded that Catholics opposed the bill for a different reason: millions of them "believe in local self-government, and the splendid individualism which Christianity inspires." He pictured the church as a bastion of states' rights. Democratic Senator Fernifold Simmons, head of North Carolina's political machine, questioned how King knew that Catholics opposed the bill on that ground. The church, after all, was not objecting to similar legislation that had to do with healthcare, like the Sheppard-Towner Maternity bill. Simmons believed that Catholics opposed the measure because they would be taxed without deriving benefit for their schools. King held fast that the church objected on states' rights.[130]

Here were Democrats arguing with Democrats, one a progressive, one a states' righter, and one a conservative machine politician. In fact, all three were correct. King gave the official party line, namely, that the church opposed the Smith-Towner bill because of the federalization involved, not a specious argument, for the NCWC increasingly objected to centralization in general.[131] Yet this was but the public reason. Simmons was correct in assuming that Catholics took exception to the measure because they would not benefit from it. Actually, some, like Bishop Walsh, feared that parochial schools would be unable to compete with a federally funded public system. Smith had his finger on the unspoken truth. The church opposed the bill because it believed that eventually a department of education would be used to wipe out Catholic schools. To judge by practical effects, King had won the day. By 16 February the Smith-Towner bill was dead, or so it seemed, and the NCWC called a halt to further protest.[132]

On the day of the Smith-King exchange, the Knights of Columbus, who had been vying with the NCWC for the position of spokesman of the Catholic church, unwittingly undid the Welfare Council's work of trying to keep religion out of the issue. Organizing its own national protest against the bill, the Supreme Council of the Knights passed a resolution objecting that the legislation was subversive of states' rights. Local councils were encouraged to endorse the resolution and send it to their congressmen and senators.[133] If the Supreme Council had stopped there, all might have been well. In an unfortunate interview with the *New York Times,* Supreme Knight Flaherty asserted that the bill attempted to place "all private and public education at the mercy of a Federal bureaucracy." Although the Knights opposed it "not so much on religious as on patriotic grounds," he added that like the Michigan amendment, "the Smith-Towner bill is ultimately aimed at the parochial schools."[134]

Flaherty's pronouncement was "just what the friends of the Smith-Towner Bill wanted," reported Burke. They circulated his words through Congress. When telegrams from knights around the country poured in, "the matter assumed very quickly the aspect of a religious issue." Senator Ransdell went to Burke and, in the name of several Catholic colleagues, begged him to do something.[135] There was little he could do, and the matter soon worsened.

Alert to the newsworthiness of controversy, the *New York Times* solicited opposing statements from Strayer and William McGinley, Supreme Secretary of the Knights. Strayer's was simply a condensed version of arguments contained in the article that the paper had printed back in 1918, as well as arguments he had since offered before both the NEA and congressional education committees. McGinley's proclaimed that "the Knights of Columbus are opposed to the Smith-Towner education bill, and supporting the Knights of Columbus in this opposition is the entire ecclesiastical and lay organization of the Catholic Church in the United States." In other words, the hierarchy and the NCWC were behind the Knights, not the other way around. McGinley argued that the bill would interfere with parental rights in education; it would also create a despotic authority in Washington. The Bureau of Education, he said, was none too friendly to the church, and if the commissioner were to become a cabinet secretary, nothing would prevent him "from declaring against the Catholic parochial school system."[136] McGinley's words confirmed the suspicions of Smith, and many in Congress now saw the matter his way.[137] Smith privately polled the Committee on Education and Labor and won the endorsement of a number sufficient to report the measure favorably. Fortunately for Catholics there were only four days left in the session, a time insufficient for the enactment of the legislation.[138]

Legislation that did make it to the president's desk in the final days of Congress was the Dillingham bill. Although politicians could find no agreement about plans for Americanization, whether embodied in the Smith-Towner, Smith-Bankhead, or Kenyon measures, they did agree on the need to restrict immigration. While the House wanted to suspend it altogether, the Senate sought a quota system that would limit the annual number of immigrants from each nation to 3 percent of its total in the American population according to the 1910 census. When the Senate's plan carried, the House abandoned the drive for suspension and supported the quota system. The idea was to enact a stopgap measure until a permanent plan of immigration restriction could be enacted. Congress sent the legislation to Wilson for signature. Never an advocate of restriction, he pocket-vetoed the measure.[139] Public sentiment for stemming the influx of eastern- and southern-European Catholics and Jews was too strong to be stayed for long.

Similarly, advocates of a national program for education would not be undone by their latest setback. Although the NEA had reworded its measure to give the appearance that it preserved local control of schooling, and by extension the right of private education, many educators supported the bill precisely because they believed that it would lead to a national school system. Nor were the bill's sponsors friendly to the church. Smith repeatedly suggested that Catholics objected to the measure because they opposed public education, while Towner thought private and parochial schools ought to be abolished. He was not alone. Klansmen and Southern Scottish Rite Masons assumed the lead in a national movement for cultural conformity. Massive immigration and modernity threatened to undermine Anglo-Saxon hegemony and Victorian mores. While local communities sought to make public schools the agencies for reinforcing traditional values, Masons and Klansmen wanted to compel all children to attend them, thus eradicating private and parochial schools, which were considered inimical to national homogeneity. Understandably, the Catholic church lobbied to protect its educational interest, linked as that was with the preservation of the faith. By 1921, then, the federal education question had become a religious issue, one in which Catholic objection to the NEA bill was viewed by Protestants as opposition to public schools, while Protestant promotion of those schools was seen by Catholics as a veiled attack on parochial ones.

3

“Those Who Do Not Answer to the Lash”

Despite setbacks at the national party conventions in 1920, the NEA had rebounded to force its bill out of the House Committee on Education during the final session of Congress. Although its Senate counterpart seemed content to sit on the measure, the impolitic statements given out by the Knights of Columbus in mid-February 1921 had played into the hands of the advocates of a department. A number of legislators felt that the Catholic church had duped them into believing that it opposed the Smith-Towner bill for secular reasons when in reality its opposition was religiously motivated, a belief that rapidly spread beyond the halls of Congress. As will be seen in the present chapter, various religious groups increasingly objected to Catholic obstructionism on the education issue and openly identified public schools as Protestant ones. With the issue clearly drawn between Catholic parochial education and Protestant public education, the question was a religious one. More to the point was a battle taking shape over which schools embodied true American values. Catholics contended that parental rights and the local control of schools—both elements of parochial education—harmonized best with the American educational tradition. Administrative progressives, evangelicals, Masons, and Klansmen argued, on the other hand, that public schools were the bulwarks of democracy and the social order.

While the NEA sought to turn this charged atmosphere to its advantage by capitalizing on the issue of religion, it faced two formidable obstacles to its legislative program: an increasing conservatism and a division within the educational trust itself. The postwar conservative backlash, evident in the previous chapter, took institutional form in the Republicans' capture of the White House with a party regular in residence. Fiscal restraint and reorganization became the watchwords of the day, neither of which boded well for the NEA's bill. The association was not, however, without political allies. In the face of an administration that seemed to favor urban industrialism and commercialism, Republican and Democratic progressives from agrarian states would form the farm bloc to promote agricultural interests. The emergence of the bloc is significant, not only because its members tended to support the NEA bill—most of them had favored it before the bloc came to be—but also because it pointed to the fact that the education issue was more or less regionally drawn, something that would become more apparent as the decade drew on. The areas that stood to gain the most from the bill were the rural South, Midwest, and West, at the expense of the commercial and industrial East and Midwest. This is not to say that the bloc's concern was merely self-interest. That would be to deny the progressive aspect of the bloc. Members believed that their agenda promoted the national well-being by seeking to bring the farm population to the same economic and educational levels as the rest of the country. They considered their constituents the primal American community: old-stock, Protestant agrarians, now disadvantaged and disvalued. They were struggling for their rightful place in American society: equality with prosperous urban capitalists; for the nation to be strong, every part must be strong. There was an increasing realization that the country was more than the sum of its parts and a growing tendency to view Americans as citizens of the nation first and of a state second. If the bloc invested this nationalism with a preference for old-stock rural folk, so be it.

More serious for the prospects of the NEA's legislative program was the deepening division within the educational trust itself. As seen in the previous chapter, the educational old guard had lined up against the idea of a department with federal aid. The only members of the new generation who had openly challenged the NEA were Charles Judd and William Burris. They were not, however, alone in their opposition; others had yet to speak out. Though most members of the educational trust desired some form of federal participation in education, their views of the type of intervention that was appropriate were widely diverse. This diversity would come to light in 1921 and 1922, shattering the illusion that the educational community was more or less of one mind about the establishment of a department of education and federal aid for a national program of schooling.

Even as the fiasco of the Knights of Columbus's crusade against a department of education unfolded in New York and Washington in the winter of 1921, the NEA Department of Superintendence met in Atlantic City to revise its legislation. The aim was to overcome conservative arguments against it, especially those of Catholics, by ensuring in the strongest possible terms the rights of states over education. In the spring of 1921, Harding had called a special session of Congress to reenact the agricultural tariff vetoed by his predecessor. On the opening day of the special session of the Sixty-seventh Congress, Horace Mann Towner introduced the new bill, cosponsored in the Senate by fellow Republican Thomas Sterling of South Dakota. Given that both houses were overwhelmingly Republican, the NEA no longer considered bipartisan sponsorship of value. Similar in the main to previous legislation, the new measure gave the governor of each state, rather than the secretary of education, the right to approve the plans for implementing its purposes. If the secretary later determined that a state had failed to comply with those purposes, the matter had to be referred to Congress; the secretary himself could not withhold funds.[1] Thus, implementation was seemingly left entirely at the local level—even the approval of plans—and only Congress had the authority to withhold further aid. As redrawn, the bill appeared to be a ringing endorsement of state control.

To win the ACE's support, the bill called for a national council on education because its recent referendum had revealed a desire for such an agency. The council was to comprise the principal educational officers of each state "to consult and advise with the Secretary of Education on subjects relating to the promotion and development of education in the United States." The secretary of education was free to appoint up to twenty-five non-educators to represent the interests of the public at large. Although members were to be reimbursed for expenses, they were to serve without pay and meet once a year.[2]

Almost immediately the NEA was served notice that, despite revisions, its program was in political trouble. The day after Towner introduced its bill, President Warren Harding went before Congress to outline his program for "normalcy." Opposed to both the New Freedom and the New Nationalism, he was pro-business. In practical terms, this meant that the government should be economical in its functions, it should refrain from competing with private enterprise, and it should ensure that regulatory commissions gave business the widest possible freedom of action. The nation faced no more important issue, said the president, than reducing the wartime tax burden, with a consequent reduction in government spending. "There are two agencies to be employed in correction," he declared: "One is rigid resistence in appropriation and the other is the utmost economy in administration. Let us have both." In fact, Harding's was a four-pronged approach to fiscal restraint: tax reduction, a hard line on new subsidies, reduction in force in the executive branch, and a regimented budget. Such belt-tightening augured ill for the Sterling-Towner bill with its $100 million aid package. If Harding's commitment to fiscal restraint were not damaging enough to the hopes of the NEA, he went on to call for the establishment of a department of public welfare to coordinate the federal government's various endeavors in education, health, sanitation, child welfare, and public recreation, the aim being greater efficiency rather than the expansion of bureaucracy.[3] This speech marked the first step in Harding's retreat from his pre-election commitment to federal aid for education. His program for restoring prosperity and the NEA's for education were at odds, the former calling for a reduction in federal spending, a decrease in bureaucracy, and the placement of education in a multidimensional department; the latter calling for an increase in federal spending and the expansion of bureaucracy through the establishment of a new, separate department to house education alone.

When the Sterling-Towner bill was unveiled in April, Edward Pace, who had opposed opposition to the NEA program since 1919, confided to Archbishop Austin Dowling, "I hope we will not be obliged to start a new fight; there is nothing to be gained by it."[4] Catholic editors did not share his view. Despite revisions to the NEA bill, they remained unconvinced that it safeguarded states' rights in education. Sounding the keynote, *America*'s Paul Blakely declared: "As far as words go, the bill does not establish federal control. As far as facts are concerned, the bill does establish federal control." It was "immaterial" that plans for implementation now needed a governor's approval. In Blakely's view, the secretary remained free to accept or reject them; the only difference was that he now could take no independent action but had to refer the matter to Congress. Moreover, Congress had the power to alter or multiply the provisions with which a state must comply to qualify for funds, and the present measure now gave it final control over disbursement.[5] The only way to ensure state control of education would be to stipulate "that in case of controversy between the Federal and State authorities, the States were designated as the court of final arbitration." As the bill stood, Congress was the final arbiter.[6] Other editors agreed (as would non-Catholic educators). The *Michigan Catholic* held that, revisions notwithstanding, the bill indirectly infringed on states' rights, paved the way for socialism, and penalized parochial schools.[7] Admitting that "some objectionable features" had been removed the *Denver Catholic Register* considered the measure "still fundamentally obnoxious."[8]

The essential difference between proponents and opponents of the bill, Catholic or otherwise as will be seen, was how to guarantee state control of education. Proponents believed that the wording of the new measure sufficiently safeguarded states' rights over the matter. Opponents, on the other hand, believed that the only way to assure those rights would be to grant, gratis, federal aid that was to be used for education as states saw fit, with no strings attached. Otherwise, if a dispute were to arise over a state's use of the funds, Congress would have the final say, and the federal government would have ultimate jurisdiction over education.

Catholic opposition to the NEA bill provoked a belligerent response. In a speech allegedly "teeming with rancor," former Senator Hoke Smith told a Baltimore audience that the Knights of Columbus were to blame for his bill's demise; he even accused them of bigotry. Protesting that the measure contained nothing harmful to Catholics, he declared hostility to any denomination that drove children into its own schools.[9] The Board of Temperance and Public Morals of the Methodist Episcopal Church held the Catholic hierarchy responsible for all opposition to the legislation.[10] Catholic rejection of the measure, commented the *Christian Science Monitor,* flowed naturally from the church's protective stance toward its own schools. To the extent that it favored parochial education, it opposed increased expenditures for the public system.[11] "It would be too bad if the interference of a bigoted priesthood would prevent the adoption of a bill which contains so many promising features," lamented the *Reformed Church Messenger.* "The all too familiar protest against 'Federal Autocracy' and the violation of 'home rule principles' is, as usual, employed to hide the 'colored gentleman in the wood-pile.' . . . It is the camouflage of vested interests and will not bear the light of day."[12] Clearly, the education question had become a religious issue. It would be unfair to blame this situation entirely on the Knights of Columbus and their injudicious statements. While the Knights certainly undid the NCWC's work of framing opposition in terms of financial concerns and state control of education, the Smith-King debate and the events that led up to it indicate that remarks by the Knights only hastened the inevitable.

In May 1921 Republican Senator William Kenyon of Iowa organized the farm bloc in the Senate. Dominated by western Republicans, the bloc collaborated with southern and western Democratic senators. Republican Lester Dickinson of Iowa led the bipartisan House wing, which tended to be less well organized and consequently less effective. An advocate of the New

Nationalism, the bloc sought government intervention in the economy to equalize agrarian opportunity with that of urban business. To alleviate a sharp postwar depression from which agriculture never fully recovered during the 1920s, the bloc not only reenacted the emergency agricultural tariff, but pushed through legislation for easier credit, readjustment of freight rates, federal regulation of stockyards and meat packers, federal regulation of future trading in grain, cooperative marketing exempt from antitrust legislation, and agricultural representation on the Federal Reserve Board. Though the Old Guard opposed these measures it was unable to prevent them, thus revealing that the farm bloc held the balance of power in Congress. It used its power to obstruct implementation of the policies of the conservative, urban, eastern wing of the party. While some historians have viewed the farm bloc as simply a self-interest group, others—and members of the bloc itself—have recognized that many of these reforms were long-deferred, dating back to the Granger and Populist movements. Resisting domination by Old Guard Republicans and the administration, the bloc was insurgent.[13]

An issue that enjoyed farm bloc support was immigration restriction. There was considerable public unrest, especially in rural areas, over the massive influx of eastern and southern European Catholics and Jews, people judged to be racially inferior and unassimilable. Although Woodrow Wilson had pocket-vetoed the Dillingham bill in the last hours of his administration, the new Congress quickly repassed the measure by overwhelming majorities. Harding then signed into law what became known to history as the Emergency Quota Act, a stopgap measure aimed at curbing immigration until satisfactory permanent legislation could be framed.[14]

Though the farm bloc also supported the Sterling-Towner bill, there was no immediate need for action by the NCWC because the president had followed up his words with action. When Dr. Simeon Fess of Ohio, a former university president and Republican chairman of the House Committee on Education, went to seek Harding's endorsement of the bill, the president rebuffed him because he wanted a department of public welfare in which education would find a home. Harding asked Commissioner of Education Philander Claxton to prepare a bill to create such a department, with the understanding that Claxton himself would become secretary of the new agency. Though Claxton did as ordered, he informed Harding that he opposed the idea and would refuse the post. After he prepared the bill, Harding sacked him. The president sent his longtime friend and personal physician, Charles "Doc" Sawyer, to the Senate Committee on Education and Labor with Claxton's bill and a message to pass it immediately. The measure simply transferred the Bureau of Education from the Interior to the proposed department of public welfare without expanding the bureau's function.[15]

Harding's bill was an undisguised blessing for the Catholic church, and both Pace and John Burke hoped that its passage would derail the Sterling-Towner measure. Still, the church needed a plan regarding the latter, and it must take into account the altered religious situation. "Catholics are now under fire," Pace wrote to Cardinal Dennis Dougherty of Philadelphia, chairman of the NCWC's Legal Department. "They are criticized as opponents of any movement linking to the improvement of public education. . . . [W]e do not want this false impression regarding our attitude to persist, and . . . we cannot afford to repeat the blunders which came so near disaster," an obvious reference to the fiasco with the Knights of Columbus. In Pace's view, the church had three choices. First, it could launch a campaign against the Sterling-Towner bill, an unpromising line because the measure had been carefully redrawn to meet Catholic objections. The proponents would argue that the church was fighting federal aid simply to prevent help from being given to public schools. Second, the church could let the bill pass and depend on Catholics in each state to stay the acceptance of federal aid. This too was dubious. Third, the church could bargain to trade its support for a clause prohibiting any state that accepted federal aid from passing a law or adopting a constitutional amendment that interfered with the freedom of private education. The third was Pace's preferred course.[16]

On 23 April 1921 he and Burke met with Dougherty to discuss the matter. Like them Dougherty favored Harding's bill and hoped its enactment would doom the Sterling-Towner measure. Until the fate of the president's bill became clear, he wanted the NCWC to do nothing, for unwarranted opposition to the NEA legislation would further antagonize proponents and be taken as more evidence of Catholic hostility to public education. Dougherty agreed to direct Supreme Knight James Flaherty to halt the Knights of Columbus's opposition; Burke was to do the same with the NCCM and NCCW. All three agreed that Pace's third alternative was the best course of action with regard to the Sterling-Towner bill. If the measure seemed likely to pass, Burke was to exchange Catholic opposition for a guarantee that federal aid would not interfere with parochial schools, a guarantee Dougherty considered unlikely to be given but worth a try.[17]

Thus, the NCWC would bide its time in the hope that Harding's bill would derail the NEA's drive and promote acceptance of Pace's clause. The backup plan to trade support for the Sterling-Towner bill in return for a guarantee of protection of parochial schools was clearly a tactical maneuver to avert religious persecution. For Dougherty, who by temperament was conservative, that was probably all it was. For Burke and Pace, it was that and probably more. Burke had endorsed the Sheppard-Towner Maternity Act, a progressive measure. While he believed that the aid feature of the Sterling-Towner bill would prove detrimental to all education by leading to federal control, he was unopposed to the idea of a department or some other expanded federal agency for education. For his part, Pace had given the bill qualified support since 1919. He had worked against the measure only in the service of the policy established by the hierarchy, not out of personal preference.

After the conference, Pace explained the situation to John Wynne, S.J., of *America*, who promised to advise Blakely not to attack the Sterling-Towner bill until it became necessary. Burke had Archbishop Austin Dowling write to the NCCW and NCCM to cease opposition to the legislation because Harding had introduced an alternative bill. Until it became clear which measure would gain ascendency, further protest "would only do harm to the Catholic interests."[18] Archbishop Edward Hanna conferred with the Knights of Columbus at their annual convention and secured their apparent cooperation with the NCWC. In fact the Knights abandoned the fight against the NEA bill altogether.[19]

On 5 May Kenyon, chairman of the Committee on Education and Labor, introduced Harding's department of public welfare bill. The NCWC received mixed reports about reaction to it. William Cochran, director of the Legal Department, recounted that sentiment in both houses ran against the measure because of a desire to keep down spending. Both Arthur Monahan and Walter Barron, Burke's lieutenant, had lunched with Towner, who considered his bill dead as a result of the agitation in some quarters for the president's measure.[20] If Towner seemed reconciled to the death of his bill, the NEA was not. Hugh Magill told the ACE that the NEA intended to fight for a separate department of education despite the fact that Harding opposed one. Without objecting to the public welfare bill as such, the NEA intended to seek the removal of education from it.[21]

When a joint hearing was held on the president's legislation, Magill asked the committee to permit educators the opportunity to express their dissatisfaction with the inclusion of education in the proposed department and argue the case for proceeding with plans for a separate department of education. Magill posed the question in black-and-white terms. "In view of the fact that the friends of public education in the United States are not satisfied with this division as provided in this bill, while the enemies of public education are satisfied with it and have so expressed themselves," he told the committee,

"I think it makes the issue pretty clearly drawn." Republican Senator Lawrence Phipps of Colorado challenged him to name the enemies. Backpedaling to say that not every opponent of the NEA bill also opposed public schools, Magill explained that he meant only to say that some who had fought the Smith-Towner bill now supported Harding's measure. The committee gave the NEA two hours on 18 May.[22]

Magill appeared before the Joint Committee with a cohort of supporters, including James W. Crabtree, secretary of the NEA; William Bagley of Columbia University, who edited the *Journal of the NEA;* John McCracken, chairman of the NEA Legislative Commission; Walter Athearn, spokesman of both the Sunday School Council of Evangelical Denominations and the International Sunday School Association; and Dr. Robert Kelly, secretary of the Council of Church Boards of Education. Though still opposed to federal aid, McCracken now considered acceptance of that feature a small price to pay for the creation of a department of education.[23] The NEA representatives argued that if the nation desired the best type of citizenship, Congress must give education a place of honor, not submerge it in another department. The Bureau of Education, whether located in the Interior or in public welfare, could never provide the leadership necessary. The importance of education cried out for the kind of recognition only a separate department could give.[24]

The religious issue was not absent. Athearn stated that approximately one hundred Protestant groups with a combined membership of 20 to 23 million endorsed the NEA bill. "The very life of the Protestant Christian churches is involved here," he told the committee. "Whatever touches the public schools touches the very heart of the Protestant church," for its adherents had no other system of education to which they might send their children. The creation of a separate department of education was necessary, said Athearn, because "the churches I represent feel that they do not have any agency or other means whereby they can get the attention due their people. They must stand for the promotion of the public school."[25] Kelly made the same point. "We have no system of parochial schools. We depend upon the American public school system for the education of our boys and girls. We believe in the American public school."[26] Plainly, for these men the public schools were Protestant schools, not in religion, but in clientele. Their remarks starkly indicate how religiously drawn the issue had become.

Athearn followed up his testimony with action. In an urgent circular, he informed his constituents that Harding's bill for a department of public welfare, chiefly supported by the Catholic hierarchy, was likely to supersede the Sterling-Towner measure. Such a misfortune would deal a direct blow to public education. A vigorous protest was necessary to let the president and Congress known where Protestantism stood. As evidence of Catholic support for the president's bill, Athearn included a copy of the circular sent by Dowling to the NCCM urging a halt to attacks on the Sterling-Towner bill until it became clear what action Congress would take regarding Harding's measure. Although hardly a hierarchical endorsement of the latter, it was all the proof that Athearn and his people needed.[27]

After the hearing, the Joint Committee's interest in the public welfare bill slackened, and Magill claimed the credit. The impression made by his witnesses threw a scare into the Catholic camp. Fearing that the Sterling-Towner movement might degenerate into a religious crusade, agents of the NCWC considered ways to forestall such a turn. Pace played with the idea of supporting a separate department of education shorn of objectionable features as a means of avoiding further religious recrimination. Such a move, however, would be inadvisable if Harding was committed to a department of public welfare, something unclear because rumor had it that he had introduced the bill simply to redeem campaign pledges. Pace asked Father Joseph Denning of Marion, Ohio, Harding's hometown, to let him know where the president stood.[28] Although Denning's response is unknown, it is certain that Harding wanted a welfare department and opposed a separate department of education.[29]

Monahan took a different tack from Pace. Possessing a distinguished career in public education, he was convinced that the hierarchy needed to state its position on public schools. To this end he drafted a declaration to be submitted to the NCWC Department of Education.[30] The mounting crisis caused him to act without awaiting approval. Making the declaration the core of a baccalaureate address delivered in June 1921 at Saint Teresa's College in Winona, Minnesota, Monahan noted that since the demise of the first Smith-Towner bill more and more people accused the church of being the enemy of public education. He called on the hierarchy to issue a statement on public, private, and parochial schools. In his view, no matter how diverse might be the opinions of bishops, all of them were in fundamental agreement on the following points: the right of parents to send children to their school of choice, the right of the state to establish a minimum school term and demand that English be the language of instruction, and the right of the state to tax all citizens for the support of public education. Although Monahan held that only basic courses need be taught in English, his wording gave the unfortunate impression of categorical opposition to foreign-language schools. The NCWC news service publicized the speech, and Monahan himself included the key elements in an article for the *NCWC Bulletin*.[31]

More controversial than the speech was his praise of Michigan's Dacey Law, which gave the state superintendent of education power to maintain standards in all schools by withholding or revoking licenses to operate. The Michigan bishops had supported the legislation in the hope that the concession of state supervision over Catholic schools would forestall a reintroduction of the compulsory, public-education amendment. What had been a tactical endorsement for them was an ideal for Monahan. When the NCWC Legal Department asked his opinion of the law, he declared that proponents of Catholic education in other states "should secure at as early a date as possible, legislative enactments similar to those provided in this bill." The Legal Department published his view as that of the NCWC Department of Education.[32]

These expressions raised a storm of protest, especially in the German Catholic community.[33] Bishop Joseph Schrembs warned Burke: "Mr. Monahan is talking too freely. There is much bitter criticism on this score on the part of a number of bishops."[34] James H. Ryan, who replaced Pace as executive secretary of the NCWC's Department of Education, also reported that many prelates objected to Monahan's words. For a while it seemed that Burke might have to issue a clarification. The Department of Education, however, advised otherwise because a published correction would do more harm than good.[35]

While there was criticism from within the church, there were kudos from without. Monahan published a revised version of the Winona speech in the Boston *Journal of Education*. The editor, A. E. Winship, reported that the article considerably helped "the better class of people," who were disgusted with the anti-Catholic element but had an unconscious anxiety about the Roman church. After lecturing in sixty cities in twelve states, he informed Monahan that almost everywhere someone had commented favorably on the article but believed Monahan would be punished for it. The fact that nothing had yet happened to him gave them faith in what Winship said about the Catholic position on education.[36] Although Monahan went unpunished, his utterances would have dangerous repercussions on the Welfare Council. Members of the episcopacy resented his speaking too freely in the name of the church, and they would take their resentment out on the NCWC.[37]

Throughout the summer the NCWC monitored the educational situation. In an article written to update Catholic educators on the status of the Sterling-Towner bill, the NCWC Department of Education reported that there were "many indications that certain Protestant bodies are seeking to create the impression that the opposition to the bill emanates chiefly from Catholic sources and are attempting to engender religious animosities, evidently hoping thereby to becloud the issue and minimize the fact that [the] Bill is opposed by many elements upon sound economic as well as fundamental grounds."[38] At the annual convention of the hierarchy in September, Dowling admitted that Catholic opposition to the Smith-Towner bill had put parochial schools "in a worse position than they were two years ago." Given the religious volatility of the situation, the NCWC Administrative Committee advised against an endorsement of Harding's bill, no doubt because it would only lend credence to the allegations made by the NEA at the recent congressional hearing that Catholics supported the president's measure simply as a means of opposing the Sterling-Towner bill. Despite the recommendation, Dowling had to quell an impulse among bishops to act contrary to the committee's advice.[39]

While the hierarchy and the NCWC secretariat privately favored Harding's bill, the latter suffered on a more popular Catholic front. Only two editors gave it notice, and their opinions were discouraging. The Detroit *Michigan Catholic* thought it whimsical that passage of the president's measure would halt the Sterling-Towner movement, especially since the proponents declared that they would put the question to a national referendum rather than see education submerged in another department.[40] *America* declared against Harding's bill on constitutional grounds and went on to call the Sterling-Towner legislation "the most dangerous measure that Congress has ever faced" because it established federal control over one of the two elements responsible for the thought and conduct of a people: education. If it carried, federal control would soon be asserted over the other element: religion. "Freedom of education and freedom of religion stand or fall together."[41] Here were Catholics contending that the educational question was really a religious issue. Dowling privately commented to Monahan: "There is a cockey [*sic*] spirit among some of our Eastern Catholics as if they were spoiling for a fight. Maybe they'll not be prepared for it when it comes."[42] And come it would.

Fortunately, throughout summer and fall of 1921 the educational question lay dormant because of reorganization. Harding remained committed to a department of public welfare, but now within the broader context of a larger plan to restructure the executive branch. The Joint Committee on Reorganization, established by the previous Congress, had held its first meeting in May and elected as chairman Walter Brown, Harding's representative. The House and Senate Education Committees set aside both the public welfare and Sterling-Towner bills until Harding submitted a proposal to the Joint Committee.[43] Still, Towner and the NEA worked hard to reach an agreement with Brown regarding the place of education in the plan of reorganization. They made no headway. Economy-minded administration leaders informed them that it was not the time to attempt raids on the federal treasury in the size contemplated in the Sterling-Towner bill. The best that education could expect in the restructured executive branch would be one of four assistant secretaries in whatever new department would be created.[44] Reorganization was aimed at streamlining bureaucracy with the purpose of increasing efficiency and economy.

During this period Harding delivered two addresses that irrevocably reversed his pre-election declaration in favor of federal aid to education. Marking the three-hundredth anniversary of the Pilgrims' landing at Plymouth Rock, he urged the nation to guard against "the supreme centralization of power at home." "The one outstanding danger today," he warned, "is the tendency to turn to Washington for the things which are the tasks or the duties of the forty-eight commonwealths."[45] The speech was widely understood as applying to the education issue. Four months later, in proclaiming American Education Week, Harding returned to the theme. Admitting that the future strength and security of the nation depended upon its overcoming the problems of illiteracy, poor physical education, and un-Americanized immigrants (all targets of aid in the Sterling-Towner bill), he urged local communities to raise the consciousness of their citizens to these issues and to "bring to their attention specific, constructive methods by which, in the respective communities, these deficiencies may be supplied."[46] The president had clearly come to view educational problems as a local matter.

On 31 October 1921 the NEA sent a delegation to Harding to appeal for a separate department of education in the reorganized executive branch. Without mentioning federal aid the petition argued that if the government was to uphold its proper role in the promotion of education, "the department at Washington must be given such dignity and prominence as will command the respect of the public and merit the confidence of the educational forces of the country." The delegates later indicated that their real hope was for a compromise, a direction in which the administration itself was moving. Both were drawn to the idea of a department of education and public welfare, without aid, and hoped to strike a bargain.[47] This did not mean that the NEA had given up the idea of a separate department with an appropriation; it was simply making sure that it did not come away empty-handed.

While faced with an administration opposed to a separate department and federal aid, the NEA witnessed widening fissures within in the educational community over the issue. Wrongly assuming that the absence of federal aid in the NEA's petition to Harding signaled that the association had abandoned the aid package and was now seeking only the establishment of a separate department, Judd considered this a step in the right direction. Nor did it matter to him if the new entity was called the "department of education and public welfare." The important thing was that the many educational agencies in the executive branch be brought together under one head, and that the purpose of the department be the investigation of educational issues. Judd sent to some two hundred educators the draft of an editorial promoting this idea, with a request for responses that could be collated into an article showing the spectrum of opinion on the issue.[48]

He was not alone in thinking that the issue ought to be aired. In December 1921 the University of Illinois hosted a two-day symposium on the federal relation to education as part of the celebration of the installation of David Kinley as the institution's new president. More than a hundred presidents and deans of colleges and universities gathered in Urbana for a frank discussion of the question, revealing considerable opposition to the Sterling-Towner bill. Objections ranged from too much federal control to not enough. At one end were men like Samuel Capen, director of the ACE, who believed that despite disclaimers, the measure would lead to the federal control of education because of its large appropriation. Worse, the more the federal government induced states to act on certain matters by offering funds on a matching basis, the more it tied up state funds for federal projects. "Let the principle . . . continue to dominate Federal legislation for a decade or two longer," said Capen, "and the major part of all state tax levies will be mortgaged in advance for the support of undertakings in Washington."[49]

Kinley made a similar point in his installation address. Considering federal aid on a matching basis a "bribe" and the "wrong principle," he argued that the plan was simply "an attempt to do indirectly what the Constitution will not allow directly," a scheme that would lead to the federal control of education. Like Blakely, Kinley argued that if a disagreement arose between a state and the federal government over use of funds, the state would have to give way. The only method that would leave intact the states' constitutional authority over education would be for the federal government to freely offer aid to be used for schooling as the states themselves determined.[50] Frederick Mumford, dean of the College of Agriculture in the University of Missouri, agreed, stating that it was "doubtful whether cooperation in any proper sense of the term can be arranged between the Federal Government and a state, because in the last analysis the Federal Government must be supreme." Dean W. S. Sutton of the College of Education in the University of Texas declared himself "opposed emphatically" to federal funds on a matching basis. Without using the word "bribe," he clearly believed the appropriation in the Sterling-Towner bill to be one, permitting the federal government to trespass on the constitutional authority of states over education. He considered the "increasing multiplication of Federal activities in matters which belong to the states . . . a real danger" and cited Harding's Plymouth Rock address in support of his view.[51]

At the other end of the spectrum of opposition stood Judd alone. Repeating his allegation of a year and a half earlier, he contended that the NEA had suppressed public discussion among educators in the hope of securing early passage of the Smith bill, which contained the apparatus for federal control of schooling. When it became clear that teachers and superintendents would not support the bill because of such control, the NEA had reversed itself, leading Judd to believe that the leaders of the association lacked principles and were willing to sell themselves for the sake of whatever might secure educational legislation. As he had earlier argued, he favored the creation of a strong department to conduct scientific research and advise Congress. He reiterated his conviction that there ought to be a great amount of federal participation in education and a certain amount of federal control, especially over those matters in which states had proven themselves incompetent. He hoped that the present discussion would prove to the promoters of the bill that "the show of unanimity [was] a forced illusion." A new bill was needed, one formulated after much discussion.[52]

The Sterling-Towner measure was not without advocates, most notably its two congressional sponsors. Admitting that Republican leaders in Congress had separated the organizational aspects of the measure from its aid package and were willing to consider only the former within the context of the reorganization of the executive branch, Towner argued the necessity of the entire bill. He did so on the basis of national security, namely, that the country must have a literate, physically fit, and Americanized citizenry suited for military service and prepared for an informed use of the franchise. Denying that Congress had either the desire or the constitutional authority to control education, he wondered why grants in support of the common schools aroused controversy and condemnation. Answering the question, he said: "It must be that such opposition is based upon a misconception of the proposed legislation. To think otherwise would be to believe that there were in our country those who really desired the destruction of our common school system. Such a belief no loyal American would desire to entertain."[53]

These last remarks were an artful way of framing the issue, one that reduced opposition either to misapprehension of the bill or to hostility toward public education. Because the former implied a certain lack of mental acumen while the latter bore the stigma of near treason, the only alternative was compliance. Yet the educators who had just voiced opposition were thoughtful men, neither un-American nor mentally weak. Their objections to the bill were based on prior experience with federal aid, especially the grants contained in the Smith-Hughes Act. It seems likely that Towner's remarks, like those of Smith before him and those of the NEA on an ongoing basis, aimed at implying that Catholics, the principal opponents of the legislation, were un-American.

Senator Sterling noted that with the First Morrill Act (1862) the federal government had used its constitutional power to "dispose of . . . the territory or other property of the United States" in order to make grants of land to the states in support of education. With the Second Morrill Act (1890), the government made the great breakthrough of considering money itself "property." Thereafter, it began granting financial aid for various aspects of education deemed important to national interest. In Sterling's view there was no more important matter of political interest to the nation than the Americanization of immigrants through education.[54]

In addition to the sponsors, several educators also supported the bill. Dean G. H. Reavis of the School of Education in the University of Pittsburgh noted that the legislation left states free to accept or reject the funds. The bill redressed a real need to equalize educational opportunity around the nation, especially in the South. He considered it a "sad commentary on democracy" if the federal government could not "make a law giving the states this help and encouragement, without controlling education." Magill assured listeners that the bill had been carefully worded to leave intact the states' constitutional authority over education. Although McCracken averred that he would like to have seen the financial features of the measure separated from the creation of a department of education, he now considered their acceptance a tolerable trade-off to win for education the national recognition it deserved.[55]

Between those who favored and those who opposed the bill were some who either sought a middle ground or wanted to bide time. Dean Charles Chadsey of the University of Illinois and a member of the Cleveland Conference thought that the symposium had clearly demonstrated that there were two sides to the issue. Because almost all of the opposition expressed against the bill related to federal aid and the control that would follow rather than to a department, he wondered if it was "possible to separate these questions in some permanent fashion" so that education could be given the national recognition it deserved. Injecting a note of caution, President R. A. Pearson of Iowa State College of Agriculture concluded: "I have felt, as this discussion has proceeded, that we are wading into deeper water, that is more and more murky. We do not know what is ahead of us, and instead of plunging along rapidly we should proceed very slowly; and perhaps we should wait awhile before we move forward further." This point was seconded by Sutton.[56]

Soon after the symposium, Judd further undermined the "forced illusion" of unanimity with a series of articles based on the responses received in reply to the draft of the editorial he had circulated among educators. The array of opinion was striking. State Superintendent J. E. Swearingen of South Carolina and President Joseph Rosier of the State Normal School of West Virginia favored the Sterling-Towner measure. Others, like Capen and Professor Henry Morrison of the University of Chicago, another member of the Cleveland Conference, wanted a department of education to conduct scientific investigations, with aid out of the question. Kinley wanted both a department and aid, but the appropriations had to be given with no strings attached. State Superintendent Thomas Johnson of Michigan wanted a department, aid, and federally established minimum standards. State Superintendent T. H. Harris of Louisiana wanted federal aid without a department of education. He would keep the bureau as it was, with the commissioner elected by a federal board of education rather than appointed by the president. President Frank McVey of the University of Kentucky thought the present bureau should simply be enlarged; in time it might eventually grow into a separate department. President A. Lawrence Lowell of Harvard opposed the establishment of a department because it would ultimately politicize schooling. Professor Alexander Inglis of Harvard and Charles R. Mann, chairman of the Advisory Board of the Division of Operations and Military Training in the War Department, desired the establishment of a federal commission of education as a means of avoiding the politicization of the issue. Frank Cody, superintendent of schools in Detroit, also wanted a commission, but for him it had to be invested with two elements of control: (1) the power to establish uniform methods and procedures of child and cost accounting, and (2) the authority to standardize definitions. With the exception of those two matters, the states should be left free to set educational policy. The foregoing fairly represents the diversity of views without exhausting the names of those who held the various positions.[57]

Judd concluded the articles by noting that proponents of the Sterling-Towner bill hoped to suppress this diversity of opinion or to label it the view of a minority. One such minority often vilified as the "enemies of public schools" were the advocates of parochial education. As counterevidence, Judd stated that he possessed a letter from an influential Catholic educator (probably Monahan or Pace), who favored the establishment of a department of education to conduct research and offer advice, but who did not care to be drawn into public expression of his views. Judd had "a considerable collection of Protestant correspondence" against the bill, no doubt much of it from Lutherans who also refused to come out into the open while the proponents of the Sterling-Towner bill were "so free with their vituperation of those who do not answer to the lash."[58] This last remark clearly indicated his feeling about the NEA and its tactics. It was using religious prejudice to whip opposition into line.

Judd's articles and Kinley's symposium were significant for several reasons. First, the symposium highlighted that a number of respected educators believed, as did many Catholics, that federal aid would lead to federal control. Second, like many Catholics, Judd thought that the proponents of the NEA bill were using the religious issue to further their cause. Third, and most important, both the articles and the symposium indicated that the educational community was deeply divided over the proper type and method of federal participation in education. There were those who wanted a department, those who wanted an expanded bureau, and those who wanted a commission. There were those who wanted federal control, and those who wanted state control. There were those who wanted aid and those who wanted advice. And there were the various combinations of all the foregoing. This divergence reached into the Cleveland Conference, with Judd, Chadsey, Morrison, and Strayer representing different positions on the issue. The only thing that members of the conference seemed agreed on was the desirability of federal participation in education, an issue that divided as much as it united them.

Despite division within the educational community, despite Harding's opposition to a separate department and to federal aid, and despite the opposition of Republican congressional leaders to the same, the NEA developed a plan of action. If the Joint Committee on Reorganization made its report during the present session, that was to be the "signal for rallying the friends of the Towner-Sterling bill to secure its enactment with such acceptable modifications as may be required to adapt it to the general plans recommended." If the committee failed to act, the NEA would mount a drive for passage of its bill in pure form. Moreover, the NEA *Journal* carried word that an anonymous organization of national scope had pledged $125,000 annually to publicize the legislation until it passed.[59]

Catholics believed that Masonry had put up the money. Burke was convinced of this, and Monahan learned from a reliable source that Masons had raised the funds to offset Catholic activity against the bill.[60] *America* too identified the Southern Jurisdiction of the Scottish Rite with the anonymous pledge. The editor considered the financial support a mixed blessing. On one hand, lodges of the Southern Jurisdiction understood the donation as a signal to launch the attack on parochial schools. On the other it unmasked "the possibilities, if not the deliberate purpose, of this Federal educational plot." Time and again, Towner and other advocates had insisted that no one could interpret his bill as an attack on parochial schools. In *America*'s view, the Southern Scottish Rite placed the proper value on such protestations because lodge after lodge saw the measure as a means for abolishing Catholic education. The Rite's financial backing of the measure clearly revealed the bill's true purpose.[61]

Then came word from Burris, a friend of Ryan, that the Masons and the NEA were working hand in glove and that the money had indeed come from them. Judd had sent to Burris correspondence between himself and Perry Weidner, secretary general of the Southern Scottish Rite. The letters showed that under the watchful eye of Magill, himself a Mason, the Rite had engaged in activities in favor of public schools in general and the NEA bill in particular. This included the Rite's anti-parochial school agitation. "In other words," Burris concluded, "Magill and the Masons have started in to put the private and parochial schools out of business, at least as far as primary schools are concerned."[62]

The NEA's acknowledgment of the pledge and the information linking it to the Southern Jurisdiction of the Scottish Rite was the first hard evidence of a connection between the Sterling-Towner bill and the emerging drive for compulsory public education. That the jurisdiction committed such a sum (roughly equivalent in purchasing power to $1.5 million in current dollars) and that the NEA accepted it, indicated that the association had hardly reconciled itself to Harding's reorganization scheme. Rather, it seemed determined to ramrod its bill through. Nor was the pledge the only concern of Catholics. An estimated two-thirds to three-fourths of all congressmen were Masons. To be sure, they were men of independent mind—as were many Catholics—and were responsible to their constituents. Still, their tie to the lodge was a card that could be played.

In the winter of 1922 the NEA's patience wore thin. Through the previous summer and fall, Harding's cabinet had stonewalled the reorganization plan. In January 1922 a frustrated Brown submitted to the president his own plan, including a department of education and welfare, sparking internecine warfare among cabinet secretaries eager to preserve their bailiwicks.[63] When members of the Reorganization Committee informed the NEA that a report would be long in coming, the association gave up waiting. Aided by the Southern Jurisdiction of the Scottish Rite, the NEA, described by Burke as "the strongest lobby in the way of political influence that Washington knows," brought crushing pressure to bear on the House Committee on Education. The Senate committee was immobilized by chairman Kenyon's departure to take an appointment as a federal judge, an adroit political move by Harding to destabilize the farm bloc by removing its leader. Burke sized up the situation as follows. Although Harding opposed the Sterling-Towner bill, he was unable to control the House committee. Although the chairman of the Republican National Committee wanted no action on the bill with mid-term elections upcoming, he was unable to prevent the bill from being reported. Although Fess, who chaired the House committee, wanted to await the report of the Joint Committee on Reorganization before bringing the Sterling-Towner bill forward, such a course was no longer tenable.[64]

No doubt the NEA's lobbying was made easier by the insurgency of House Republicans in the farm bloc. An important piece of non-agricultural legislation favored by the latter was the Sterling-Towner bill, whose aid package proposed to use tax dollars generated from commercial and industrial areas to equalize educational opportunity in rural states. As historian Don Kirschner points out, farmers in areas that were consolidating schools bore an almost unbearable economic burden as they witnessed their purchasing power decrease by about half while their property taxes skyrocketed to support this educational reform.[65] The bloc supported the Sterling-Towner bill for more than monetary concerns. It believed that the federal government had a role to play in the eradication of illiteracy, the Americanization of immigrants, and the improvement of education by promoting better teacher training and increasing teachers' salaries. Seven of the ten Republicans on the House Committee of Education hailed from midwestern states (Fess, Towner, Albert Vestal of Indiana, Edward King of Illinois, John Robsion of Kentucky, Adolphus Nelson of Wisconsin, and Samuel Shelton of Missouri) and voted for nearly every piece of farm-bloc legislation that came before the House in 1921 and 1922.[66]

On 28 February Republican leaders of the House committee met informally with representatives of the NEA. Certainly present were Fess and Towner, as were others of the farm bloc who favored the Sterling-Towner bill. No doubt, Magill represented the NEA. Despite reluctance to report the bill out of deference to Harding, Fess felt compelled by his colleagues and the NEA to do so. This was a bold, insurgent move on his and Towner's part, especially in light of their respective positions, Fess as chairman of the committee and Towner as chairman of the Republican caucus. Here were two of the party's progressive leaders preparing to go head to head with the president and the Old Guard.[67]

On receiving word that the committee intended to report the bill, Burke urged Fess to hold a hearing before bringing the measure forward, but the chairman refused on the ground that there was no demand for one. If Fess needed a demand, Burke would provide it. He wired Hanna and probably others on the NCWC Administrative Committee to urge that they get prominent men, preferably Protestants, to telegraph Fess that public opinion demanded a hearing.[68]

The NCWC received help from within the NEA Department of Superintendence that was meeting in Chicago during these crucial days. The Welfare Council had agents there working on friendly but up to then timid representatives. The session devoted to the Sterling-Towner bill revealed division in the ranks as several prominent educators spoke against the measure. Inglis of Harvard led the attack by declaring his categorical opposition to federal aid. Without denying either the evils that beset education or the delinquency of some states in remedying them, he argued that the NEA had failed to make the case that states were unwilling or incapable of doing so. In his view, federal aid was reprehensible because of the dilemma it posed, namely, that it be given freely and without supervision, thereby leaving the money subject to misappropriation or squander; or that it be granted with supervision, inevitably leading to federal control, as had happened with the Smith-Hughes Act. Though unalterably opposed to aid, Inglis argued that the federal government should have some agency to play an advisory role in education, be it a department, bureau, or commission (his preference being the last). Capen and E. C. Broome, the superintendent of schools for Philadelphia, agreed. Both opposed aid to education, though Capen remained open to its theoretical, future possibility, but not on a matching basis as proposed in the Sterling-Towner bill. Like Inglis, both favored the creation of a federal agency, in their views a department, to conduct research and offer advice. President W. A. Jessup of the State University of Iowa, on the other hand, questioned the idea of a department out of fear that it would politicize education. Without addressing the Sterling-Towner bill which he supported, Winship spoke about the NEA itself, which he suggested was self-serving and "snaky," using the latter adjective no doubt because Monahan had convinced him that the Catholic church was unopposed to public education while the NEA kept insinuating that the church did oppose it. Winship challenged the NEA to give disinterested service to education. Only two speakers raised no objection to the bill: Strayer, who, without mentioning aid, confined his remarks to the need for a federal department; and Olive Jones, principal of P.S. 120 in New York, who embraced the Sterling-Towner program in toto.[69]

Praising Inglis for having the courage to beard the lion in its den, Judd reported that listeners preferred Strayer's eloquent plea for national participation in education in the belief that federal subsidies were just over the horizon. During the days that followed, however, "sober thinking was beginning to take the place of boisterous applause and vague platitudes," especially in view of indications that the Republican-led Congress would endorse no subsidy. Denying urgency for action on the question, Judd again called for reconsideration: "The time has come for careful, close thinking on essentials. . . . Education is not begging for financial support; it is not overlooked in American public life, but it is in need of a new and broader organization. . . . This new and broader organization ought to be of a form that will appeal to all the people of the nation . . . as something more than a plea on the part of teachers for political recognition."[70]

Apparently, the rift in the Department of Superintendence and the numerous telegrams of protest sent by prominent educators and others as a result of Burke's campaign convinced Fess to alter course. On 3 March he sent word through his son to the NCWC that the Sterling-Towner bill would remain in committee. Three days later, Republican leaders of the House Committee on Education met again informally with representatives of the NEA. The grist for their discussion was Harding's firm commitment to reorganization and his advice that the committee take no action on the Sterling-Towner bill until he could submit a proposal for a reshuffled executive branch, almost certainly to include a department of education and welfare. Afterward a chastened Towner reported: "We have decided not to do anything until we get the reorganization report. We . . . decided that it was not well to stir up muddy waters." Before the week ended, House majority leader Frank Mondell of Wyoming reiterated his determination to oppose all new subsidy bills, which included the NEA measure.[71] A congressman for almost twenty-five years, he and other party regulars would pay at the polls in November for their support of the administration on this and other matters.[72]

If Towner was willing to bow to the wishes of his party's leadership, William Bankhead of Alabama, the ranking Democrat on the Education Committee, was not. As mentioned in the first chapter, educational reform was an element of southern progressivism that perdured throughout the 1920s. On 9 March Bankhead urged Fess to call a committee meeting "for the purpose of considering the Towner-Sterling bill at this session of Congress." Fess replied that such action was impossible until the reorganization plan was submitted to Congress.[73] On the floor of the House, Bankhead accused the Republican leadership, namely, Fess, Mondell, and those on the steering committee, of a conspiracy to prevent the bill's passage during the current session. "The method of administering to it the coup de grace is very simple," he said. "It can not be reported . . . without a meeting of the committee, and no meeting of the committee will be called by its chairman, Doctor Fess. It is one of the oldest and yet most effective of all methods of giving the sleeping potion to a bill." Bankhead had spoken up, he declared, so that the "friends of education," the NEA in particular, would know why the legislation would receive no action.[74]

Here was the farm bloc divided against itself along party lines. Both Bankhead and Fess favored the Sterling-Towner bill, though the chairman was willing to compromise his belief in service to the administration. The fact that the Republican committee members had frozen their Democratic counterparts out of the conferences with the NEA, even though some like Bankhead supported the legislation, offered mute testimony to the disorganization and factiousness of the House wing of the farm bloc. As will soon be seen, Fess would issue his own ultimatum to Harding, one that would prove as toothless as Bankhead's to the Republican leadership.

The decision to keep the Sterling-Towner bill in committee was encouraging to Catholics. Burke was confident that no education measure carrying a federal subsidy would be enacted "for a generation to come." Encouraging, too, was Kenyon's replacement as chairman of the Senate Committee on Education and Labor: William Borah of Idaho, an ardent anti-statist and self-proclaimed progressive. As historian George Mayer observes, for Borah "progressivism meant limited campaign expenditures, direct primaries, high taxes on business, isolationism, and anything that his Idaho constituents wanted at that particular moment." Equally pleasing to Catholics was the rift in the NEA Department of Superintendence, which Burke hoped to exploit. On the other hand, he confided to his friend Bishop Peter Muldoon that the church would have to yield on the idea of elevating the Bureau of Education to department status. "The feeling for that is very strong throughout the country," he wrote, "and I don't see on what ground we could effectively oppose it; nor do I see any harm that it would do to the cause of Catholic education."[75] To Bishop Edmund Gibbons of Albany, another member of the Administrative Committee, Burke made a similar point: "Our situation will be hopeless if we as Catholics oppose all federal remedial legislation simply because it is federal. We must show that in a particular instance federal aid will do more harm than good as for example in the control of general education by the Sterling-Towner Bill." He foresaw no danger whatever in the creation of a department of education and welfare, likely to be proposed by the Reorganization Committee, so long as there was no subsidy attached.[76]

Alarming was Masonic involvement in the recent campaign. Worse, anti-Catholicism was spreading. In March 1922 a group of ministers organized the Evangelical Protestant Society "to defend American democracy against the encroachment of Papal Rome." The organization's manifesto resurrected the accusation of a Catholic plot against Protestantism in general and Protestant America in particular. The conspiracy's aim was "to undermine our public school system in the interest of parochial schools, where Romanism may be taught; to re-write American history in the interest of the papacy and thus to poison the minds of even Protestant children."[77] On the basis of this indictment, the National Patriotic Council, an offshoot of the Evangelical Society, adopted a policy close to that of the Southern Jurisdiction of the Scottish Rite Masonry, namely, to prevent the sectarian use of public funds, to wipe out parochial schools, and to bar Catholics from public office.[78]

As anti-Catholicism spread and evidence of Masonic collusion with the NEA mounted, disaster struck. Without warning, Rome suppressed the NCWC. Word reached the American bishops in the final days of March 1922. The meeting of the hierarchy in the previous September had revealed dissatisfaction with the Welfare Council, especially over pronouncements made by NCWC agents like Monahan and the social activist John Ryan. More significant, the meeting was to have been the occasion for repudiating the leadership of Cardinal William O'Connell of Boston, who had turned a blind eye to the secret marriage of his nephew-chancellor, James O'Connell. The plan misfired, and the cardinal, who had never been friendly to the NCWC, engineered its dissolution with the support of Dougherty, who himself had recently grown wary of the organization and resigned the chairmanship of the Legal

Department. In Rome too late to participate in the election of Pius XI, O'Connell had his friend Cardinal Gaetano De Lai, secretary of the Consistorial Congregation, draw up the decree of suppression, which the unsuspecting new pontiff signed as if it were routine business. Quickly mobilizing against this disaster, the NCWC Administrative Committee appealed for a suspension of the decree until it had an opportunity to present its case. Pius granted the appeal.[79]

On behalf of the Administrative Committee, Burke wrote a report for the Vatican arguing the necessity of the NCWC. In the belief that the most cogent argument against the suppression was the need for national, united action by Catholics, he recounted the origin, structure, and purpose of the organization, at the heart of which was the school question. Because the spirit of America was non-Catholic, and because it permeated public education, Catholics had established parochial schools for the prosperity of the faith. This course had a serious drawback. Americans prided themselves on their public schools, while Catholics disapproved of them and considered their own superior. This attitude aroused deep hatred among many Protestants. "The greatest desire of our enemies is to cripple or to destroy the Catholic schools system," said the report. "The great national organization of teachers [NEA] is unfriendly to us. The Masons have publicly announced their campaign for making attendance at public schools compulsory all over the United States. A great fear has seized the Catholic body." The report pictured the Sterling-Towner bill "as the first step towards the harmful control, if not the suppression, of Catholic schools and colleges." Thus far the political influence of the NCWC had enabled it to parry the threat. Those in the public eye believed that the Welfare Council was the authoritative voice of a united hierarchy and represented the vote of a united Catholic body.[80]

Three bishops went to Rome to defend the NCWC: Schrembs, Henry Moeller of Cincinnati, and William Turner of Buffalo. Eventually eighty-one members of the hierarchy signed a petition asking the pope to reverse the decree. From April through June 1922 the NCWC labored as a lame duck, crippled by increasing public knowledge of its plight.[81]

During this period, the battle with Masonry became more apparent, in part through the efforts of Burris. Undeterred by Republican stonewalling of the Sterling-Towner bill, Magill and Strayer circulated a petition among prominent educators asking for declarations in favor of the Sterling-Towner bill.[82] Burris admonished Strayer for not respecting Harding's desire to keep the legislation in committee until the reorganization plan was proposed. In Burris's view, the committee would never report the measure because of the $125,000 Masonry pledged in its support. Since the Scottish Rite had "some old scores to settle with a very important element of our population which is opposed to the bill," neither party would risk injecting religion into the upcoming election.[83] Burris expanded on these thoughts in writing and Judd published them in the editorial section of the *Elementary School Journal.*[84]

In the pages of that magazine, Judd confirmed that the Southern Jurisdiction had made the pledge, for he had been so informed by Weidner. He also learned from Weidner that there were inaccuracies in the NEA's announcement regarding the money. The Southern Jurisdiction had earmarked approximately $150,000 over the next two years for support of the Rite's educational program, which included passage of the Sterling-Towner bill, the creation of a national university, and the enactment of laws for compulsory public education. Judd was concerned about the ethics of the NEA. Why, he wondered aloud, had it claimed that the money came from an organization of "national scope," when in fact it came from the Southern Jurisdiction of the Scottish Rite? Why had the NEA announced the amount to be $125,000 annually until the bill was passed, when in fact it was about $150,000 over two years? Why had the NEA said nothing about the other educational purposes that the money was to support? He thought that these would make appropriate questions for the members of the NEA to ask their representatives to the annual convention.[85] In effect, he accused the NEA of exaggerating the scope and amount of support while glossing over the other purposes for which the money was to be spent, like the suppression of private education.

For his part, Strayer turned Burris's letter over to John Cowles, grand commander of the Southern Jurisdiction. Cowles admonished Burris for not verifying that the Rite was the organization behind the pledged $125,000. Playing the legalist, Cowles denied the allegation because the annual fund for support of its educational programs amounted to $200 per month for each state in the jurisdiction (or $76,800 a year, roughly the annual amount Weidner had admitted to), not the $125,000 claimed by the NEA. As for having "old scores to settle" with the Catholic church, Cowles denied that religion was at issue because Lutherans were equally opposed to the bill. The real concern was "class distinction" and "what is best for the interests of the country." He accused the Catholic church of separatism and discrimination. Although the Federal Council of Churches was open to all, Catholics refused to join. The same was true of American fraternal organizations, including the Masons, the

Knights of Pythias, the Boy Scouts, and the Girl Scouts. In each case Catholics separated themselves and set up parallel organizations. This was especially apparent with the parochial school system, which was the church's equivalent to the public school system. "If there is any discrimination in this country," concluded Cowles, "it is on the part of others, and not us. . . . We want all the people of the United States to work and live together in peace and harmony—no classes or class distinctions, or discrimination against anyone because of his religion . . . that is why we want all the children who will be the future rulers of this country, to go to the Public Schools."[86]

This letter reveals how the same reality might be interpreted differently by different observers. Catholics had, since the turn of the century, embarked on a separate-but-equal path. Their purpose was to preserve traditional Catholic values from corrosion by the acids of modernity. They viewed themselves as the true Americans, faithful to traditional ways and charged with the responsibility of preserving them and reinserting them into the larger society. As Cowles's letter indicates, their view of themselves was not shared by many outside the church. Masons, the Ku Klux Klan, Evangelical Protestants, and other similar organizations viewed Catholic separatism in a negative way. They believed that Catholics held themselves apart from the rest of society because they were discriminatory, class-conscious, and un-American. Given that each side viewed itself as truly American, there was little common ground on which—let alone a common forum within which—to resolve differences. Catholics viewed Masonic, Klan, and evangelical opposition as religious prejudice. Masons, Klansmen, and evangelicals considered Catholic separatism undemocratic and un-American.

Burris forwarded Cowles's letter to the NCWC, which used it in its fight for survival in Rome. Burke sent a copy to Bishop Schrembs, who was spearheading the NCWC's defense at the Vatican. The letter created a "veritable sensation" among the first Vatican officials who saw it, causing them to proclaim that the decree of suppression would be reversed. Given their reaction, Schrembs had the letter translated into Italian and printed as supplementary material in support of the NCWC's case.[87]

In fact, Masonic rhetoric grew more belligerent. The next issue of the *New Age* carried an editorial attacking an article Blakely had published months earlier in a Washington State Catholic paper. He had argued that because public education excluded religion, Catholic parents were conscience-bound to withhold taxes for the support of common schools and to keep their children out of them. Reluctantly, he admitted that taxes could be paid, albeit under protest. The Masonic editor replied that the first duty of Americans was to obey the law whether they agreed with it or not. Religion was excluded from public schools, he said, because Catholics had protested against the teaching of any faith but their own. "A lot of yap there has been from the priests and jesuits [*sic*] about what the consciences of Roman Catholics forbid," declared the *New Age*. "We all know very well . . . that the priests and jesuits [*sic*] *do not allow the Roman Catholic people to have any consciences of their own—these things are prescribed for them by the hierarchy!*" The solution to this problem was the Rite's legislative program comprising the Sterling-Towner bill, compulsory public education, and the requirement of matriculation in public schools as a prerequisite for holding any "office of trust or profit."[88] At the end of May, two thousand Masons gathered in Atlantic City for the annual convention of Masonic clubs. Delegates endorsed the Sterling-Towner bill and urged their brothers in Congress to hurry the measure through.[89]

The NEA also kept up the pressure. In early May it sent a delegation, including Magill and Strayer, to ascertain Harding's attitude. The president assured them that his reorganization proposal would not be long in coming and would give education distinctive recognition. Later Sawyer informed Magill that the reorganization plan called for a department of education and welfare, a compromise suggested earlier to the president by Fess, which Magill finally agreed to support. At least that was what Sawyer told Burris.[90] In reporting to the annual convention of the NEA, Magill was more reserved. "Just what attitude the National Education Association and other allied organizations . . . should take toward the proposed recommendation for a Department of Education and Welfare cannot be determined until the plan has been submitted and its exact provisions known," he said. "In the meantime we should continue to work unitedly for the Towner-Sterling bill."[91] Fess hoped that the reorganization plan would soon come forward and "meet the approval of the best judgment of the teachers." If the plan was not presented within a "reasonable time," he would feel compelled to have his committee consider the Sterling-Towner bill, which it would favorably report.[92] Despite the declarations of Fess and Magill, the NEA ceased congressional agitation, and both Sawyer and the NCWC believed that the measure was finished.[93] In fact it would die a quiet death in the Sixty-seventh Congress.

While things were going well for the NCWC on the Washington front, they also went well on the Roman front. Realizing that he had been taken advantage of, the pope began the process of reversing the decree of suppression. It took time, however, because the change in pontificates had spawned a power struggle within the Vatican between the conservative De Lai and the progressive Cardinal Pietro Gasparri, papal secretary of state. The problem was compounded by the need to effect the reversal in a way that saved face for De Lai's Consistorial Congregation. Ultimately, Pius returned the case to that congregation for a vote, while arming Gasparri with a papal veto in case the decision went against the NCWC. Fortunately, Gasparri never had to play that card. The congregation reversed the decree of suppression, though it fell to De Lai to write the new instructions under which the resurrected NCWC could operate, instructions that Dowling described as "full of broken glass."[94]

One of the shards had to do with the nature of the NCWC. The instructions stated that the NCWC was not to be identified with the hierarchy. Legally, however, it was. The charter of incorporation named every bishop individually and conveyed all rights to their successors. Yet opponents of the NCWC would always have De Lai's instructions to fall back on in support of their view. Still, the NCWC continued to function as it had in the past, with only a change of name from Welfare Council to Welfare Conference to reflect the understanding of most bishops that it was not the hierarchy sitting formally in a canonical council, but the hierarchy sitting informally as an organization in conference.[95]

Clearly, as one of those who did not answer to the lash, the Catholic church sought to minimize the impact on itself of the education question's transformation into a religious issue. The NCWC muted opposition to the Sterling-Towner bill among its own members, the Knights of Columbus, and the Catholic press. Its bureau of education publicized the fact that while the church upheld parental rights in schooling, it acknowledged the state's right to levy taxes on all for support of public education. The NCWC received unexpected help from the Republicans in maintaining a low profile. The Harding administration's promotion of governmental economy and its advocacy of local control enabled the church to hide beneath party coattails on the Sterling-Towner issue, though proponents of the measure accused it of cloaking its opposition with support of the Republican legislative agenda. Another welcome turn of events was the widening division of opinion within the educational trust over the proper mode of federal participation in education. When the NEA and the insurgent farm bloc sought to force the Sterling-Towner bill out of committee, the NCWC fostered opposition to the legislation among educators, indeed within the NEA's Department of

Superintendence itself, thus aiding the administration and congressional leaders in holding the line. The Welfare Conference was even able to turn the religious issue to its benefit during the suppression crisis. When the Vatican understood that the church in America was under attack, it reinstated the NCWC. Yet the church would soon lose its ability to object by indirection, especially with the NEA in open alliance with the Scottish Rite Masons.

4

A Battle on Two Fronts

Having survived the attack on its existence, the NCWC emerged from the fray intact and in time to face one of the most serious threats to Catholic schools in American history. Implicit in the NEA's open alliance with the Southern Jurisdiction of the Scottish Rite was the identification of the Sterling-Towner bill with the eradication of parochial schools. The temper of the nation did not help. In the wake of the war, the public increasingly abandoned the idea of the melting pot in favor of reducing immigration to a trickle. With the Emergency Quota Act in place, Congress set to work on permanent legislation to ensure maintenance of the cultural status quo, namely, the dominance of the northern and western Europeans, the Nordic race. The flip side of this immigration policy was the proper method of dealing with aliens already in the land. Many old-stock Americans found the answer in mandatory public schooling to inculcate in children, immigrant and hyphenate, the values of democracy and Anglo-Saxon culture.

This situation portended a battle on two fronts: a drive for compulsory public education at the state level and another for a national program of education at the federal level, each a different facet of the same movement. These two issues pitted Catholic Americans against old-stock Protestant Americans, with the most treasured possession of each at the struggle's center: its school system, viewed as prototypically American by each side. For old-stock Protestants, the public school was the mainstay of democracy because it was the instrument for instilling patriotism and traditional social values in segments of the population viewed as alien. For Catholics, the parochial school was the descendant of the traditional religious school of the nation's colonial and early republic heritage, established by the founding fathers. It was the place where the next generation could safely be schooled for citizenship while preserving cherished religious and cultural values. With both sides championing their own institutions as truly American while suspecting those of the other, the situation was explosive. As Lynn Dumenil points out, "They were in effect at war over the right to define what it meant to be American."[1]

In the summer of 1922 the Southern Jurisdiction of the Scottish Rite launched a campaign in Oregon to implement the compulsory public schooling element of its educational program. Masons sponsored an initiative requiring all children between the ages of eight and sixteen to obtain their primary education in the public school, effective 1 September 1926. A parent was to be fined $5 to $100 or be imprisoned for two to thirty days for each day a child failed to attend. At first glance the state seemed an unlikely spot for the drive. Ninety-five percent of the children of primary-school age were already in public institutions.[2] It was the state's very approximation to the American ideal that drew the Southern Jurisdiction to it. Oregon was chosen, said a national spokesman of the Rite, "because she has no foreign element to contend with and is, more than any other state, purely and fundamentally American." A leading Portland Mason expressed the hope that success there would serve as an example for the rest of the nation.[3] This was the course of action some Catholics had warned about. As early as 1919 and as recently as May 1922, first Edward Pace and then Bishop William Russell cautioned that the proponents of compulsory public education might employ the tactic of the prohibitionists, namely, enact the law in states where resistance was light; when other states followed suit, demand a constitutional amendment.[4]

Although the Rite claimed responsibility for the measure, Oregon Masons were divided over the issue. Those brothers who backed it were either members of the Ku Klux Klan or belonged to the radical wing of Masonry. Historian David Tyack has argued that the Klan was in fact the prime mover behind the Oregon initiative and used the Scottish Rite as a front. Indeed, Fred Gifford, one of the brothers responsible for the Masonic endorsement of the initiative, headed the Portland Klavern. While the initiative divided Masonry, it had no such effect on the Klan. In fact the Invisible Empire had sponsored a similar initiative in Ohio the year before and was conducting a concurrent drive in Oklahoma.[5]

Klan unity and Masonic division probably had more to do with the nature of the two organizations. The cement that held the Klan together was nativism, racism, anti-Catholicism, and anti-Semitism. This is not to deny, as recent historians have pointed out, that many Klansmen joined for reasons of local concern. Still, the issues that gave the Klan national cohesiveness were the aforementioned, often translated into local action. Both the Sterling-Towner bill and compulsory public education were settled Klan policies and corollaries of its nativism and anti-Catholicism. That, however, was not the case with Masonry. Men joined the order for a variety of reasons: a quest for truth and meaning in life, a need for ritual, a sense of belonging, or simply because membership was expected of those in a given profession or locale. More to the point, decisions of the Supreme Council of the Scottish Rite were non-binding on members. Hence, the greater unity on the issue in the one organization rather than in the other.

More significant than sponsorship of the law was the way that proponents and opponents of the initiative framed it. Each side pictured the issue as a struggle for true Americanism. For Klansmen and Masons, the matter had nothing to do with religion as such; rather, it was a campaign for 100 percent Americanism and traditional Anglo-Saxon culture. Viewing political Romanism as tyranny and therefore antithetical to democracy, supporters believed that parochial schools were seedbeds of foreignism, nurturing children on principles antagonistic to true Americanism. To safeguard national institutions, all children must be "Americanized" in public schools.[6] Catholics, on the other hand, viewed their schools as the bulwarks of democracy and traditional values. For them, the secularization rampant in public education represented the abandonment of the founding fathers' vision of America as a Christian nation and the forsaking of their understanding that religion formed an integral part of schooling. As Catholics saw it, the nationalist philosophy of education propounded by the NEA, the Southern Jurisdiction, and the Klan made the child a creature of the state, a theory that undermined both parental rights and the freedom of education. Considering this philosophy un-American—out of harmony with America's educational past—Catholics viewed themselves as defenders of the nation's educational tradition.[7]

James H. Ryan articulated the Catholic view. He warned the annual convention of the NCCW of "a titanic struggle" impending between the "nationalist point of view" held by the Klan and Masons and the "traditional American attitude" held by Catholics. In the balance hung the orientation of American education, "the perpetuity of American institutions, the continued welfare, if not the existence, of the Church itself." The nationalist philosophy was ever the same: "the child is the ward of the state." The traditional attitude upheld the right and duty of the parent to provide an education for the child, while admitting the state's right to insist on education and even provide schools for that purpose. In Ryan's view, the nationalists were ignorant of American educational history. "Private education existed in this country at least a full century before public education," he noted. "Protestants founded religious schools before the Revolution." Practically all of the nation's founders received their education in private schools. These same founders "conceived of America as a cradle of religious liberty and as a preserver of the religious rights of every man, no matter what his beliefs may be," continued Ryan. For Catholics, the preservation of their religion depended on the parochial school. To rule such schools out of existence "would be to invade the realm of conscience." It would also be the deliberate destruction of "one of the greatest supports which democracy has here in America": religion. "Democracy needs religion," declared Ryan. "Religion is not only the sure basis of good morals, but it is the sure basis of good citizenship as well. . . . It is time we get back to the old allegiances, to the honored and respected principles which have always guided this nation." Catholics were the keepers and guardians of these ideals. If the church were to explain its position and defend its rights, Ryan was sure the public would thank it for "maintaining the religious school, one of the greatest contributions to the welfare of America as a nation and to the preservation of democracy in the world."[8]

Clearly, each side viewed itself as quintessentially American. For one, the public school was the mainstay of democracy because it was the instrument for inculcating patriotism and contemporary, Anglo-Saxon, Protestant social values in segments of the population viewed as alien. For the other, the public school had abandoned the path carved out by the founding fathers. Only religious schools had remained true to the American educational tradition. Only the religious school was capable of maintaining the welfare of the nation and democracy worldwide. Viewed from the vantage point of history, the issue at stake was a desire for conformity and homogeneity versus a desire for diversity—uniformity against pluralism.

Beaver State Catholics mobilized for defense. Archbishop Alexander Christie of Oregon City established the Catholic Civil Rights Association. Catholics were forced to maintain a low profile, letting other groups conduct the fight so as to avoid further antagonizing the proponents by confirming their suspicion about Catholic political power. They found allies in the Portland Committee of Citizens and Taxpayers, the Lutheran Schools Committee, and the Non-Sectarian and Protestant Schools Committee. These groups held that the initiative was un-American and denounced it as an abridgment of parental rights and the first step toward state absolutism, one that would substantially increase taxation.[9]

When the hierarchy met in annual convention in late September 1922, the bishops considered the steps to be taken in defense of Catholic educational rights. In addition to ordering the NCWC to oppose the Sterling-Towner bill, they decided to issue a pastoral letter stating the church's position on education, something that Arthur Monahan had advocated more than a year earlier. Archbishop Austin Dowling, Bishop William Turner, and Bishop Edmund Gibbons were commissioned to draft the pastoral and submit it to each prelate for criticism and approval, an unworkable plan that doomed the letter from the start.[10] The command to oppose the Sterling-Towner bill marked an end to the Burke-Dougherty-Pace strategy of trading support for the measure in return for a guarantee of the right to maintain parochial schools. That right was now in jeopardy, and it was unlikely that the NEA would consider granting such a concession because many of the backers of its bill, particularly the Southern Jurisdiction of Scottish Rite Masons and the Ku Klux Klan, also supported compulsory public schooling.

The NCWC warned Catholics about the formation of a national conspiracy against parochial education, with the NEA at the center. Already allied with the Rite, the NEA openly promoted two of three elements in the Masonic educational program, namely, the establishment of a department of education and the foundation of a national university. NCWC officials believed that at least some authorities in the NEA also covertly supported the third element: compulsory public education. In 1922 the American Legion, which, in historian John Higham's words, was "the supreme embodiment of the 100 per cent philosophy" and whose educational program was avowedly nativist, linked itself with the NEA through the establishment of a joint advisory committee on cooperation.[11] The NCWC was "not surprised" that the NEA "would initiate and support, either in whole or in part, an educational program of this character." Surprising it was, however, that behind the NEA was a host of organizations like the Masons, the American Legion, feminist groups like the League of Women Voters and the Women's Christian Temperance Union, church bodies, and anti-Catholic societies whose purposes were "only remotely connected with the progress of education." This naturally led to the conclusion "that there exists in the United States a formidable union of forces directed against the Catholic school and, by consequence, against freedom of religion as well." The NCWC urged Catholics to explain to open-minded friends the nature of parochial schooling. "We want religious peace in the

United States. We do not want religious war. . . . If, after having exhausted every argument to justify our right to the maintenance of a separate system of schools, that right is curtailed or threatened . . . , the only alternative left us is to fight—and fight we will. This is not a warning, a piece of bravado, an open threat to the N.E.A., American Legion, or the Masons. It is a simple statement of set purpose which we mean to go through with to the bitter end."[12]

In fairness to the NEA, Ryan did not hold the entire organization responsible for its policy. The enemy was a clique of educators centered in Columbia University's Teachers College, principally George Strayer and William Bagley, who operated through the NEA, women's clubs, and fraternal organizations, particularly the Scottish Rite Masons.[13] According to John Burke, these men were not anti-Catholic themselves, but nationalists who sincerely believed that the country's well-being demanded federal control of education and compulsory public schooling.[14] Still, they distanced themselves from no one who supported their program. Because Catholics considered themselves to be as American as their fellow citizens, it was difficult for them to understand the hostility toward their schools as anything other than religious prejudice. Hence, the depth of feeling expressed by the NCWC that religious war was about to be engaged.

On 7 November 1922 voters in Oregon passed the compulsory public education law, confronting the Catholic church with the abolition of its schools in that state, effective September 1926. A month later, Dowling, Burke, Pace, and Admiral William Benson, retired chief of naval operations and president of the NCCM, met in Washington to discuss the situation with influential Catholic Democrats: Senators Joseph Ransdell, David Walsh, and Thomas Walsh, Representatives Nicholas Sinnot of Oregon and Vincent Brennan of Michigan, and Constantine Smyth, chief justice of the United States Court of Appeals for Washington, D.C. Dowling began by expressing the dashed hopes of Catholics that the church's war record would render religious prejudice negligible by proving the essential Americanism of the church. Both Ransdell and David Walsh believed that the church had brought the present disaster upon itself. Catholics, especially many in the clergy, had been too critical of public education and had opposed progressive legislation like the Sterling-Towner bill and other measures that favored public schools. Worse, they publicly boasted about defeating such legislation. Walsh believed that the church should attack no "wise" legislation that promoted public education and that Catholics ought to be willing to bear their share of taxation to enhance such schools. Ransdell thought a Protestant misconception about the aims and purposes of the Knights of Columbus had contributed to the situation. In the view of many, the knights were engaged in a religious joust with the knights of the Ku Klux Klan. According to Thomas Walsh, Democratic Senator Joseph Robinson of Arkansas had recently been invited to join the Klan on the basis that it was just a Protestant version of the Knights of Columbus. Smyth believed that the separate-but-equal style of Catholicism, wherein the church established its own organizations to parallel those already in the larger society, had created a great deal of opposition in the minds of Protestants. All agreed that the church should make a forthright statement in favor of both public and private education, something that Dowling assured them that the bishops would do.[15]

Convinced that the Oregon Law was not a local issue but the spearhead of a national movement, the bishops of that state asked for the NCWC's help. Archbishop Edward Hanna invited Christie to meet with the Administrative Committee in January 1923.[16] Present at the meeting were Burke, Ryan, Pace, Christie, Bishops John Carroll of Helena and Michael Gallagher of Detroit, and three Catholic judges, including John Kavanaugh of Portland. Kavanaugh explained that Christie wanted the matter resolved by a decision of the U.S. Supreme Court. In Kavanaugh's view, the law was unconstitutional and should be tested on the basis of equity. His fellow jurists advised otherwise. In view of the present Court's strong advocacy of religious, family, and parental rights, they thought that the attack should be made on those grounds rather than on property. So too did William Guthrie, a former professor of constitutional law at Columbia University, whose advice Burke had sought. In Guthrie's opinion, expressed by letter, the statute was unconstitutional because it violated the spirit of personal liberty contained in the Fourteenth Amendment. Although it also interfered with property rights, he warned that a decision made solely on that ground "might not adequately protect the Catholic Church in the future, and might not be a satisfactory basis upon which to have any decision stand." The committee then asked the jurists to depart. With only clerical advisers present, the board considered the case. Given the national Catholic interest at stake, the committee agreed to lead the fight against the Oregon School Law and entered an agreement with Christie.[17]

The NCWC rose to the defense none too soon. Around the country the 100-percenters began seeking proscriptive or regulatory legislation against church schools. In December 1922 *Colonel Mayfield's Weekly* (Texas), believed to be a Klan mouthpiece, decried the disloyalty of parochial institutions, which inculcated allegiance to a foreign dictator and took the Bible out of the hands of children. It announced that Representative A. D. Baker would introduce a bill into the state legislature calling for compulsory public education. Once Catholic grade schools were closed, those "slave pens" that went by the name of Houses of the Good Shepherd (Catholic homes for wayward women) would be next. The Baker bill was less drastic than expected. It empowered the state superintendent to give quarterly examinations to children in private schools to ensure that their performance equaled that of their public-school counterparts. If it failed to do so, he was required to place the pupils in public institutions. The bishops of Texas brought such pressure to bear on the legislature that the Klan vote split, defeating the bill.[18]

In California, Masons and Klansmen joined forces as they had in Oregon. Golden State Masons boasted a long tradition of devotion to public education and were successfully concluding a ten-year campaign for the removal of the Catholic Joseph Scott from the Los Angeles City School Board.[19] In January 1923 Assemblyman Hugh Pomeroy introduced several bills aimed at controlling Catholic education by regulating the building of new private schools, empowering the state superintendent to inspect them and approve all courses taught in them, and stipulating that the principals of these new institutions must be 100 percent Americans. Several months later Assemblyman George Cleveland sponsored a compulsory public education bill. In the belief that these measures were Klan-backed, and out of fear that the Invisible Empire would tear the state apart with a school fight, the Assembly's Education Committee killed some of the bills and forced the withdrawal of others. Denying any Klan connection, Pomeroy claimed his legislation had the support of San Francisco merchants, who were determined to bring the Oregon Law to California even if they had to do so by initiative. Nothing came of his threat.[20]

Elsewhere around the country 100-percenters tried unsuccessfully to enact Oregon-type legislation. The Arkansas legislature defeated such a bill. In Michigan James Hamilton tried yet once more to amend the state constitution by initiative. Thanks to the timely passage of a new law regulating this procedure, he failed to secure the requisite number of signatures to place the measure on the ballot.[21]

The Hearst papers, too, argued the educational case for 100 percent Americanism. Arthur Brisbane, a syndicated editor of the chain, styled the public school the "LABORATORY OF DEMOCRACY" and considered "NEGLIGIBLE" the accomplishments of men who had attended private schools. Fortunate were the children who matriculated in public institutions, for "they alone receive[d] a really AMERICAN AND DEMOCRATIC education." According to Brisbane, state schools were the great levelers of society wherein all met without regard to class distinctions. He warned that the public school had enemies: "selfish tax-payers that choose to send their children to private schools" and those who constantly opposed public education because they did "not believe in the American principle upon which the public school is founded." Americans, declared Brisbane, would brook no interference with their institutions.[22] Here was a classic statement of the old-stock American side of the battle.

The NCWC tried to curb the anti-parochial school campaign in several ways. In February 1923 its Department of Education established the Catholic School Defense League to disseminate information among non-Catholics about the nature and aims of parochial schools. Free monthly publications were to be sent to congressmen, state legislators, professionals, and businessmen to acquaint them with the Catholic viewpoint. By September 1923 the League had raised more than $5,000 and distributed more than 100,000 copies of its literature.[23]

The NCWC also entered into direct negotiations with the NEA. At the January meeting of the Administrative Committee, Bishop Louis Walsh reported that the NEA's annual convention in the summer was to be an international affair, hosting representatives from many countries. The committee asked Burke to contact the NEA about the possibility of ensuring that Catholics were included in the foreign delegations so that they might explain the harmonious working of church and state in their national systems of education.[24] This he did during the spring, while Ryan and his friend William Burris worked to persuade NEA officials to introduce a resolution in support of private education. Their efforts failed. While professing belief in the right of private education, William Owen, president of the NEA, refused Catholics a place on the program. Nor would he seek a resolution upholding the right of private schools. He promised, however, to introduce and support one for the appointment of a commission to define the limits of state absolutism in education, a panel that was to include public, private, and parochial educators.[25] As late as mid-June, Catholic officials held on to the hope that Owen would redeem his promise.[26] He failed to do so. Realistically, how could he have done otherwise while the Scottish Rite backed the NEA's program with liberal amounts of money? It would be another year—after Masonic largesse had evaporated—before the NEA reluctantly supported private schools.

In April the Administrative Committee made adjustments with regard to the ill-starred pastoral letter on education. Dowling considered his mandate impractical. Given the urgency of the situation and the impossibility of composing anything that would please a hundred bishops, the committee scrapped the idea and ordered him to have the NCWC Department of Education write a statement on Catholic educational principles to be released through the CEA at its annual convention.[27] The CEA adopted the statement in the form of a lengthy resolution. Affirming that the welfare of the nation demanded an enlightened populace, it stressed the long-standing Catholic belief that "religion, no less than secular knowledge is necessary for the worthy discharge of the duties of citizenship." Because Sunday schools were an inadequate means of imparting religious instruction, Catholics had founded parochial schools, the type of educational institutions that antedated their public counterparts. To forbid attendance at church schools was tantamount to religious persecution and struck a blow at parental rights and personal freedom.[28]

Catholic hopes received a boost in June 1923 with the Supreme Court's decision in the case of *Meyer* v. *Nebraska,* litigation bearing on the Oregon Law. Two years earlier, the Nebraska legislature had passed the Reed-Norval Act mandating that all instruction in public and private schools be given in English.[29] Attorneys for the NCWC filed an amicus curiae brief pointing out the constitutional principles of the case without siding with either party lest an adverse decision affect the coming Oregon suit. It apprised the Court of the Oregon Law and the assumption on which it rested: "that the police power of a state over the education of minors is virtually unlimited." While upholding a state's right to set "certain minimum standards" in education and to "prohibit certain species of additional instruction duly found to be inimical to the public welfare," it denied that a government had authority to arrogate to itself the education of all children or to forbid any and all instruction above the prescribed course of study. To allow a state such power would be to ape Soviet Communism.[30]

On 23 February the Court heard the case. Burke reported that the justices were unusually interested, asking many pointed questions. Chief Justice William Taft inquired about the history of education and the rights of the state therein. One associate justice asked if a state had the right to compel children to attend public schools. "It is evident," Burke confided to Bishop Peter Muldoon, "that they are expecting the Oregon case. If the Nebraska case is decided in favor of Nebraska, we will have a difficult time, indeed."[31]

Fortunately for Catholics, it was not. The Court ruled that the Reed-Norval Act violated Fourteenth Amendment liberties. Writing the majority opinion, Justice James McReynolds argued that on one hand, parents had the duty to provide an education for their children; on the other, the state had the right "to compel attendance at some school and to make reasonable regulations for all schools." While the Court conceded that the state might go very far to improve the quality of its citizens, it must respect the rights of individuals, especially parents. "It is the natural duty of the parent to give his children education suitable to their station in life," noted McReynolds. Although the state might consider it advantageous for everyone to be competent in English, it could not promote this by unconstitutional means. The Reed-Norval Act did just that.[32] The decision vindicated personal and parental rights in education and virtually ensured that the Court would overturn the Oregon law.

In December 1923 attorneys for the NCWC took steps to undo that legislation by employing a special judicial procedure for expediting constitutional challenges. Acting on behalf of the Society of Sisters of Jesus and Mary, which operated a number of parochial schools in Oregon, church attorneys filed an amended bill of complaint asking that a federal district court of three judges sit en banc to grant an injunction against enforcement of the law. Appeal from such a tribunal was immediate to the Supreme Court, which could not refuse to hear the case. Arguments were heard in January 1924, and in March the three-judge panel granted the injunction on the ground that "parents possess a natural and inherent right to the nurture, control and tutorship of their offspring that they may be brought up according to the parents' conception of what is right and just, decent and respectable." Like the high bench in *Meyer* v. *Nebraska,* the district court upheld parental rights. Governor Walter Pierce, who had been elected with Klan support two years earlier, appealed to the Supreme Court.[33]

While working to secure the rights of private education, the NCWC had to hold the front against the Sterling-Towner bill, legislation openly identified by elements of Masonry with the attack on parochial education. For example, on 17 February 1923 the editor of the Masonic *National Observer* of Minneapolis declared, "The Jesuits and other public school enemies have seen 'the handwriting on the wall.' They realize that a national department of education, with a secretary in the President's Cabinet, as provided by the Sterling-Towner educational bill, spells the eventual doom of their alien, imported parochial school system, and these subverters and underminers of American ideals and liberty are fighting to the last ditch the creating of a national department of education." A fortnight later the *Fellowship Forum* (Washington, D.C.), an organ of the Klan wing of Masonry, reprinted the Minneapolis statement in whole.[34]

In the battle against the Sterling-Towner bill, help came once again from President Warren Harding. In February 1923, scarcely a month before the lame-duck Congress expired, he submitted the long-awaited reorganization plan. As anticipated, and much to the chagrin of the NEA, it proposed a combined department of education and welfare rather than a separate department of education. One of four assistant secretaries was to head the Bureau of Education. This officer also had responsibility for the Bureau of Indian Schools, the Smithsonian Institution, Howard University, and Columbia Institution for the Deaf (present-day Gallaudet University). The legislation also abolished the Board of Vocational Education and transferred all its duties and authority to the assistant secretary of education.[35] The plan pleased Catholics because it simply unified and coordinated existing educational agencies without expanding them.

Harding then removed from the legislative scene one of the sponsors of the rival Sterling-Towner bill. He appointed Horace Towner to replace E. Montgomery Reily as governor of Puerto Rico. As chairman of the House Committee on Insular Affairs, Towner was an ideal choice, the more so because of his tact and patience, which Reily lacked. Though the appointment probably had little to do with the education question, it robbed that cause of an ardent proponent, just as William Kenyon's appointment to the federal bench had previously robbed the farm bloc, which favored the NEA's bill, of an able leader.[36]

While the Reorganization bill and Towner's removal played into the NCWC's hand, the Catholic church also gained an ally in the United States Chamber of Commerce. In November 1922 the chamber had appointed a special committee to study the creation of a department of education with federal aid. Intended to provide information for a referendum among local branches, the committee's report took a negative stand on both issues. From December through February, the chamber conducted balloting, and the NCWC Department of Education sent literature against the Sterling-Towner bill to each local branch. The result was gratifying. On separate questions, local chambers resoundingly rejected the creation of a department and federal aid by a three-to-one margin.[37] Editorializing on this resounding defeat, the NCWC expressed hope that the NEA and its supporters would "profit by this stinging rebuke on the part of the most representative business group in the United States and . . . devote their energies to saner more useful activities."[38] The editor of the *New York Times* disagreed. Breaking editorial silence on the education question, he asked who knew more about schooling, the NEA Department of Superintendence or the business community?[39] This criticism, however, did not mean that the *Times* endorsed the Sterling-Towner bill.

Opponents of the measure themselves suffered a setback in June 1923. The same day that the Supreme Court handed down the decision in *Meyer* v.

Nebraska, it settled *Massachusetts* v. *Mellon,* which had far-reaching implications for the Sterling-Towner bill. At issue was the Sheppard-Towner Maternity Act (1921), which offered aid to states on a matching basis for needy mothers and their newborn, a perfect parallel to the federal funding in the Sterling-Towner bill. Endorsed by the NCWC, the Sheppard-Towner Act now came back to haunt it. The State of Massachusetts had refused the aid and sued Secretary of the Treasury Andrew Mellon for violation of states' rights. Attorneys for the state argued that the Constitution granted Congress no power to legislate regarding healthcare, a Tenth Amendment right reserved to the states. The act imposed on a state the unconstitutional option of either yielding a portion of its sovereignty or forfeiting its share of the appropriation, which the taxation of its citizens helped raise. The tax burden too fell unequally on the states.[40] These were the arguments that Catholics and others marshaled against the Sterling-Towner bill. Thus, the Court's decision in this matter would have tremendous repercussions on the federal education question.

The Court found itself in the position of being asked to settle what it considered a political question. Expressing a unanimous opinion, Justice George Sutherland declared invalid the argument that the burden of taxation fell unequally on the states. Congress taxed the citizens of the United States, not the states themselves. Nor had the act violated states' rights because each state remained free to accept or reject the federal money. Nothing was "to be done without their consent." "It is plain," concluded Sutherland, "that the question, as it is thus presented is political, and not judicial in character, and therefore is not a matter which admits of the exercise of judicial power. . . . We are called upon to adjudicate . . . not quasi sovereign rights actually invaded or threatened, but abstract questions of a political power, of sovereignty, of government." The case was dismissed for want of a justiciable issue.[41] This landmark decision made it impossible to test the constitutionality of any similar legislation. Federal aid offered on a matching basis for Tenth Amendment matters was not unconstitutional as long as states were free to accept or reject it. As such, the issue remained a political matter, a question of government. If the Sterling-Towner bill passed, the Court would provide no recourse.[42]

The midterm election of November 1922 had altered the political landscape in Congress. Democrats made sweeping gains, especially in urban areas of the Northeast and Midwest, greatly reducing the Republican majority in both houses. The election also replaced a significant number of Republican party regulars with anti-administration, western Republican progressives who bolstered the ranks of the farm bloc. The latter's gain in strength was offset by its lack of effective leadership after the departure of Kenyon. Bloc members distrusted old-time Republican progressives like Senators Robert La Fallotte of Wisconsin and George Norris of Nebraska, who could have provided the needed guidance. Instead, Republican Senator Arthur Capper of Kansas became the titular head of the movement and failed to give it any cohesive agenda.[43]

In the wake of that election, the Democratic party was one of extremes. In the words of historian David Burner, it comprised "the most Jeffersonian or populist of the farmers, particularly in the South, and the most powerful of the urban immigrant machines." In other words, it was an amalgam of states' righters, nationalist progressives, and urban immigrants. Rural Democratic progressives shared in the moralism of that movement, which expressed itself in a commitment to Prohibition and Protestant, Victorian values. Urban, Democratic, political machines, on the other hand, tended to be dominated by Irish-Catholic politicians attentive to the needs of the immigrant and supportive of a cosmopolitan lifestyle, including the culture of drink. For these reasons, the cultural tensions evident in society as a whole played themselves out with a vengeance within the Democratic Party.[44]

At its annual convention in July 1923, the NEA terminated its moratorium on agitating the Sterling-Towner bill. A year earlier, it had promised not to press for a separate department of education until Harding submitted his reorganization plan. With that plan before Congress, the NEA proceeded. Strayer reported that senators and representatives had persuaded the Legislative Commission that a "Department of Education may be realized in the next session of Congress," no doubt because of the weakening of the Old Guard in the midterm election. In defense of the bill's hefty appropriation, one of its principal drawbacks, he argued the justice of federal aid on the basis of the nationalist philosophy. Government statistics showed that the per capita income of states varied from $345 to $850. In four of the six poorest states, citizens devoted a larger percentage of their income to the support of public education than did those in the wealthy State of New York. "It is unjust to require the people of one State to tax themselves twice as heavily as do the people of another State for the accomplishment of a National purpose," declared Strayer. ". . . The safety of the nation, the perpetuity of our form of government . . . depend upon the education provided for all of the children of the Nation." Therefore the Sterling-Towner bill would be reintroduced in the next Congress.[45]

Charl O. Williams, who replaced Hugh Magill as field secretary of the NEA, outlined the plan of campaign. She asked for several hundred volunteers for service in a speakers' bureau to boom the bill in towns and cities around the country. Addresses on and discussions of the measure were to be held in every summer school program for teachers. Directors of these programs had given enough support for her to predict that 200,000 teachers would be informed about the legislation. She urged the NEA to employ a professional writer to prepare attractive literature on the bill. "There are 433 congressional districts in the 48 States," declared Williams, "and there the battle will be largely, if not wholly won."[46] A later addition to the campaign was a weekly radio program broadcast by Radio Station WRC in Washington, D.C., to drum up support.[47]

The Klan continued to harp on its anti-Catholic theme. Outlining the Empire's immigrant restriction policy, Imperial Wizard Hiram Evans stated that a nation could not long endure if it allowed a temporal allegiance higher than the one to its own government. In his view, the Catholic hierarchy believed that the president was "subordinate to the priesthood at Rome," and the parochial school alone was "sufficient proof of a divided allegiance, a separatist instinct."[48] Putting the stamp of Klan approval on the NEA bill, he accused the Catholic church, "one of the oldest and most powerful special interests in the world today," of keeping it from becoming law: "The hierarchy of the Roman Catholic Church stands against America on this issue. The public school, in its every phase, aspect and result is repugnant to the Pope and all his priesthood. . . . [T]he Roman Catholic hierarchy is the one influence that is successfully obstructing adequate public school education in America."[49]

As the Klan gained a reputation for intolerance and bigotry, the Southern Jurisdiction of the Scottish Rite tried to dissociate itself from it. At the October 1923 meeting of the Supreme Council, the Rite forced George F. Moore, a former grand commander, to resign from the board because of his connection with the *Fellowship Forum*, an anti-Catholic organ that voiced the views of the Klan wing of Masonry. His resignation came in response to accusations that he used the journal to further Klan aims.[50] Although the jurisdiction tried to sever any connection with the Invisible Empire, the two organizations backed identical educational programs and considered the Catholic church their principal adversary in the promotion of a department of education. Moreover, the Klan chose Masons as recruiters and considered their lodges, especially those of the Scottish Rite, prime hunting grounds for members because of Masonry's anti-Catholic attitude.[51]

While distancing itself from the Klan, the Rite organized a drive for the Sterling-Towner bill. Although denying any anti-Catholic sentiment, Reynold Blight, editor of the *New Age*, reported to a brother Mason: "The only foes we have to fear are the parochial school interests. . . . From all over the country comes the information that the Catholics are fighting it [the Sterling-Towner bill] bitterly and will move heaven and earth to defeat our purposes." Blight had attended many meetings to plan the congressional campaign for the measure. The Masonic lobby could prove formidable, for by a Klan count Masons filled nearly two-thirds of the seats in the Senate of the Sixty-eighth Congress and almost half of those in the House.[52]

Though the NEA, the Southern Scottish Rite, the Klan, and others were determined to push through a department of education, both they and the Catholic church had to fathom the position of a new principal, President Calvin Coolidge, who in August succeeded the late Harding. No one knew where Coolidge stood on education, so representatives of the NEA paid him a visit in November. Fearing that Coolidge might think that the NEA reflected national sentiment on education, Ryan asked high-level educators to urge the president not to endorse a department of education. Among those who complied were Nicholas Murray Butler, president of Columbia University, and A. Lawrence Lowell, president of Harvard.[53] There was actually little cause for fear. Even more pro-business than his predecessor, Coolidge was what historian Paul Johnson calls a "minimalist politician," one who viewed the prime function of government as the enforcement of law. Government was essentially unproductive; the private sector generated wealth. The less the government interfered in business and the more it reduced the tax burden, the more it would promote prosperity. Like Harding, Coolidge was economy-minded. A wizard at public finance, he wanted to pay down the national debt through the creation of budget surpluses.[54] In short, he disfavored an expansion of federal power and further federal spending. In his first state of the union address, he declared education to be a "peculiarly local problem," an endeavor that must "always be pursued with the largest freedom of choice by students and parents." Opposed to federal aid, Coolidge did believe that education could benefit by the counsel of the federal government. In his view, schooling belonged in a joint department of education and welfare, part of Harding's Reorganization proposal, a grand scheme that he felt morally obligated to support while filling out the term of his deceased predecessor.[55]

While the new president set forth his view on education, the ACE's Committee on Federal Legislation considered all alternatives in the fall of 1923. Concluding that the ACE could support only the department feature of the Sterling-Towner bill, the committee urged the NEA to separate the measure into two pieces of legislation, one for a department and the other for federal aid, before resubmitting the proposal to Congress.[56] Disregarding this advice and the wishes of Coolidge, the NEA had Republican Senator Thomas

Sterling reintroduce its bill unchanged. Perhaps in a bid to woo support of the commercial and industrial states that would pay for the programs contained in the legislation, the association chose Republican Representative Daniel Reed of upstate New York, a Northern Scottish Rite Mason and supporter of the bill's predecessors, to replace Towner as cosponsor of the measure in the House.[57]

Because the Sterling-Reed bill was an unchanged copy of the Sterling-Towner bill, Catholic editors marshaled old arguments. Most charged that it would result in federal control of education with all the evils attendant on a bureaucracy, cataloged by Catholic Democratic Senator David Walsh of Massachusetts, as "an orgy of incompetency, neglect, delay, procrastination, and graft."[58] A few papers warned of the measure's anti-Catholic connections. The *Denver Catholic Register* pointed out that the Sterling-Reed bill was only one part of the Southern Scottish Rite's educational program, which included compulsory public schooling. "Proof that all these schemes are fitted into one grand plot will be found in any issue of the Scottish Rite clip sheet."[59] The Chicago *New World* also pointed to the company that the bill kept. The Ku Klux Klan supported it and also campaigned actively for the abolition of parochial schools, a campaign that the NEA refused to condemn. But then, asked the editor, how could it censure an organization favoring its bill?[60] In fact, the Klan backed it in the hope that federal aid would strengthen public schools and weaken Catholic ones.[61]

Ryan argued to the Catholic clergy that the church was at war with Masonry over parochial education. "A *status belli* [state of war] already exists," he declared. ". . . A war to extinction has been decreed, and only the complete wiping-out of every private school the country over will satisfy the purposes of the enemy." The real opponent was not the Klan. The latter was simply being used by the Southern Jurisdiction of the Scottish Rite Masonry, said Ryan. "Another fact of the utmost significance . . . is that the Scottish Rite has joined hands with the National Education Association . . . in the promotion of the Towner-Sterling bill [*sic* for Sterling-Reed]." The Rite, the NEA, and its supporting organizations boasted a following of 25 million people. Not only was this a formidable phalanx but also it did "not understand the word 'compromise,' and has served notice again and again, upon Congress, the public, and, just recently, on no less a person than the President of the United States, that its program must go through without change."[62] In Ryan's view, the Sterling-Reed bill was one element of a grand scheme to destroy parochial education. His contention that the Southern Jurisdiction was using the Klan as a pawn is understandable. Their educational programs were identical, and the Rite had established its policy at least a year or more before the Klan. Yet, the reality of who was using whom was probably the other way around, as indicated by the Rite's recent attempt to distance itself from the Klan.

In Congress, the Sterling-Reed bill picked up an erudite critic, Democratic Representative Henry St. George Tucker of Virginia, a new member of the House Committee on Education and a former professor of constitutional law who adhered to strict construction. Because the Court's decision in *Massachusetts* v. *Mellon* left it to Congress to argue the constitutionality of such legislation, he tutored the House in a speech that covered twenty-four pages in the *Congressional Record.* In a lengthy historical, legal analysis of the intent and interpretation of the General Welfare clause, he argued that Congress could act under it only in matters that touched the people as a whole, not a part, not a class, but the public. This, he said, was the clear intention of the founding fathers, who at the Constitutional Convention specifically rejected any wider view. Tucker disagreed with arguments that the Morrill Acts, granting aid to agricultural colleges, had set a precedent for federal intervention. He contended that because laws enacted under the Articles of Confederation remained in force under the Constitution, the First Morrill Act (1862) had simply enlarged upon the Land Ordinance of 1785 by extending land grants for education. The Second Morrill Act (1890), however, presented a different matter because it granted direct monetary aid to the land grant colleges. The question remained, had Congress acted constitutionally in passing that law? The constitutionality of the Second Morrill Act had never been tested, and despite his own belief that it should be, Tucker doubted that it ever would be because of the Court's decision in *Massachusetts* v. *Mellon.* Hence the need for Congress to have its constitutional principles in order when considering the Sterling-Reed bill. His lecture demonstrated not only the serious constitutional questions involved but also that the bill would have a fierce opponent on the Education Committee.[63]

In January 1924 the Senate Committee on Education and Labor held hearings on the bill, while the Joint Committee on Reorganization did the same with the educational aspects of its proposal. Representing the NCWC, Ryan appeared before the former and sent a copy of his statement to the latter. The NCWC, he said, would favor the government's every effort in the promotion of education so long as the efforts were constitutional. Unfortunately the means of achieving the laudable purposes of the Sterling-Reed bill were not and would lead to the federal control of education. In support of this contention, he had Tucker's speech printed in the record of the hearing. Then, for the first time, the NCWC made a positive recommendation on the issue. Recognizing the gravity of the American educational situation, it called for the appointment of a national education commission "to conduct a scientific inquiry into the educational needs of the country and, as a result of this survey, to propose ways and means of correcting defects which now exist." Moreover, the NCWC urged Congress to create a Federal Board of Education to direct the various educational agencies scattered throughout the government. The board's purpose would be to investigate problems, disseminate information, set standards for efficiency, and "stimulate in every way possible the people and educational agencies of each State to better efforts in behalf of education." In Ryan's estimation, such an agency would have all the advantages and none of the disadvantages of a department of education. It would stimulate without controlling.[64] Thus, the NCWC threw in its lot with educators like William Burris, Frank Cody, the late Alexander Inglis, and Frank McVey, all of whom wanted a federal commission on education and/or a federal board of education.

The questions asked of witnesses who appeared before the committee indicated that Democratic senators from the urban industrial states that would be expected to foot the bill for the aid in the Sterling-Reed measure—men like Walsh of Massachusetts, Royal Copeland of New York, and Woodbridge Ferris of Michigan—opposed either the funding feature or the entire legislation. Joining their ranks was Republican Senator James Couzens of Michigan.[65]

Shortly after the Senate hearing Representative Frederick Dallinger of Massachusetts, Republican chairman of the House Committee on Education, introduced a bill that the NCWC could support. It provided for the better definition and extension of the Bureau of Education. The bureau's field of investigation was to be expanded to include the areas covered in the Sterling-Reed measure: illiteracy, Americanization, physical and health education, and teacher preparation. Absent, however, was federal aid for these features. Perhaps borrowing from Ryan, the bill proposed a federal council on education, comprising the commissioner and one representative from each executive department, to coordinate the various educational agencies throughout the government. In imitation of the Sterling-Reed bill, the measure empowered the commissioner to appoint a national council on education to advise him.[66]

The NCWC heralded the Dallinger bill as legislation "which educators can support without stultifying themselves." Ryan predicted that if enacted, it would "accomplish the alleged purposes of the Sterling-Reed Bill in a shorter time and in a more efficient manner than would that much-discussed educational measure."[67] Faced with a choice between the two pieces of legislation, the editor of the Catholic *New World* said he would gladly choose the Dallinger bill because it would "assist and not strangle, and . . . direct without controlling."[68] Without giving it the stamp of editorial approval, *America* ran an article in its favor. The author concluded that the measure would "give educators what they have long desired—an active, fearless, intelligent and effective Bureau of Education."[69] Thus, for the first time, Catholics gave a piece of federal educational legislation favorable notice. Though not an official endorsement, the NCWC's positive stance was a demonstration of Catholic support for public education and an indication of its strategy to take a benign view of acceptable alternatives to the NEA's measure.

The School of Education of the University of Chicago also gave the bill qualified support. Recognizing that Dallinger had proposed the legislation as a way around the logjam caused by the Sterling-Reed bill, the editor of the *School Review,* Charles Judd, highlighted the positive features of the measure: it moved the Board of Vocational Education into the Bureau of Education, it retained the idea of an interdepartmental council on education to coordinate the educational activities of the various executive departments, and it greatly increased the appropriation for the Bureau of Education so that it could conduct scientific investigations. While urging educators to petition their congressmen to have Dallinger alter his bill to elevate the bureau to department status, Judd concluded that if such an effort were to fail, "let us by all means have an expanded Bureau of Education."[70]

From February through June 1924, the House Committee on Education held intermittent hearings on both the Dallinger and Sterling-Reed bills. Dallinger advised the NCWC to absent itself, no doubt because it favored his measure, and an appearance would charge the atmosphere with the religious issue. So the NCWC solicited the help of others. Ryan got the United States Chamber of Commerce to send a delegate to present the results of its 1923 referendum. He also had the Lutheran church send three representatives: Mr. C. M. Zorn, chairman of the School Committee of the Evangelical Lutheran Conference; Reverend J. C. Baur of Fort Wayne; and Reverend F. J. Lankenau, editor of the *Lutheran Pioneer.* Among Protestant denominations, Lutherans would suffer most from the enactment of compulsory public education laws. Like Catholics, they had built a system of parochial schools, many of them conducting instruction in German. The three Lutherans opposed the Sterling-Reed bill because it was supported by the Scottish Rite Masons of the Southern Jurisdiction, the Ku Klux Klan, and various patriotic groups, all of them organizations that also favored compulsory public schooling. As Lutherans saw it, the logical outcome of enactment of the Sterling-Reed bill would be federal control of education and the end of parochial schooling.[71] By calling on the Lutherans, the NCWC was able to make its point by indirection, without fueling the flames of anti-Catholicism.

After the hearing, Catholic Democratic Congressman Loring Black, Jr., of New York, a member of the House Committee, attacked the legislation from the floor. The bill was too expensive, and its apportionments would, in reality, do little to equalize educational opportunity. Industrial-commercial states like New York, which already trained and paid its teachers well, would receive more in aid to promote these very functions than would a rural state like Mississippi. The only way real equalization could be achieved would be "by ultimately wiping out territorial lines in education, so that teachers would get uniform pay regardless of habitats and would also be required to teach where sent by Washington, regardless of residential preferences." Black believed that the bill would necessarily lead to federal control of education. In support of this contention, he noted that during the hearing, John A. H. Keith, president of the Pennsylvania State Normal School, had admitted that the secretary of education would set the standard for Americanization to be met under terms of the bill. More to the point, Dr. J. O. Engleman of the NEA had asserted that the secretary would determine standards for all schools to meet, despite the measure's many protestations against federal interference. "Whether they want it or not," concluded Black, "the sponsors of this bill are going to see Federal control if the bill becomes law. The whole scheme and logic of the legislation means that."[72]

As John McCracken, chairman of the ACE's Committee on Federal Legislation, summed up the situation, the hearings had "developed no new aspects of the question." Rather, they revealed that interest in the matter had increased and that feeling was "growing intense and bitter, both among supporters and opposers of the measure." It was generally conceded that there was no legislation before Congress packed with as much "political dynamite" as the Sterling-Reed bill. Consequently, neither senators nor representatives were eager to be forced into taking a stand on the measure.[73] In other words, there was little likelihood that the education committee of either chamber would report the bill.

While the Sterling-Reed bill remained safely in committee, a different piece of legislation easily made it through Congress: the permanent solution to the immigration problem. A patently discriminatory measure, the Reed-Johnson bill excluded all Asians and continued the present quota system until 1927, though reducing each country's allowance from 3 to 2 percent of the number of its natives resident in the United States according to the 1890 census, rather than the 1910 census. Because the 1890 census antedated the massive influx of eastern and southern European immigrants, the bill virtually eliminated future migrations from those areas. After 1927 total European immigration was not to exceed 150,000 annually, and quotas were to be assigned on a proportional basis according to national origin. The number of newcomers from each nation was to bear the same relationship to 150,000 as the number of Americans of that national ancestry bore to the total population of the United States in 1920. This feature essentially slammed the door on the undesirable eastern and southern Europeans. Many Catholics correctly viewed the legislation as aimed at them. During committee hearings, the NCWC protested the measure as discriminatory and unjust. After an orgy of rhetoric about preserving a "distinct American type," meaning old-stock, Nordic ancestry, western and southern congressmen rammed the bill through the House. The Senate passed it with only six dissenting votes, and Coolidge signed it into law with the terse statement, "America must be kept American."[74] With the influx of undesirable foreigners stemmed, it remained to be seen if the country would use either a national system of education or compulsory public schooling (or both) to Americanize those already in the land.

As the first session of the Sixty-eighth Congress ended, the educational situation at the national level seemed secure for Catholics. The Joint Committee on Reorganization had reported a bill that included a department of education and relief, virtually identical to the earlier proposed department of education and welfare. While the NCWC considered it less preferable than the Dallinger bill, it was a minimally acceptable alternative to the Sterling-Reed measure, one to be endorsed only out of necessity.[75] News from the congressional Education Committees was encouraging. The Senate committee voted five to three against reporting the Sterling-Reed bill; the House counterpart refused to vote on either the latter or Dallinger's measure. Dallinger was running for the Senate and probably wanted to keep a volatile issue from damaging his chances.[76]

With the Sterling-Reed issue at bay, the NCWC focused attention on the national party conventions where the educational question would be complicated by pressure within the Catholic community to have both parties adopt an anti-Klan plank. As general secretary, Burke considered it his duty to recommend "some definite policy" for the NCWC to accept. "My own judgment," he wrote to the Administrative Committee, "is that it would be a very serious and perhaps a fatal error to work for the insertion of such a plank, to demand it, or to lay public stress and approval upon the statement and position of those who do favor it." His concern was that if the church sought or supported such a plank, either or both parties would demand Catholic support in return. In his view, the Klan was not a Catholic issue, but an American one and the best way to solve it "as an American issue is to leave it alone." If Catholics began playing politics, they would become tools of particular interests and get mired in "political hypocrisy and political chicanery." "The day that it [the NCWC] is termed a political body," concluded Burke, "that same day its usefulness will cease."[77] As he saw it, the NCWC was a Catholic lobby, a Catholic voice. It articulated the Catholic position on issues and attempted to sway Congress and the government to its point of view. What kept it from being a political machine was its unwillingness to trade Catholic support for party favor. The Administrative Committee endorsed his view completely.[78]

Burke drove the point home in May 1924 when the Indianapolis diocesan chapter of the National Council of Catholic Men demanded that the state convention of the Democratic party adopt an anti-Klan plank. The chapter acted without the permission of either the local bishop or the NCWC. Informed of the matter while convalescing from surgery in Hot Springs, Virginia, Burke ordered Ryan to try to reverse the action or at least to have the Indianapolis chapter make clear that it had acted on its own and not in the name of the NCWC.[79] The mind of the Administrative Committee was definite: the Welfare Conference would never trade political support for concessions. The NCWC could not back one party or another simply because it stood for or against the Klan. The same held true for the education question. Although he believed that the NCWC would be criticized by coreligionists for being insufficiently Catholic on what was supposed to be a critical Catholic issue, the NCWC and the church had to remain free from partisan politics. "As the charge against the Church is that she is a political power or aims to be such," concluded Burke, "so I think the church here and her attitude towards and trust in American institutions will be on trial, as perhaps never before. We to whom has been entrusted her public national position must sacredly strive to keep her position intact, above suspicion."[80]

Even as Burke articulated these principles, the NCWC was engaged in negotiations with the Republican party regarding its platform. In March, party officials had contacted him to inquire what the church would like to see in the party's platform. He thought that the NCWC ought to ask for a plank against the Sterling-Reed bill. Handling the negotiations for him, Ryan met in May with James B. Reynolds, executive secretary of the Republican National Committee. He encouraged the party to oppose both the nationalization of education and federal aid. Although women's organizations were pressuring it to endorse the Sterling-Reed bill, Reynolds was not intimidated by them and promised to arrange for Ryan to meet with the chairman of the platform committee once one had been appointed. In early June, Ryan reported that Republican leaders knew the church's position on the bill and had given reasonable assurance that they would not endorse the measure.[81]

Negotiations with the Democrats went less smoothly because of the Klan issue. The party was divided: urban North against rural South and West. In the North, its membership consisted mainly of wet, Catholic immigrants and hyphenates. In the South and West, it consisted of dry, Protestant, old-stock Americans. The Klan had gained sufficient strength within the party that it hoped to nominate the presidential candidate. The two main contenders for the nomination were Georgia-born William Gibbs McAdoo of California, representing progressive agrarian Democrats, and New York–born Catholic Al Smith, representing urban, machine Democrats. Senator Oscar Underwood of Alabama was a dark horse lacking credible southern credentials. Born in Massachusetts, he was a wet, anti-Klan Democrat whose father had been a colonel in the Union army. So, the Klan was forced to back McAdoo whether he wanted its support or not. Planning to work directly with Democratic officials, Ryan arranged a conference with the National Committee chairman. When Underwood announced his intention to fight the Klan openly at the convention, Burke ordered Ryan to cancel the meeting, for a battle against the Klan would directly or indirectly involve the church. So Ryan had John Wynne, a New York Jesuit, make a strong presentation of the NCWC position to Elizabeth Marbury and Caroline Ruutz-Rees, National Committee women from New York and Connecticut, respectively. Both promised to keep out of the platform a plank endorsing the Sterling-Reed bill.[82]

Although assured that neither party would endorse the measure, Ryan took no chances. Both he and William Cochran, director of the NCWC's Legal Department, went to Cleveland to be on hand at the Republican convention in case pressure for the Sterling-Reed bill became so intense that they would have to remind the platform committee of its commitment to the NCWC, a precaution that proved unnecessary. The convention endorsed the department of education and relief recommended by the Reorganization Committee and desired by Coolidge. Ryan considered the plank "more or less a gesture" and minimally acceptable to the NCWC, which preferred expansion of the Bureau of Education to the creation of a mixed department. In any case, the endorsement was harmless because Congress seemed likely to scrap the Reorganization bill.[83]

As for the rest of the convention, the urban business wing of the party was firmly in the saddle. Coolidge's nomination was a foregone conclusion. The platform praised the protective tariff, demanded economy in government, called for more tax cuts, opposed the cancellation of war debts, and recommended policies to give limited aid to farmers. It was a platform that Coolidge heartily endorsed: "I am for economy and after that for more economy."[84]

Ryan and Cochran left Cleveland for the Democratic convention in New York City, where the political atmosphere was explosive. With the Klan in control of several state delegations, Underwood allied himself with the backers of Smith to do battle by asking the party to condemn the secret empire by name, a request that would lead to an acrimonious fight that split the convention down the middle. Despite the delicacy of the situation, Ryan had to maintain a high profile because the NEA, rebuffed by the Republicans in Cleveland, came to New York in force to press for an endorsement of the Sterling-Reed bill. He saw Homer Cummings of Connecticut and other members of the platform committee and explained the Catholic position. They were receptive and, as he later expressed it, "stood behind us loyally."[85] That was an understatement. In fact, the party adopted an education plank written by Ryan and Pace. Upholding the "sovereign right" of states in education, it declared that each was responsible for the instruction of its citizens and "for the expenditure of the moneys collected by taxation for the support of its schools." The plank added that the federal government should offer "such counsel, advice and aid as may be made available through the Federal agencies."[86] Although Ryan and Pace had aimed at strict construction of the Constitution, the final clause was as elastic as the Constitution itself, leaving dangerously vague the kind of aid that the government was to offer. Four years later the NCWC would try to rectify this mistake at the 1928 convention. Looking ahead to the NEA's annual convention, Ryan remarked to Burke that the association's field secretary would "not have a very glowing report to make.... I sincerely hope that in this report they do not drag us in."[87]

Broadcast over the radio, the Democratic convention was long and contentious, at times requiring police intervention. It deadlocked over the presidential nomination. Backers of McAdoo and Smith went head to head for ninety-six ballots before the two contenders withdrew. On the 103rd ballot, the convention selected John W. Davis, a prominent Wall Street attorney, who was a home-rule advocate nearly as conservative as Coolidge. Though Davis tried to please both wings of the party, his support in the campaign came mainly from urban Democrats. The public display of disharmony and division within the party sealed its fate at the polls.[88]

The NEA convention opened in Washington, D.C., on the day that the Democratic convention closed. Frustration ran high among delegates, though officials kept it in check for the most part. Neither Strayer nor Williams mentioned the political conventions. The former reported only that the House and Senate Education Committees had held hearings on the Sterling-Reed bill, and he urged that the NEA use "every honorable means to bring the bill to the favorable consideration of representatives in Congress." Williams reported that the Reorganization Committee favored a department of education and relief, an unacceptable proposal. In her view, the association still had to answer the objection of federal control raised by those who misunderstood the Sterling-Reed bill.[89] James W. Crabtree, secretary of the NEA, spoke directly about religious objectors. No longer were they viewed as enemies of public education, but as "catspaws [*sic*]" being used by a "class . . . composed of the money-hearted rich and an arrogant aristocracy, which [was] violently opposing the adequate support of the [public] schools." The real enemy was "big interests" that were trying to keep taxes to a minimum. "There is no fight directed toward the private or religious schools," said Crabtree. "There should be none. These institutions have their place and receive the encouragement of the Association. There are thousands of teachers in private schools in the membership of the Association. These schools should have the respect of public-school authorities. . . . Our children and youth need these schools and many more. The big interests work great injury to private schools in attempting to prejudice them against the public schools."[90] In Crabtree's view, Catholics were being used by the same elements that supported the policies of the Harding-Coolidge administrations. Echoing progressive denunciations of big business, his words marked a belated and diplomatic rapprochement of the NEA with parochial education.

The NEA backed up Crabtree's remarks with action. Having suffered setbacks in Congress and in the party conventions, the NEA seemed willing to make a bid for Catholic support, grudgingly and a year late. The association adopted a resolution giving lukewarm support to private schools. After rhapsodizing public education, the resolution concluded, "The National Education Association, while recognizing the American public school as the great nursery of broad and tolerant citizenship and of democratic brotherhood, acknowledges also the contributions made to education by private institutions and enterprises, and recognizes that citizens have the right to educate their children in either public or private schools."[91]

The barometer of frustration came in the report of Bagley, editor of the NEA's *Journal.* Abandoning the discretion or conciliation of previous speakers, he blasted both parties for kowtowing to "strategic minorities" to win votes, minorities that protected their own ends at the expense of national interest. "Contrasted sharply with the treatment that our representatives have met in Congress is the treatment of quite the other sort at Cleveland and New York," declared Bagley. "At those two conventions the great questions have been, not what is right and just and best for the broadest and most enduring interests of our country, but how can we placate the minorities that hold the balance of power? Organizations that could be counted on to deliver the vote of these minorities had no trouble reading their planks into the platforms, or in suppressing planks they did not like." Bagley was accusing the NCWC of playing politics, of trading votes for concessions, the very thing that Burke and the Welfare Conference had avoided doing. Whether Bagley was unfairly reading this interpretation into the NCWC's recent efforts or genuinely believed it to be the case, the lesson seemed clear to him. If opponents were willing to cast votes for whoever promised to promote selfish interest, the NEA ought to do the same. "Should there not be a group that will throw its votes *en masse* to the party that promises to do the most for the basic interest or our national life," asked Bagley, ". . . and should not our profession be the nucleus of this group?"[92] His words were greeted with thunderous applause.

Interpreting them as a call for the formation of a third party, news reporters tried to contact NEA officials to sound the depth of support for one, but executives were unavailable for comment. Members of the association, however, admitted to deep hurt over the lack of backing that their program received from the major parties.[93] Given Crabtree's progressive-like denunciation of the big interests as enemies of public education and Bagley's call to cast votes for a party that would promote schooling, perhaps the two were suggesting that the NEA abandon its role as a lobby in order to play politics in the upcoming convention to nominate a progressive presidential candidate, an event that held the promise of the formation of a third party.

In any case, Bagley's speech raised doubts about the sincerity of the NEA's endorsement of private education. Despite assurances from Dr. J. A. C. Chandler, chairman of the Resolutions Committee, that the endorsement was meant as "an expression of friendship to private schools and loyalty to education everywhere," the rapprochement was probably an act of realpolitik.[94] At least the NCWC interpreted it so. Ryan confided to Burke that the NEA adopted the resolution only to mollify Catholic members and win the support of the NCWC for a department of education.[95] If the NEA was making a bid for Catholic support, it would be sorely disappointed. The hierarchy alone possessed authority to alter policy regarding the Sterling-Reed bill, and it was highly unlikely that the bishops would consider an NEA resolution sufficient protection for Catholic education.

Swallowing its frustration over the setbacks handed to it by both parties, the NEA was polite enough to cheer Coolidge's address at the closing session of its convention on 4 July, even though his words were not what the association wanted to hear. He reiterated the points made in his 1923 state of the union address, namely, that education was "peculiarly the function of the several states," that the federal government would grant no financial aid to schools, and that his administration would support only the creation of a department of education and relief.[96] In effect, the president told the NEA to back off; there would be no federal control, no federal aid, and no separate federal department of education, at least not on his watch. Although the NEA might applaud him, it was not about to appease him, as it had attempted to do with Catholics.

Coolidge's words won the endorsement of both the *New York Times* and Judd. In an editorial parsing the speech, the *Times* pronounced Coolidge's views "sound" on all counts. More acerbic, Judd considered the NEA "sorely in need of guidance." Echoing Upton Sinclair, who contended that the leaders of the NEA had engineered a top-down takeover of the association, Judd asserted, "There is a grave responsibility resting on those who have assumed to dictate, so far as possible, the policies and the officers of the Association." The NEA had recruited teachers after the war with the promise of federal subsidies, only to have them watch the organization exhaust its funds—their dues—"in politics and lobbying." The wisdom of Coolidge's words, said Judd, "might very well serve the officers of the National Education Association as the keynote of a new policy which would lead the organization out of the morass into which it has been drawn by internal jockeying." Making a final appeal to heed the president's policy, he concluded, "It is recognized on all sides that what this country really needs is a national agency for the scientific investigation of education and for the widest kind of publicity on educational matters." He begged the NEA to give up the false notion that public schools could not succeed without federal money. Schooling needed scientific research, not federal funds.[97] The fundamental difference that pitted Judd against Strayer and his allies in the NEA was this: Strayer and his colleagues believed that they had done the scientific research to show that federal aid was warranted, and they sought the creation of a department to give schooling the political recognition and the money they believed it deserved; Judd, on the other hand, viewed the creation of a federal educational agency—preferably a department, but an expanded bureau at least—as an apolitical organ for scientific research in education.

Heartened by progressive gains in the 1922 elections, the Conference for Progressive Political Action (CPPA) had called for a nominating convention to meet in Cleveland on 4 July, the day that Coolidge addressed the NEA. Most conventioneers wanted to form a third party, but Senator Robert La Follette forestalled the creation of one because it would place at risk the congressional seats of progressives who had won election two years earlier as Democrats or Republicans. In his view, the CPPA should simply run an independent candidate with the support of progressives in both parties. If the election bid were successful, a third party could be formed later. This was realistic advice because Republican members of the farm bloc had refused to bolt from the party. The convention nominated La Follette and chose Democratic Senator Burton K. Wheeler of Montana as his running mate. Republican Senator Smith Brookhart of Iowa was the only member of the farm bloc to support the progressive ticket. La Follette wrote the platform, which made no mention of education.[98] If Bagley and Crabtree had intended the NEA to trade teachers' votes for an endorsement of the Sterling-Reed bill, nothing came of it.

In the end, the election was a Coolidge landslide, returning impressive Republican majorities to both houses of Congress, while at the same time depriving the farm bloc of its position of holding the balance of power. Conservatives gained complete control. Many Senate Democrats began to adapt themselves to the business mentality of their Republican counterparts, as did House Democrats. As historian David Burner observes, they "were slipping into an ill-defined conservatism that presented no clear alternative to the well-articulated conservative policy of their Republican opponents.[99]

Whatever the motivation behind the NEA's support of private schooling, Catholics welcomed it because the movement for compulsory public education showed no sign of easing. In March 1924, 100-percenters brought the matter to an unsuccessful vote in two states: the Mississippi legislature turned down an amendment for compulsory public schooling, and Missouri voters did the same in a special election.[100] One of the most concerted drives took place in Washington State. Luther Powell, excommunicated founder of the Oregon Klan, went north to establish another domain for the Invisible Empire. A Klan front known as the Good Government League filed a petition in January 1924 for an initiative based on the Oregon law.[101] The Invisible Empire in Washington differed from that in Oregon in several respects. Probably larger than the Beaver State Klan, it suffered from indolence. This alone might have ensured defeat of the compulsory public schooling initiative, but the situation was made more difficult by the Klan's political isolation. Whereas Masons in Oregon had helped to put the law across, most in Washington repudiated the Klan. Clarence Reames, former U.S. attorney for Washington and a member of the Southern Scottish Rite, told the Knights of Columbus that hundreds of his brothers would consider it an honor to help Catholics defeat the measure. As mentioned earlier, the policies developed by the Rite's Supreme Council were non-binding, and Washington State Masons chose to establish the Friends of Educational Freedom. Initially, the Invisible Empire hesitated to collect signatures to place the initiative on the ballot out of fear that the petition would amount to little more than a Klan roster. Setting caution aside, it proceeded and gained the requisite number.[102]

The Klan also carried the fight to Michigan, where James Hamilton, the redoubtable opponent of parochial education, had obtained a writ of mandamus to force the amendment for compulsory schooling onto the November ballot. Imperial Wizard Evans visited the state to rally Klansmen to the cause. In a rousing Fourth of July speech, he bluntly, and rather crudely, championed the Invisible Empire's educational platform, which pictured the common school as the engine for the Americanization of alien elements: "We'll take every child in all America and put him in the public school of America. . . . We will build a homogenous [*sic*] people, we will grind out Americans like meat out of a grinder. This question is just as vital as the question of slavery—we can't be half in the public school and half in the parochial school." The Michigan Klan backed the Imperial Wizard to the hilt. It raised funds to garner votes and planned to have some 5,000 of its members posted near the state's 458 polls to taut the measure.[103]

While the Klan pressed forward its campaign for compulsory public education, Catholics and the Scottish Rite sparred over what it meant to be American. Charles Lischka, a researcher in the NCWC Bureau of Education, contended that in a public crisis, people cloaked their motives in patriotism, advancing pretexts for their cause that in normal times would be recognized by the majority as patently false or erroneous. Admitting that the patriotism of the proponents of compulsory public education was laudable in itself, he asserted that it was being put to partisan use in an errant cause. The issue was whether the state may abridge the educational rights of the individual, the parent, or the church. He argued that it was "*the natural duty*" of parents to instruct their children. The parental right over education, said Lischka, "does not amount simply to a privilege or a prerogative; it is not a secondary right, not an acquired right, not a right conferred by human authority, . . . but it is a primary and direct right, an innate and inviolable right, a permanent and inalienable right." The last two words placed his argument in line with the Declaration of Independence. Although the state might have an educational right, that right was not inherent in the nature of the state, and the parental right was both anterior and superior to it. Lischka noted that the federal district court in Oregon and the Supreme Court in *Meyer* v. *Nebraska* had specifically recognized this parental right. In light of that, he asked: "Can anyone affirm that the principle of state monopoly of popular education is an American principle? It never has been, and, God willing, never shall be, recognized as an American principle, for it is foreign, abhorrent, revolting to the true American spirit."[104] In other words, those who under the cloak of patriotism sought compulsory public schooling were foreigners in spirit in their attempt to adopt the statist type of education current in some European countries. In upholding individual and parental rights, Catholics and others like them were the true Americans, adherents to the principles of the Declaration of Independence.

Refusing to accept this argument, the Supreme Council of the Southern Jurisdiction of the Scottish Rite reasserted its belief in the propriety of compulsory public schooling. Considering it a policy "harmonious with American ideals and institutions," the Supreme Council declared that it was impossible "to maintain a homogeneous democratic government if a heterogeneous population is permitted to acquire elementary instruction from diverse sources, some of them unsympathetic or hostile to the American system of government."[105] The *New Age* accused the Roman church of separatism and "a supercilious, holier-than-thou attitude toward other sects and religions that has provoked misunderstanding, suspicion and resentment." The magazine found this kind of behavior utterly repugnant to the spirit of democracy, which demanded tolerance, sympathy, and a desire for the common good. Parochial schools were incapable of inculcating these virtues because they instilled a contrary spirit in their students. The editor declared that the Rite was not so much opposed to private schools as in favor of public ones, the only real bulwarks of democracy.[106] In short, Masons and those like them were the real Americans, the true upholders of democracy.

Despite the Supreme Council's renewed affirmation of compulsory public education, the prospects for the Washington initiative and the Michigan amendment were not good. The Washington measure was opposed by the majority of papers, organized labor, the Catholic church, and even many Masons. The Michigan amendment faced a united phalanx of editors and the hostility of Catholics, Lutherans, Episcopalians, and Dutch Reformed. In the end, both measures were soundly defeated, the Washington initiative by a margin of 3 to 2 and the Michigan amendment by almost 2 to 1.[107] Although these reverses marked the last time such proscriptive legislation came to a vote, one last compulsory public education bill was introduced in Indiana just months before the Supreme Court struck down the Oregon law.[108]

While the Klan, the Masons, and other 100-percenters failed to secure additional legislation of the Oregon variety, they were successful on a more personal front. The NCWC bureau of education had a teacher registration service for the placement of Catholic instructors in parochial schools throughout the country. Beginning in 1924 this office began receiving numerous letters from Catholics recounting how they had been fired from or turned down for teaching positions in public schools because of their religion. Perhaps the tersest example is a two-sentence rejection notice to Mary Monahan from an official in the Kansas Department of Education: "Your qualifications make you the very teacher we want. But I am sorry, for my board are averse to hiring a Catholic as a teacher in our schools." All the letters originated from the American heartland, the farm-bloc states of Iowa, Kansas, Oklahoma, Illinois, Indiana, Missouri, Kentucky, and Minnesota. In many instances, the offended party accused the Klan of being behind the deed.[109]

The Oregon case was argued before the high bench in mid-March 1925, and the outcome seemed certain, at least to the Court. On the morning of the trial, one of the NCWC's attorneys, William Guthrie, asked Chief Justice Taft for additional time to make an extended argument. "I don't see why you want any more time," replied the jurist. "In principle, this case is simply the *Meyer* case over again."[110] Lawyers for the governor claimed that parents had no more control over their offspring than they had over themselves. Since the state could reasonably limit the rights of adults for the sake of the common good, it could do the same with children. One lawyer went so far as to claim that the state stood as *parens patriae* [parent of the country] to minors and therefore had unlimited control over them. The Fourteenth Amendment did not abridge a state's right to legislate for the common good, which demanded that all children be educated in the public school. The latter was the only safeguard against foreignism, religious distrust, and disloyal teaching. Both Guthrie and Kavanaugh argued that the Oregon law aimed at the destruction of private education. To accomplish this, it negated the liberties of private schools, of the teachers who taught in them, of the parents who sent their offspring to them, and of the children who attended them. It was a gross assault on Fourteenth Amendment rights.[111]

On 1 June 1925 the Supreme Court handed down a unanimous decision upholding the lower tribunal. The opinion, written by McReynolds, admitted that to ensure the public welfare, the state had the right, within reason, to regulate, supervise, inspect, and examine all schools and teachers. However, the practical effect of the Oregon law was the destruction of private and parochial schools without sufficient warrant for such drastic exercise of the state's police power. More significant, McReynolds condemned the law as an abridgment of parental rights and the beginning of state absolutism. The theory of liberty on which all American government rested, he wrote, "excludes any general power of the state to standardize its children by forcing them to accept instruction from public teachers only. The child is not the mere creature of the state; those who nurture him and direct his destiny have the right, coupled with the high duty, to recognize and prepare him for additional obligations."[112] The decision was a vindication of personal, parental, and property rights and established the outer boundaries of state control over education. The government could reasonably regulate private schools; it could not abolish them.

At the same time that the church successfully concluded the Oregon litigation, it faced the Sterling-Reed bill for the last time. Even though the Senate Committee on Education and Labor had voted against reporting the measure and the House counterpart seemed willing to let it slide into oblivion, thereby leaving the Reorganization bill in sole possession of the congressional field, both the Scottish Rite and the NEA were determined to push the Sterling-Reed bill through in the final session of Congress. In October 1924 the Supreme Council of the Southern Jurisdiction pledged to continue efforts for its passage.[113] The *New Age* claimed that objections that the measure would standardize parochial schools or assert federal control over them were simply "canards and deliberate and malicious misrepresentations." The bill never mentioned church schools; its sole purpose was the promotion of public ones. "Only invincible prejudice can find anything to oppose in such a program and purpose," declared the editor.[114]

For its part, the NEA asked its members to inform congressmen about the unacceptability of the reorganization bill and to urge them to pressure the House Education Committee to report the Sterling-Reed legislation. As chairwoman of the Education Committee for the Women's Joint Congressional Committee (WJCC), comprising representatives from twenty-one distaff organizations, Williams made sure that the bill was high on their list of priorities. In addition to supporting the Sterling-Reed bill, the WJCC promoted such other progressive measures as the Sheppard-Towner Maternity Act and the Child Labor Amendment. By fall, Charl Williams reported that twenty-seven national associations of women and men supported the measure as the only way of improving the schools of the nation.[115]

The NCWC countered her efforts by working on its own female branch. At the annual convention of the NCCW, Archbishop Dowling denied that the Sterling-Reed bill was of any concern to Catholics as Catholics. Rather, it was "a bonus bill," a "political bill," which the church, like many non-Catholic organizations, feared lest "this great federalized institution should be developed." The measure was a "dangerous thing educationally" and might eventually be directed against parochial schools. Dowling encouraged his listeners to become members of the women's groups that favored the bill in order to fight the measure from within.[116] Ryan was even more strident. He told the New York Council of Catholic Women that those behind the legislation were "either knowingly or ignorantly advocating deliberate treason to our highest national ideals." The bill was "dangerous, vicious, and a wolf in sheep's clothing," for federal control always followed federal money. Appealing to emotions, he concluded, "We don't want nationalized children who will develop into machine men as a result of standardized education."[117] Agnes Regan, executive secretary of the NCCW, asked each unit to familiarize itself with the Catholic position to offset the support of non-Catholic women's organizations.[118]

Opposition by inaction came from the ACE. At the annual convention in May 1924, representatives instructed the organization's Committee on Federal Legislation to take a referendum on the measure. Given the complex nature of the ACE's membership and the fact that only a third of the member organizations had sent representatives to the convention, John McCracken, the committee's chairman, and Charles R. Mann, the director of the ACE, decided to poll the membership on the question of conducting a referendum. Member organizations panned the idea and with it any support for the Sterling-Reed bill.[119]

In the end, all efforts for or against the legislation were academic because senators and representatives thought the Reorganization bill contained the only viable plan for education. The Senate Steering Committee, which set the legislative agenda, considered that bill one of the few measures other than necessary supply bills that ought to be enacted during the lame-duck session. The House even placed it on the calendar for action. Reports on the prospects of the measure were contradictory. Old Guard Republican Senator Reed Smoot, who had signed the majority report in favor of a department of education and relief, believed that the bill would encounter no serious opposition. On the other hand, farm-bloc Democratic Senator B. Pat Harrison of Mississippi, who had turned in a minority report in favor of a separate department of education, gave the bill no chance. With Congress prepared to take up reorganization, the ranks of the NEA broke. Some officials decided to take what was within reach and back a department of education and relief.[120] This caused the NCWC to rethink its own position on that legislation. Catholic authorities had previously considered it marginally acceptable because it avoided a separate department with federal aid, though going beyond the more desirable expansion of the Bureau of Education. The Executive Committee of the Department of Education now decided that the NCWC should remain neutral and let fate run its course.[121]

As fate had it, the Reorganization bill died. When Smoot asked for its consideration by the Senate, no one supported him. Sterling spoke against the measure because he wanted a separate department of education. Two ardent anti-statists, Republican William Borah of Idaho, chairman of the Senate Committee on Education and Labor, and Democrat William King of Utah opposed it because it increased bureaucracy by creating a department of education and relief. Others did not want to devour the final days of Congress with a debate on such monumental and controversial legislation. The motion to consider was defeated by a substantial margin.[122] Although the bill had been placed on the House calendar, when the day came, the House passed over the legislation. On 4 March 1925 the Sixty-eighth Congress expired and with it the Reorganization bill, the Sterling-Reed bill, and the Dallinger bill.

The Catholic church had successfully negotiated a battle on two fronts. On the state level, it parried a direct thrust at parochial education. The warring parties had squared off over the issue of whose schools were truly American. Proponents of compulsory public education contended that state schools were the only democratic ones capable of bringing all children together on a level playing field that inculcated tolerance of differences while instilling traditional Anglo-Saxon, Victorian values. Catholics argued that parish schools harmonized best with the religious tradition of the nation and the heritage of the founding fathers, which respected the individual and his or her beliefs, thereby sustaining democracy. Without deciding the matter of whose schools were the more American, the Supreme Court held that the Constitution supported parental rights in education and prevented the state from blanket standardization of schooling. On the national level, the two sides fought the same fight in different form over the Sterling-Reed bill. Given the alliance of the NEA with the Scottish Rite, whose educational program was identical with that of the Klan, it seemed certain, at least to Catholics, that the measure ultimately aimed, like the Oregon law, at compulsory public education. Forced to take a higher profile than in the previous Congress to show that it was unopposed to public education, the NCWC espoused the establishment of a federal board of education and endorsed the Dallinger bill for expanding the functions of the Bureau of Education. Still, the federal education question was so religiously charged that Congress chose not to address it, even indirectly through the Reorganization bill. Having lost the battle on both fronts, the NEA was forced to reconsider its strategy.

5

Guardians or Betrayers of Catholic Interests?

For seven years the NEA had tried unsuccessfully to establish a department of education with federal aid in support of a national program of education. The proposal had run headlong into administrations committed to economy in government and local control of schooling. The significance of its latest frustration was not lost on the NEA. It now understood the warning voiced as early as 1918 by Philander Claxton and as recently as 1923 by the ACE that the Smith bill and its successors contained two measures: one for a department and another for aid. Many prominent educators and the majority of institutions comprising the ACE favored the former but opposed the latter. So, the wisdom of separating the two finally sank in. If the NEA were to achieve any success with its legislative program, it would have to minimize opposition and maximize support. It never again attempted comprehensive legislation of the Sterling-Reed type. The two proposals contained therein would now be sought piecemeal.

As for Catholics, they were faced with the prospect of reexamining the education question in a way that had not occurred since 1919 when progressive elements in the church had toyed with the idea of accepting some form of federal participation in education. Events after that had caused the church to fear that a department of education with a large subsidy might lead to the destruction of parochial schools, so intertwined had become the drives for a department and compulsory public schooling. To be sure, many Catholics opposed the NEA bill in the belief that education belonged to the states, not the federal government. Their concern for parental rights and parochial schools served as the bedrock on which the states' rights argument rested. They were certain that federal funds would lead to federal control. The Supreme Court's decision in the Oregon school case afforded parental rights and parochial education a measure of safety. This altered situation and the NEA's new bill would cause the bishops to reassess their stand on a department, especially one shorn of aid.

While providing security to parochial schools, the Oregon decision dazed the Southern Jurisdiction of the Scottish Rite Masony and the Ku Klux Klan, but left them unshaken and undeterred in their belief that public schools ought to be the only ones in the land. They continued to view compulsory public education as the means for Americanizing children and instilling in them the values of democracy and Anglo-Saxon culture. The problem was how to accomplish this end in the wake of the Court's ruling. Although the Rite would flirt with the idea of a constitutional amendment, both organizations quickly viewed the new NEA bill as the best means of achieving the goal.

When it became apparent that the lame-duck Sixty-eighth Congress would enact no education legislation, the NEA decided to formulate a new measure. Because there was "no chance of securing additional National support for education at this time," its Legislative Commission decided to drop federal aid from the new bill and seek simply the creation of a department of education. George Strayer, the chairman, was authorized to invite interested educators to a conference in Washington to draft the legislation.[1] On 7 March 1925 Strayer's ad hoc committee met. Present were John McCracken, chairman of the ACE's Committee on Federal Legislation; Charles Judd of the University of Chicago; Samuel Capen, former director of the ACE and now chancellor of the University of Buffalo; William Owen, past president of the NEA; Dr. Thomas Finegan, president of Eastman Teaching Films; and Strayer. Judd and Capen, who had long championed the establishment of a department for the scientific investigation of education, carried the day. Though both men were open to the possibility of federal aid if serious scientific study of schooling indicated that it was warranted, federal grants to education were not their aim. The meeting resulted in a bill embodying the Judd-Capen perspective.[2]

John Burke and James H. Ryan informed the NCWC Administrative Committee that the NEA had drafted new legislation, devoid of federal aid, to create a department of education for scientific investigation and research. Both thought that the idea would win the approval of everyone. The Administrative Committee thought that the NCWC ought to advocate something positive, perhaps the forthcoming bill itself. At the very least, it ought to be represented at the conference that the NEA was planning so that interested educators could discuss the new measure. The committee decided that Edward Pace, who would attend the conference in his capacity as president of the ACE, should represent both the NCWC and CEA. He should not, however, commit the hierarchy to the new bill.[3] By late April, the NCWC began to have suspicions about the NEA's real intent. Ryan reported that although the new bill was to provide no aid, some thought it was the first step to that end.[4]

The NEA unveiled the measure at its annual convention in June. The legislation called for the establishment of a department of education, housing both the Bureau of Education and the Board of Vocational Education. Through research and dissemination of information, the secretariat was to help states establish and maintain better school systems, devise better methods of financing education, design improved school buildings, enhance teaching methods, and develop more adequate courses of study. Borrowing from the Dallinger bill, the NEA included a federal conference on education, consisting of a representative from each executive department, to coordinate the educational activities scattered throughout the government. The proposed department was to function on the sum of $1.5 million.[5] Essentially, the bill was the first half of the old Smith-Towner/Sterling-Reed measures, though now remedying the ills of education through research rather than federal aid.

The interpretation given to the bill by the two parties who created it was instructive: Judd and Capen on the one hand, and Strayer on the other. The twist that each side gave to it revealed that they had agreed on common legislation but remained immeasurably apart on its aim. Judd pronounced the NEA's deletion of the $100 million aid package a "great gain to the cause of education," a concession that opened the way for "the right kind" of department. "The department should be, as is provided in the bill," he commented, "a federal agency for investigation and publication, not an auditing department for the disbursement of federal subsidies." The new legislation made it possible for all those who had opposed the original Smith-Towner bill and its successors to join in support of the new measure.[6] In Strayer's view, those at the drafting conference remained "convinced of the legitimacy" of the NEA's total program, namely, the establishment of a secretariat with a large appropriation for federal aid. "In the present situation, however," he reported, "it seems wise to bring all of our strength to bear in support of a bill creating a Department of Education." The new bill would gain the backing of educators who wanted a department but opposed government grants. It would also accord with President Calvin Coolidge's desire that the federal role in education be limited to research and the dissemination of information, the functions of the president's own proposed department of education and relief. This line of action would enable the NEA to win new backers for its project without alienating those proponents who also favored federal aid.[7] Given thoughts like these, the NEA's dropping of aid was a maneuver of realpolitik.

That the NEA had simply deferred seeking government grants was made apparent by the remarks of Charl Williams: "The consensus of opinion of those who conferred on this important question," she told conventioneers, "was that it would be unwise to include the Federal aid appropriations in the new bill. It seemed best at this time to build upon the sentiment for a Department of Education which had been created throughout the country and in Congress by the supporters of the Education Bill as well as by the report of the Reorganization Committee."[8] During a session of the Department of Classroom Teachers, Mabel Wilson, President of the Seattle Grade Teachers' Club, delivered an address in favor of both federal aid to and federal control of schools. After rehearsing the history of national grants to education, she concluded, "States can no longer live unto themselves alone. . . . When the good of the cause as a whole suffers, education . . . must not only accept Federal aid but also submit to a certain amount of national guidance when such guidance comes to the rescue in the unselfish interest of a higher and broader type of nationalism."[9]

The intention of the NEA was not lost on the NCWC. After the convention, Ryan reported that although the association had deleted the objectionable $100 million aid package, it had really yielded nothing. It still sought the federalization of education through a department dispensing federal funds, but what it had previously attempted in one step, it now sought in two.[10] Burke informed the hierarchy that the NEA was seeking a department first and would ask for aid later. Given the strong sentiment for a department, and since aid was not an issue in the present measure, he proposed that the bishops reopen the educational question at their annual convention and weigh the wisdom of intransigent opposition to the new bill.[11]

The day before the hierarchy's convention, the Administrative Committee discussed the course of action it would recommend. On one hand, the position of Catholic schools seemed secure: the Supreme Court decision in the Oregon case offered unprecedented safety. The new NEA bill contained no subsidy, the fulcrum of federal control according to the axiom that government supervision followed federal money. The proposed department would be as helpful to parochial schools as to public ones because it was to conduct studies and disseminate information. On the other hand, there was reason for caution. Many Catholics suspected the new bill foreshadowed federal control of education. It also had the support of the same anti-Catholic groups that backed its predecessors. The committee decided to recommend that the hierarchy take a neutral position regarding both the NEA's bill and the Reorganization bill with its department of education and relief.[12]

On 16 September 1925 fifty-six bishops assembled for the meeting. Curiously, instead of advising neutrality, Archbishop Austin Dowling introduced the issue as an open question. Apparently, he was concerned about Catholic obstruction of public educational endeavors and its potential backlash on the church. Noting that neither the NEA bill nor the Reorganization bill proposed any federal control, he remarked that they were not directly hostile to parochial schools. Cardinal Patrick Hayes of New York moved that the hierarchy endorse the Reorganization bill. Bishop John Carroll of Helena, Montana, went further, advocating support of the NEA bill itself. Believing that aloofness in such matters courted unfortunate consequences, he reminded his colleagues that as pastors, they had the responsibility of watching over not only those within their folds but also those without. Though firmly opposed to federal control, he thought that the duties of the proposed department of education could be circumscribed to avoid that possibility. Under proper safeguards, a department might even prove beneficial to parochial schools. Dowling noted that in the Oregon case, the church's attorneys had admitted the state's right to assert a minimum of control over education. In states where legislatures had contemplated asserting control over all education, Catholics had deemed it wise to accept a certain amount of regulation in order to avoid the evil of total control.[13] His implication was clear: better a department with the possibility of minimal control over education than risk the chance of complete federal control of schooling. While Carroll seems to have been motivated by a combination of progressivism and genuine pastoral concern for the good of all, Dowling and Hayes seem to have been motivated by a concern that the church advocate something positive lest a worse evil befall parochial schools.

In the face of this willingness to compromise, voices of opposition were raised. Bishops William Turner of Buffalo, Francis Howard of Covington (still secretary general of the CEA), and Hugh Boyle of Pittsburgh stated that the NEA was the church's principal opponent in educational matters. If a separate department were created, the association might push to enlarge the secretariat's scope and work to gain control of education. Since public sentiment opposed every form of centralization, why should the bishops take a contrary position, especially because education belonged to the parent first, to the state second, and to the federal government not at all? Any step to enlarge the federal role in education was a step in the wrong direction. Given that Catholics were a minority, home rule was the correct policy. Boyle went so far as to assert that the bishops should oppose any federal educational legislation even if it were clear that the legislation would pass despite their objection.[14]

Sensing the drift, Dowling inquired if the hierarchy wished to counter the measures informally or openly. The latter course, he warned, would ensure the passage of one bill or the other, no doubt because Catholic opposition would be interpreted as hostility to public education. Hayes made a final push for the Reorganization measure but was overcome by nay-sayers. Still, he pleaded that the bishops give Dowling's Department of Education discretion in handling the matter. The hierarchy decided to take no public stance, while opposing informally any legislation tending toward federalization. Cardinal William O'Connell of Boston, chairman of the assembly, emphasized that Dowling's department was "not to be hampered in any way in its activity, but may use its discretion informally to direct its course of action."[15]

Thus, the hierarchy arrived at the same decision as the Administrative Committee, namely, that public neutrality toward the new bill was the best course; there was, however, a twist. The bishops clearly considered both bills dangerous, but did not wish to make their opposition public at the time. So the Administrative Committee directed Burke to oppose neither measure openly, but to counter informally—quietly—federal control. If public action became necessary, he was to consult with the committee.[16] There was a problem: neither Congress nor the Catholic press, especially *America,* knew that the hierarchy had muted, not abandoned, its opposition to federalization. This situation would bring the Washington secretariat of the NCWC into disrepute among some Catholics and cause them to question the nature and authority of the Welfare Conference itself.

Even before the NEA bill was introduced in Congress, Catholic editors took an almost uniformly negative stand toward it. Unaware that it lacked federal aid, a few papers, like the Hartford *Catholic Transcript* and the New York *Catholic News,* argued foolishly that it would result in federal control because that always followed government money.[17] With a clearer notion of the bill's details and the NEA's intent, other editors, especially Paul Blakely of *America,* decried the legislation as the first step toward federal control. Its advocates were simply seeking a department first and federal aid later. In Blakely's view, the new measure, lacking as it did a subsidy, was like the camel that just wanted to get its nose under the tent. Soon the whole animal would be inside.

Switching analogies, he compared the bill to the Trojan horse and concluded, "Why throw away what we have securely gained by seven years of hard fighting, by volunteering to bring the hateful thing within our walls?"[18]

For diverse reasons, two church papers favored some kind of positive action. The Cleveland *Catholic Bulletin,* although opposed to the NEA bill for the same reasons as its counterparts, echoed Dowling's and Hayes's fear that continued Catholic antagonism might be construed as opposition to public education, something proponents would be able to exploit. Consequently, the editor urged the church to support the Reorganization bill with its department of education and relief.[19] Without any reservation, the Milwaukee *Catholic Citizen* endorsed the NEA measure as a positive step in the growth of the president's cabinet. No doubt emboldened by the Supreme Court's endorsement of the right to private education, the editor dismissed Catholic objections about federal control: "We apprehend that this fear is somewhat exaggerated. Federal encroachment in education would be restrained because of constitutional limitations. But to recognize education by the dignity of a cabinet place might be altogether in the upward course of evolution."[20] This was a forward-looking, progressive opinion that placed the paper in the camp of educators like Judd, Capen, and McCracken, and in the company of Catholics like Pace, Carroll, and Arthur Monahan. Exaggerated or not, however, fear was too great and danger too real for most Catholics to agree with the Milwaukee editor. He remained a lone, Catholic voice for some time.

The *New York Times* also gave the new bill a half-hearted endorsement—half-hearted because the editorial was actually a paean to the self-sacrifice of teachers. Only in the final paragraph did the editor address the department of education. Apparently, the absence of federal aid from the new NEA measure inclined him to prefer this legislation to the Reorganization bill. Affirming that the "maintenance of school systems is the function of the States," the editor asserted that the establishment of a separate department to consolidate federal educational activities and to serve as a voice for the nation's teachers was consistent with the theory of states' rights. Supporting "at least a minimum standard of school provision in every community," the *Times* echoed the nationalist argument put forward by representatives of the NEA: "What is done by the several States is of concern to the whole people. No State educates for itself alone. And no State suffers alone from a neglect of its children."[21] While the editor seemed to think that a department would serve as a voice for teachers, the NEA's new bill generated little interest in their ranks. If established, the department would probably have served as a voice for the educational trust.

As the NEA continued its drive for a department, an unrepentant Southern Jurisdiction of Scottish Rite Masonry kept up its drive for compulsory public education. The *New Age* paid only lip service to the Supreme Court's decision in the Oregon case. Admitting that the duty of all loyal Americans was to abide by the ruling, the magazine carried an article by "a Special Contributor," perhaps Grand Commander John Cowles himself, who averred that the Rite had supported the Oregon law, not out of opposition to parochial schools, but in the belief that the best interest of democracy was served by forcing all children to attend public schools. The Oregon law was one way of achieving that end, and "the Supreme Court simply said: 'You can't do it this way.'" The author suggested reorganizing the campaign along the following lines: (1) strengthen the public schools to such a degree that all parents will feel compelled to send their children to them; (2) encourage increasingly larger appropriations for public education; (3) seek strict regulatory legislation for private and parochial schools; (4) combat state aid to private education; (5) maintain this program until all parents are convinced that the public school is the only patriotic place for their children.[22] In the next issue, the editor committed the magazine to this policy.

According to *America* the Rite went further, asserting in its mid-June Bulletin that if the majority of citizens wanted all children educated in public schools, mandatory public education would become a reality. "The right of the individual ceases where the right of the state begins. . . . He must submit his personal preferences and predilections to the common sense of the whole. . . . His liberty is circumscribed by the best interest of the community at large."[23] The editor of *America* commented that the members of the Southern Jurisdiction had for so long "fed upon a rank diet of State idolatry that they are unable to appreciate the American spirit of the [Supreme Court's] decision." The Rite, *America* averred, claimed that a minority had no rights a majority was bound to respect. Such a view was out of harmony with the Declaration of Independence, which held that certain rights were inalienable and inviolable by any majority.[24] The *New Age* responded that the fundamental principle of American government was majority rule. The aim of the Constitution was not to frustrate the will of the people but to ensure democratic government. That was why there was an amending process. Hinting that the Rite intended to use this means to achieve its educational goal, the editor declared: "We yield to none in our respect for the Supreme Court. . . . But we also remember that there is a higher tribunal even than that of the Supreme Court—the people of the United States."[25]

Other Catholic journals aggravated the situation. Commenting on the NEA bill in *Commonweal*, Mark Shriver remarked that legislation, like a person, was known by the company it kept. Because the same groups that backed the Oregon law were lining up behind the new NEA measure, he detected a sinister attempt to use this apparently innocuous legislation to achieve the same goal. No friend of civil liberty would deny, wrote Shriver, "that the only good educational bill is a dead educational bill, if it be tainted by the flavor of the Oregon law, the insidious policies of the Scottish Rite Masons and the policies of the National Education Association."[26] The *Tidings* (Los Angeles), too, warned that the NEA bill aimed at curtailing educational freedom. Why else, asked the editor, should the Scottish Rite, a sectarian group in the commonly accepted meaning of the term, be so interested in it? The Protestant fear of Roman domination of public schools was "nothing more or less than an expression of belief in their distinctively Protestant character and a corollary of the freely spoken ministerial dictum that 'America is a Protestant country.'"[27] A host of other papers echoed the warning.[28]

For the *New Age*, the question was not a sectarian religious issue. The issue was the separation of church and state. The schools in America were to be secular, so the Southern Jurisdiction was fighting to ensure that they were under the domination of no sect. In Catholic countries around the world, argued the editor, the Roman church denied freedom of conscience and coerced state support for its own institutions. The *New Age* vowed to combat the extension of Roman political domination over the United States.[29]

In October the Supreme Council pledged the Southern Rite's support for both compulsory public schooling and the NEA bill. In so doing, the resolution trumpeted the message of children's, as opposed to parents', rights: "Now is the time to arouse the latent forces of the Scottish Rite in behalf of the greatest of all our principles, the right of every child to free education in the public schools. We cannot . . . but insist upon the existence of the principle that the right of the child to avail himself of the educational opportunities of the public school system is superior to the right of the parent or any corporation, secular or religious." John Cowles was empowered to take every action in support of the NEA measure. Editorializing on the resolution, the *New Age* explained that by endorsing the NEA's new bill, the Supreme Council intended in no way "to question the validity of the decision handed down in the Supreme Court with regard to the Oregon School Law case, but still remains unshaken in its firm convictions as to the value of compulsory education of all children in the public schools."[30]

The Rite's championing of children's rights was decades ahead of its time. To be sure, courts were making concessions toward children "in slow, subtle, halting fashion," as historian Morton Keller notes. But as the decisions in

Meyer v. *Nebraska* and the Oregon case made clear, they were nowhere near mitigating the custodial rights of parents to the extent contemplated by the Southern Jurisdiction.[31] More to the point, it is doubtful that the Rite was championing such rights in any altruistic or comprehensive way that even present-day courts would recognize. This was a novel argument advanced in favor of a partisan position. Rather than launch a drive for a constitutional amendment in favor of compulsory public education, the Supreme Council threw its weight behind the NEA bill. The Rite's contentiousness about public schooling, combined with its backing of the NEA's measure, enlivened Catholic suspicion that the bill's real intent was the control, if not abolition, of parochial education.

Like the Southern Jurisdiction, the Ku Klux Klan remained undeterred in its conviction that public schools were the agents for Americanization. Writing in the *North American Review*, Imperial Wizard Hiram Evans declared that the Klan was committed to preserving the country "for the benefit of the children of the pioneers who made America." As such, it stood for the white, Protestant, "old pioneer stock," that is, the "Nordic race" (Anglo-Saxons). Nordic Americans had become strangers in the urban areas of their own country, overrun as those parts were by aliens. It was not enough to close the gates to immigration; "America must also defend herself against the enemy within, or . . . be corrupted and conquered by those to whom [it had] already given shelter." While nodding to the threat posed by Jews and blacks, Evans considered the real internal enemy to be the Roman Catholic church, which was "fundamentally and irredeemably . . . actually and actively alien, un-American and usually anti-American." That it was in politics was proven at the 1924 Democratic convention wherein "Catholic politicians . . . seized upon Catholicism as the cement for holding the anti-McAdoo forces together." The Klan was the defender of "a Protestant and an American 'vote.'" "One of the Klan's chief interests is in education" said Evans. "We believe that it is the duty of government to insure to every child opportunity to develop its natural abilities to their utmost. We wish to go to the very limit in the improvement of the public schools; so far that there will be no excuse except snobbery for the private schools." The Klan considered parochial education to be a propaganda system in the hands of aliens who could not "possibly understand Americanism or train Americans to citizenship." Church schools should be "closed" or at least subjected to state control.[32] The Klan's position on Catholic education had not altered a whit. It would have parochial schools shut down and would accomplish this end by improving public education to such an extent that there would be no alternative to it.

In the fall of 1925 the NCWC received information that the Southern Scottish Rite had begun establishing statewide educational leagues throughout its jurisdiction. To assess Masonic strength and activity, Burke wrote to all the bishops in that territory.[33] Responses were generally encouraging. In the South, only two leagues were reported, one in Kentucky consisting of two people (a president and secretary) and the other in Georgia. Although the latter was rabidly anti-Catholic and bent on the federalization of education, it did not enjoy united Masonic support. Reports from Texas and Oklahoma said that Masons in those states were two-faced, privately stirring up anti-Catholicism among Klansmen while publicly lamenting the antagonism toward the Catholic church. With regard to education, Texas Masons had long maintained an Education Program Committee that sent out speakers to stump for compulsory public schooling regulated from Washington. In Oklahoma, the brotherhood was divided over the education question. From the West came word that California, Oregon, Washington, Idaho, and Wyoming each had an educational league, but all were dormant.[34]

This information convinced Burke that Masonic statements were inflated, not an improper conclusion to draw.[35] The Rite itself had referred to its forces as "latent," and the fact remained that the Southern Jurisdiction had never presented a united front on the issue of compulsory education, now linked to support for the NEA bill. No doubt the fecklessness of the jurisdiction's membership to the cause of mandatory public schooling gave Burke a false sense of security about the federal educational question, a misconception soon to be dispelled.

While the Southern Jurisdiction labored to drum up support among its membership for both compulsory public education and the NEA bill, champions of the latter sought the support of Coolidge and the Catholic church. In early September 1925 a troika of Massachusetts public-school patriots—Payson Smith, state commissioner of education; A. Lincoln Filene, chairman of the National Committee for a Department of Education; and Mrs. Frederick Bagley, representative of the Women's Division of the Republican State Committee—called on the president to win his endorsement of a separate department of education. Coolidge stood firm for a department of education and relief. Against objections that the head of such a department would have to be a jack-of-all-trades rather than a nationally prominent schoolman, the president responded that just because a university had schools of medicine, law, and theology did not mean that its president had to be a doctor or a lawyer or a divine. He had to be an able administrator.[36] Coolidge felt no sympathy for the argument that education deserved the dignity of a separate cabinet post occupied by an educator. If any new department were to be added to the executive branch, it would be manifold and under the direction of an able executive. Moreover, Coolidge was committed to economy in government. He wanted to reduce the national debt and lift the tax burden off citizens.[37] Neither of these goals could be accomplished by increasing government expenditures for the investigation of education.

Some weeks later, Joy Elmer Morgan, editor of the NEA *Journal,* conferred about the new bill with Frank Crowley, director of the NCWC's Bureau of Education. Morgan wanted to know the church's position on the bill and whether the NCWC could get a representative group of Catholic educators to support it. Crowley sidestepped the question by answering that Catholic opinion had yet to crystallize.[38] Although faithful to the letter of the hierarchy's instruction that the NCWC take no public position on the measure, this response was not altogether truthful. The prevailing wind in the Catholic press blew decidedly against the bill. The only thing that had yet to crystallize was the willingness of the bishops to oppose the measure openly. Events would soon force them to take that step.

The NEA apparently pulled out all the stops in its effort to put the new bill across. A week before the Sixty-ninth Congress convened, John J. Cochran, secretary to Missouri congressman Harry Hawes, reported to Frederick Kenkel, head of the Central Bureau of the Central Verein, that mail was pouring into the Capitol in favor of the legislation, even naming the bill's congressional sponsors before its introduction.[39]

On the second day of the new Congress, Republican Senator Charles Curtis of Kansas introduced the legislation. The NEA probably chose him as sponsor to replace the retired Thomas Sterling because he was a party regular with strong ties to the administration. He also hailed from a farm-bloc state, thus seemingly straddling the divide between agrarians and Coolidge. That Curtis was also Senate majority leader probably appeared to be an additional asset, though he never gave the measure the support or encouragement that Sterling had. In the House, Daniel Reed of New York, chairman of the Committee on Education, continued to serve as sponsor for the legislation.[40] The NCWC gave the bill little chance of passage, at least during the first session. Noting that the proposed department of education was to be limited solely to research, Ryan wrote that even this new style of agency would fail to gain the support of a public grown weary of the proliferation of federal offices and the encroachment of the federal government on states' rights. Having become more vocal about the need for a resurgence of local self-government, Coolidge himself was prepared to abandon the reorganization scheme. Though the president remained committed to a department of education and relief that would conduct investigations and disseminate the results, he was willing to forego it if it remained embedded in the reorganization plan. Without Coolidge's support, which certainly would not be forthcoming, the prospects of the Curtis-Reed bill were poor indeed.[41]

Apparently, something in Ryan's article caused Bishop Howard concern that agents of the NCWC might be softening in opposition to the measure. He asked Archbishop Michael Curley of Baltimore for information about Ryan's and Burke's attitudes toward the bill. In Howard's view, although the bishops had considered it unnecessary to commit themselves to formal opposition, their position was "emphatically opposed to measures of this character."[42] After meeting with Ryan, Curley reported to Howard that the priest's article was in fact an objective "exposition" of the Curtis-Reed bill without expression of opinion. The NCWC Department of Education was, however, "up in the air" about the action it should take regarding the measure because it had received no instructions from Dowling and no direction from anyone else. Having missed the convention of the hierarchy, Curley criticized the bishops' lack of spine at the meeting. In his view, they should have simply ordered the NCWC secretariat to oppose the Curtis-Reed bill.[43] His pointing the finger at Dowling was understandable because the minutes of the meeting specifically stated that he and his department were to act at their discretion regarding the bill. Curley was unaware that the Administrative Committee had ordered the secretariat to take no public stand on the measure without specific authorization from the committee itself.

Ironically, it was the combination of the hierarchy's instruction, the Administrative Committee's order, and the bill's poor prospect that soon embroiled the church in bitter internal recriminations. In mid-December 1925 Burke attempted to dampen the zeal of *America*, arch foe of federal education legislation. He met with Blakely and Wilfrid Parsons, S.J., the editor, to inform them about the unlikelihood of congressional action on any school legislation. He counseled that Catholic opposition was unnecessary at the moment and would have greater impact if delayed until warranted. A week later, Burke had an interview with Coolidge, who confirmed the priest's assessment that Congress would pass no education bill.[44]

Just before Christmas, Curley summoned Burke to Baltimore to inform him of his intention to publish a statement against the Curtis-Reed bill. Asked for an opinion about such a step, Burke responded that he had no instructions from Dowling. The minutes of the meeting of the hierarchy made it clear to him, however, that the bishops wanted to take no public position on the bill, but to pursue a policy of "watch and wait." Any opposition was to be "of the quiet kind." Heartened by Howard's definite attitude, Curley replied that his reading of the minutes convinced him that the bishops wanted the NCWC to oppose the Curtis-Reed bill. There was no room for doubt. Burke insisted that the bill had no chance of passage by the present Congress. So, too, thought Coolidge. At that, Curley relented, agreeing for the present to make no pronouncement.[45]

The NCWC's official silence on the Curtis-Reed bill soon began to raise suspicions about the secretariat's intention regarding it. On Christmas Eve, Parsons wrote to Curley that he had just learned from a trustworthy source that "some sort of a deal is on foot at Washington by which the N.C.W.C. is favoring, or to favor, the Federal Department of Education, either publicly or privately." Parsons added that Burke had recently urged *America* to abate its opposition to the bill. Curley responded that at their meeting, the bishops had been unwilling to go on record as formally opposed to the measure, but "individually" (rather than "informally" as the minutes stated) they opposed it. "There is no question about the minds of the Hierarchy on the matter," he told Parsons, "consequently I cannot understand Father Burke's attitude."[46] Burke, on the other hand, understood the minutes of the hierarchy and the explicit instructions of the Administrative Committee to mean that the secretariat was to take no public stand for or against the bill, but to work against it informally. To him, that meant checking on its prospects with the president and Congress, actions he had taken. Everything indicated that the matter was quiescent. He could hardly have guessed that proponents of the measure would take official Catholic silence as consent. Indeed, word in political circles in Washington seems to have been circulating to that effect, no doubt giving rise to Parsons's suspicions.[47]

Parsons must also have expressed his concern about the NCWC to the dean of the hierarchy, Cardinal O'Connell of Boston. Writing for the latter, Monsignor Richard Haberlin, chancellor of the archdiocese, informed Burke that O'Connell had heard that "the N.C.W.C. is about to come out in favor of the Federal Department of Education or, at least, has given assurances in legislative circles that the [NCWC's] Department of Education favors the Curtis-Reed Bill." O'Connell wanted the truth.[48] Burke quickly replied that neither the NCWC nor any of its departments had endorsed the measure or planned to do so. The Washington office was abiding strictly by the instructions of the Administrative Committee.[49] Although Burke had no idea how the rumor had started, his suspicion was soon raised.

When he called on the apostolic delegate, Archbishop Pietro Fumasoni-Biondi, to wish him a happy New Year, Monsignor Filippo Bernardini showed

Burke the lead editorial in the current issue of *America*. Entitled "An Alarm and A Warning," it began: "As this edition goes to press an ugly rumor is circulated referring to 'a political deal to put the Curtis-Reed Federal education bill across.' The details, which would tell a curious story, need not now be stated. Are the guardians of our interests betraying them?" Bernardini, a nephew of the papal secretary of state and a defender of the NCWC during the suppression crisis in 1922, feared that the piece was aimed at the Welfare Conference.[50]

Burke returned home and fired off a letter to Parsons. Referring to O'Connell, he said that "one of the highest ecclesiastical dignitaries in the country" had recently been informed that the NCWC favored the Curtis-Reed bill. Now came the editorial in *America* with a similar rumor. Burke wondered if the two were connected and asked if "the guardians of our interests" referred to the NCWC.[51] When the editor replied that the information was confidential, there followed a fruitless exchange of letters among Burke, Parsons, and the Jesuit vice provincial. Burke wanted a retraction and an apology. Parsons and the vice provincial were unwilling to oblige.[52]

To his confidant Bishop Peter Muldoon, Burke bitterly lamented Parsons's willingness to believe such a thing of the NCWC, especially in view of all the money and energy the Welfare Conference had expended in defeating the Smith-Towner bill and its successors.[53] To Dowling, Burke questioned the wisdom of continued silence on the Curtis-Reed bill. The problem was not with Congress, because the bill lay dormant in committee. The problem was with the Catholic body itself, "many of whom are beginning to think we really favor the bill, because we are doing nothing in the way of protest." Burke admitted that the NCWC's silence gave some justification for drawing such a conclusion, particularly in light of the conference's previous antagonism toward federal education legislation. Presciently, he concluded: "It may be that through silence we would be jockeyed into a false position. That would be grossly unfair and unjust to the work of the N.C.W.C." Because the situation was "fraught with dangerous emergencies," he suggested that the Administrative Committee meet soon and invite other bishops, like Curley, who were interested in education, in order to lay a plan of action. Admitting that a meeting might be a good idea, Dowling encouraged Burke to be less anxious and keep a sense of humor.[54] In this instance, Burke, an inveterate worrier, was more on the mark than Dowling, who grossly underestimated the situation of both the education question and the NCWC itself.

America picked up powerful support from Curley, who told Pace that the hierarchy definitely wanted a public protest against the Curtis-Reed bill but that Burke had asked the magazine to soft-pedal opposition. Curley agreed with *America.* No politician could be trusted. Passage of the Curtis-Reed bill would be but the first step toward a federal appropriation for education and then complete federal control. Curley was quite upset. Pace responded that for decades the government had had a Bureau of Education to do research on schooling and disseminate the results. The Curtis-Reed bill would simply expand that function, aiding education in much the same way as the Department of Agriculture did farming, without exerting any federal control. In Pace's view, opposition to such a measure at this time would be interpreted as opposition to education in general. Informed of this interview, Burke confided to Muldoon that the temper of those insisting on active opposition by the NCWC was not good. "Indeed, it is angry: unwilling to be deliberate: or calculating: or thoughtful." Again, he recommended that the Administrative Committee meet together with the Executive Committee of the NCWC Department of Education in Chicago to discuss the situation.[55] Again, his advice was not taken, which, in view of subsequent events, was a mistake.

Curley's mounting frustration caused him to lash out privately in all directions. He continued to blame the hierarchy for the NCWC's lack of opposition to the bill because the bishops had failed to take a corporate stand at their last meeting, even though they "individually" (*sic* for "informally") opposed the measure. "Just what that distinction means, I do not know," he complained to Howard; "and the men in the N.C.W.C. have not the slightest idea of where the Bishops stand." This last remark was not to excuse agents of the Welfare Conference. It was sarcasm, revealing Curley's pique at their apparent blindness to the obvious. After this jibe at Burke and Ryan, Curley absolved them of responsibility—for the moment. In his view, the hierarchy "should have said one thing or the other." He held Dowling responsible too for his lack of direction of the Washington office. Both Burke and Ryan were "perfectly willing to do what they are told by those in authority," but their immediate superior had left them "up in the air."[56]

For his part, Howard had little use for the NCWC secretariat. He viewed himself and the CEA, over which he presided, as the true guardian of Catholic educational interests. The secretariat was a meddlesome wartime appendage of the church. "It is my opinion that many would be pleased to have the annual meetings [of the hierarchy]," he confided to Curley, "if we could dispense with some of the bureaucratic adjuncts that have been bequeathed to the Bishops as a heritage of the war."[57] He considered the War Council's peacetime counterpart unnecessary. A conservative recently elevated to the episcopacy, Howard apparently felt that bishops alone should look after the church's interests without help or interference from priests and lay persons, bureaucratic adjuncts as he called them. Arthur Preuss, conservative editor of the *Fortnightly Review*, would echo the sentiment.

Amid this internal suspicion and recrimination came word that the Scottish Rite was vigorously promoting the NEA bill in North Carolina, Texas, and Wyoming. Masons in Wyoming were using a leaflet entitled *Education Bill Not Abandoned*, printed courtesy of the Nebraska Education League. Its purpose was to explain why the Sterling-Reed bill had given way to the Curtis-Reed bill. "If the form of the original Smith-Towner measure has been laid aside, its principle has not been abandoned," reported the tract. "For the present it is not possible to secure the appropriation contemplated." The tone of the leaflet was 100 percent American, branding opponents "enemies of the public school," hence enemies of the nation's best interest.[58]

With the Rite becoming increasingly supportive of the measure, Burke checked the pulse of Congress. In mid-January he sent William Montavon, successor of the late William Cochran as director of the NCWC Legal Department, to investigate the bill's chances. Montavon learned that the House had just passed a revenue measure that would considerably reduce the federal purse, thus making the economy-minded Reorganization bill far more attractive to Congress than the Curtis-Reed bill. If the latter reached the floor during the current session, members in both houses intended to seek delay of consideration until the Reorganization Committee finished its work. There was still reason for caution because Congress had received numerous appeals in favor of the Curtis-Reed bill but no protests against it. This was dangerous, warned Montavon, for silence bespoke consent. Heedless of his own warning, he concluded that there was "little reason to expect that this matter will reach a critical stage at an early date."[59] He was dead wrong.

A week later Burke did some checking himself, contacting Democratic Senator Peter Gerry of Rhode Island, who also commented on the bill's poor prospect. The measure had been entrusted to a Senate subcommittee, which seemed unwilling to report it back. Moreover, Senator Lawrence Phipps of Colorado, Republican chairman of the Committee on Education and Labor, opposed it. Gerry commented on the lack of interest that Curtis himself showed in the bill. It seemed he would be happy if it never came out of committee. In fact, Curtis had introduced the bill only as a favor to the NEA and was content to let the committee do as it pleased. Gerry advised against complacency, however, and urged Burke to keep in touch with Congress.[60]

Thus, in an apparently cloudless sky, the storm broke. On 26 January Burke received unofficial word that the Education Committees of the House and Senate would conduct a joint hearing on the Curtis-Reed bill to coincide with the meeting of the NEA Department of Superintendence to be held in Washington at the end of February. The Senate Committee on Education and Labor had been informed that Catholics accepted the Curtis-Reed bill because the appropriation had been removed. Curtis averred that he had received hundreds of letters in favor of the legislation, but none opposed. The hierarchy's decision against a public stand had backfired. Interpreting the lack of opposition as approval, Congress decided to proceed. Fortunately, the Executive Committee of the NCWC Department of Education was meeting the next day in Chicago. Burke telegraphed the news to Ryan with a request that it consider the matter.[61]

At the meeting Ryan read Burke's telegram, and Dowling explained that at its last meeting the hierarchy had opposed the Curtis-Reed bill but had opted against a formal, corporate expression of this position out of fear that its opposition might be misunderstood. While some of the Executive Committee believed it was still unnecessary to make such opposition, others felt that it was no longer possible to remain neutral. Pace argued persuasively (against what he personally believed) that the committee should base its decision on the attitude of the bishops and of the country in general: both opposed centralization. Still, he urged against taking a completely negative stand. The church should offer a substitute measure for the expansion of the Bureau of Education along the lines of the Dallinger bill (1924), which the NCWC had favorably reviewed. The committee accepted Pace's counsel, and Dowling wired the decision to Burke.[62] The latter received conflicting advice from Muldoon, who reminded him that the hierarchy wanted no public opposition and wondered why the Administrative Committee should not carry out that order. Burke telegraphed these opposing views to the committee's chairman, Archbishop Edward Hanna, with the explanation that the NCWC's silence was construed as consent. Hanna favored an active protest but wanted Burke to seek the advice of Curley. If he agreed, the NCWC would openly oppose the bill.[63] There is no doubt about Curley's opinion.

On 11 February 1926 Phipps officially informed Burke that the joint hearing would be held beginning on the twenty-fourth. Complaining that because it was to coincide with the meeting of the NEA Department of Superintendence in Washington, D.C., "it looks very much to me as if the 'game' were not being played fair," Burke demanded a just share of the time for the NCWC. Phipps gave him half the hearing.[64]

While the NCWC geared up to fight the Curtis-Reed bill, Andrew Haley, Washington correspondent for the Tacoma *Ledger* and a Catholic, sounded out the members of the Joint Education Committee. As a reporter for a non-

Catholic daily, his presence in Congress raised no suspicions, and so the NCWC used him to snoop. Regarding the House members, Haley reported that Reed favored the bill and held that opposition to it came from the same groups that objected to women's suffrage and Prohibition, namely, Catholics. The congressman believed that in time they would be won over. Loring Black, a New York Catholic Democrat and an intransigent opponent of the bill, was just as confident it would be killed. In his view, the hearing was simply a delaying tactic to keep it in committee. He informed Haley that Tammany Hall had made the committee assignments for all New York congressmen and had purposely earmarked him only for the Committee on Education so that he would be a prime candidate for appointment to any special Joint Committee, like the present one. Haley was unable to discover much about the other three House members. Republican John Robsion of Kentucky was unavailable, but most thought him in favor of the bill, which he was. Republican Florence Kahn of California refused to commit herself, and, as it turned out, was replaced on the Joint Committee. Democrat Bill G. Lowrey of Mississippi said he would issue a press statement later.[65]

As for senators, Haley learned that Phipps favored increased recognition for education but opposed the growth of bureaucracy. Republican Senator Hiram Bingham of Connecticut, a former university professor, staunchly opposed the bill, as did fellow Republican James Couzens of Michigan. Democrat Royal Copeland of New York remained a question mark. He was out when Haley visited the office, and his secretary refused comment. Haley finally persuaded her to let him look through Copeland's files to see what he had written to constituents about the bill. The correspondence revealed him to be noncommittal. Still, Haley thought that Copeland could be counted on to oppose the measure because, according to Black, Tammany Hall had put the word out to fight it.[66]

All Catholic parties set differences aside to present a united front in opposition to the Curtis-Reed bill. At Howard's urging, Curley asked Burke to have Blakely appear at the hearing. He was reluctant to attend, so Parsons came instead. Curley claimed to have secured the testimony of Governor Albert Ritchie of Maryland and President Frank Goodnow of Johns Hopkins University, though Burke reported that Curley had let the matter slip and he (Burke) had to secure their services.[67] The NCWC headquarters worked feverishly to mobilize its forces. Agnes Regan, executive secretary of the NCCW, sent out a circular urging local units to protest against the bill to Phipps and their own congressmen, and Ryan did the same with 133 presidents of Catholic colleges. Through the good offices of Mark Lally, a Knight of

Columbus in Dunkirk, New York, Reed's hometown, Montavon mounted a campaign of criticism in the congressman's district. Burke himself lined up forty-three witnesses, both Catholic and non-Catholic, to appear before the Joint Committee.[68] O'Connell sent a representative and Cardinal Dennis Dougherty sent a written protest. "So," concluded Burke, "it will be a great historical question as to who saved the day—if the day is saved. There won't be any question as to who lost it, if it is lost."[69]

To publicize the church's position, Ryan wrote a pamphlet entitled *The Curtis-Reed Bill—A Criticism,* which was sent to every congressman, to the Catholic and secular press, and to all branches of the NCCW and NCCM. It opened with harsh words for the NEA and its legislation. Alleging that "educational opinion favoring the bill has been manufactured by a politico-educational machine known as the N.E.A.," Ryan called the legislation "a lobby measure pure and simple . . . whose sole genesis as well as the development of favorable support depends on an organized advertising campaign of the worst type." Not only was the bill objectionable on constitutional grounds, namely, that education was a Tenth Amendment matter; if enacted, it would eventually lead to federal control because the NEA would next press for the aid package, deleted from the present measure simply to improve its chances. Agreeing that education was a national concern, Ryan thought that the matter was best left to the people, for they were concerned about their schools and would do what was proper for them. A federal bureaucracy would only kill local interest. In his view, the government's role would be adequately served by an enlarged Bureau of Education, like the one proposed in the Dallinger bill of 1924.[70]

The NEA was as active as the NCWC. It sent out 93,000 pieces of literature, much of it in Curtis's franked envelopes, and cooperated with allied organizations in visiting members of Congress. Williams interviewed sixty-nine congressmen herself, while representatives from other groups saw most of the rest. Moreover, the NEA invited all the legislators to attend a banquet of the Department of Superintendence during the convention. Three days before the hearing, Williams even called on Montavon to convince him that the Curtis-Reed bill was in keeping with the American form of education and government and posed no threat to parochial schools.[71] Montavon and the NCWC remained unmoved.

While the Department of Superintendence met, the NCWC received help from an unexpected quarter. As Burke expressed it, "the Washington Post threw a bombshell into the N.E.A. camp through an editorial condemning the Curtis Reed Bill [*sic*]."[72] Dismissing the idea of a department of education, the paper declared: "The government of the United States should be stripped of extraneous bureaucratic growths and restored to its true functions set forth in the preamble of the Constitution. No new Federal departments or bureaus should be established except for the execution of the powers already granted to the government."[73] In short, the paper had come out in favor of states' rights and home rule.

The joint hearing opened on 24 February. The NEA was given the first day, the NCWC the second, and the third was split between them. Proponents argued that although education was traditionally a local concern, the federal government had always involved itself in schooling. The national interest now demanded that there be some kind of national leadership in and unification of the educational enterprise. All denied that federal control was desirable. In their view, the Curtis-Reed bill offered an acceptable solution of the dilemma between national and local interest in education. The department would exercise influence through research and information, which states could accept or reject. Thus, the federal government would promote uniform solutions to problems without being able to enforce them. The cabinet status of the secretary would also give him the prestige to exercise national leadership in his field. Although a few proponents thought that an expanded Bureau of Education might fulfill some of these functions, none held that it was an adequate substitute for what was needed: a federal department of education. It should be noted that both Judd and Capen were present to endorse the establishment of a department, but flatly opposed federal aid.[74]

The NEA tried an end run around the NCWC. After spending an entire day presenting its case, the association demanded the second day as well. Accession to such a request would have deprived the NCWC of all its out-of-town witnesses, who had planned to be present only for that day of the hearing. Fortunately, the Joint Committee denied the request.[75] In all, thirty-one of the forty-three witnesses lined up by the NCWC came to the hearing, fourteen of them representing Catholic organizations or colleges. Never had the church appeared in such force before a congressional committee. Opponents, both educators and citizens, Catholic and Protestant, argued against the bill on two grounds. Some contended that although the measure appeared to be innocuous, it aimed directly or indirectly at federal control of education. Those who had no fear about federal control held that Congress lacked the constitutional authority to enact such legislation.[76]

Only once did the issue of parochial schools come to the fore. Appearing on behalf of the CEA, Reverend George Johnson, a professor of education at Catholic University, told the committee that it was "vain to argue that because the present bill does not provide for federal control of education, or even disavows it, that it will not make for such control." The threat was veiled. The purpose of the bill, research and information, could be a dangerous step toward standardization because education lacked the objective quality of exact sciences and educational philosophies differed widely. The secretary of education would impose his or her philosophy on the nation by dictating how and what research should be done. Johnson thought that essential investigations could be conducted more safely by the Bureau of Education.[77]

Broaching the religious question, Republican William Holaday of Illinois asked how the secretary of education might strike at church schools. The first way, answered Johnson, might be by prescribing a curriculum based on a philosophy that was non-Catholic. A second way might be by establishing standards for teacher training based on a contrary philosophy. Even if the latter were unobjectionable, the fact that a standard was set would preclude the church's seeking new modes of operation based on a different philosophy. Johnson apprehended an ultimate danger in the bill: secularization. In his view, many teachers were enamored with the French system of education, which elevated the secular almost to the status of religion, complete with tenets, dogma, and creed. Would scrapping the bill really deter such a movement? asked Republican Woodbridge Ferris of Michigan. Johnson replied that it would keep education close to the person on the street and the farmer in the field, both of whom believed in God. As long as schooling rested in those hands, it would be free from the atheism of the university.[78]

The testimony of two NCWC officials was important, not for their comments on the bill, but for the description of their portfolios. A new employee, Montavon lacked mastery of the organization's structure. He represented the NCWC, he said, which acted "for the body of Catholic citizens of the United States in matters of public interest and of national importance." When Phipps asked how many members it comprised, Montavon replied 20 to 25 million Catholics. Rephrasing, the senator inquired how many comprised the working arm of the NCWC and how they got their positions. Montavon answered that the Administrative Committee had twenty members (actually seven) elected by the hierarchy. When asked how many people were in his department, his response was four. "So," summarized Phipps, "through these four representatives, of which you are the chairman, and the committee of 20 of the hierarchy, you represent the Catholic men and women of the country." Montavon replied affirmatively.[79] His description of the NCWC would occasion a fierce attack from the *Fortnightly Review.* So, too, would the testimony of Charles Dolle, executive secretary of the NCCM, who claimed to represent "all

National, State, and local Catholic lay societies" as well as "all Catholic laymen of the United States who do not belong to any of these separate groups."[80]

These descriptions of their portfolios were simply wrong. The secretariat of the NCWC represented the view of the Catholic hierarchy. Though some prelates, like O'Connell, Curley, and Howard, might try to minimize its authority by contending that it represented only the Administrative Committee, the NCWC was according to civil law the hierarchy of the United States. The charter of incorporation named each bishop individually and specified that all rights passed to their successors.[81] While the NCWC represented the hierarchy, it did not represent all Catholics in the United States. It did have two lay organizations, the NCCW and the NCCM, both voluntary and, in a measure, independent of the hierarchy. If the bishops took no public stand on an issue, each organization remained free to do so. Their freedom was limited only by official action of the hierarchy. Like the hierarchy, neither had officially taken a stand against the Curtis-Reed bill. Moreover, the NCCM, for which Dolle spoke, existed largely on paper. Unlike its distaff counterpart, it remained a small organization in the wake of the crisis of the suppression of the NCWC.[82] Montavon's and Dolle's exaggerated claims were certain to provoke a response from within the Catholic community.

Outside the committee room, the Curtis-Reed bill came under attack on the floor of the House. John Philip Hill, a Baltimore Republican, argued to his fellow representatives that a department of education was unnecessary because the present bureau was able to perform all necessary functions. "If it [the bureau] needs more money for its legitimate purposes[,] appropriate it," he said, "but do not create a new department." Though the Curtis-Reed bill authorized an annual appropriation of only $1.5 million, he was convinced that within five years of its passage "pressure would be brought to bear upon Congress to make annual expenditures of $100,000,000 or more," the amount in the previous NEA bills. Moreover, a cabinet-level secretary would politicize education and probably lead to federal control. Concluded Hill, "The Bureau of Education can minister to the general educational welfare. We do not need a new Cabinet post of secretary of education. It is another case of attempted usurpation of local State rights, and I am against it." His words drew applause.[83]

Washington papers also hammered at the legislation. Like many Catholics, the editor of the *Post* considered the bill's real aim to be federal control exerted through an eventual subsidy. "The plan adopted by the educators is to advance cautiously, behind a mask," he told readers. "They should throw away the mask and march forth boldly against the Treasury, under a banner bearing the glorious motto, 'Down with home rule! The States are a failure!' "[84] A week later the *Star* ran a lengthy editorial quoting negative opinions of the bill from some two dozen papers around the country.[85] In imitation of the *Star,* Ryan put together a pamphlet reprinting editorials against the measure from ninety-eight journals in every part of the nation. Major papers objecting to the legislation included the Providence *News,* the New York *World,* the Brooklyn *Eagle,* the Buffalo *Courier,* the Rochester *Democrat and Chronicle,* the Newark *News,* the *Philadelphia Inquirer,* the Philadephia *Public Ledger,* the Toledo *Times,* the *Columbus Dispatch,* the Louisville *Times,* the Indianapolis *News,* the *Chicago Tribune,* the *Chicago Daily News,* the Minneapolis *Journal,* the Saint Paul *Pioneer Press,* the Omaha *World-Herald,* the Baltimore *Sun,* the *Charleston Daily Mail,* the New Orleans *Times-Picayune,* the Houston *Post,* the Dallas *News,* the Los Angeles *Record,* the Sacramento *Bee,* and of course both the *Washington Post* and *Star.* The nearly one hundred editors, whose pieces were reproduced, opposed the bill because its enactment would either exceed Congress's constitutional power, violate home rule, or increase bureaucracy.[86]

While many papers panned the Curtis-Reed legislation, Lally and his allies campaigned against it in Reed's home district and throughout the state of New York. Initially, he had planned to hold a rally against the bill in Dunkirk, with President Nicholas Murray Butler of Columbia University, an arch foe of the bill, as the featured speaker. The district, however, was heavily Catholic, causing Lally to fear that the rally would look like little more than a church function. Instead, he launched a massive region-wide mail and telegraph protest against the measure to bring Copeland into open opposition and perhaps neutralize Reed. Apparently, the campaign had the desired effect. The congressman and senator received hundreds of telegrams and letters opposing the measure. After the hearing, Montavon found a chastened Reed "in a rather apologetic mood regarding his bill."[87]

Within days Black reported to the NCWC that the House committee would kill the Curtis-Reed bill if only by a narrow margin.[88] According to the *New York Times,* the contest was hotter than Black let on. The NEA and affiliated organizations were exerting tremendous pressure on Congress to bring the measure forward. The Ku Klux Klan and the Southern Jurisdiction of the Scottish Rite were also working earnestly on the bill's behalf. All sought its immediate passage by the House so that there would be ample time for the Senate to approve the measure. At the center of the storm were congressional leaders, in a quandary over whether to let the bill come out of committee, with a consequent disruption of the legislative agenda for the remainder of the session. Proponents claimed to have 170 representatives committed in writing to the measure. Opponents claimed to have more than the 218 votes needed for its defeat. The *Times* itself remarked on a decided shift in the political wind against federalization. Coolidge had called a halt to such schemes, and his spokesman in Congress said that the president would be relieved if the Reorganization bill were "pigeonholed." Indeed, Coolidge had cooled toward that measure, the only feature of which he would again advocate was a department of education and relief. Influential senators also noted a growing antagonism among congressmen toward further federalization. All the movement needed was a champion in Congress.[89]

There were at least two contenders for that position. In the *Nation's Business,* the journal of the U.S. Chamber of Commerce, Republican James Wadsworth, Jr., Old Guard senator from New York, denounced the drift toward federalization. Asserting that the strength of the nation rested on its history of local initiative, he lamented the growth of matching grants of federal aid, like those in the Smith-Hughes Vocational Education Act, the Sheppard-Towner Maternity Act, and those proposed in the Sterling-Reed bill. Such measures had led to the creation of a vast federal machinery across the nation for their enforcement. "If we continue this centralization of power and this assumption of governmental functions," he argued, "we shall most certainly smother the ability of our people to govern themselves in the several states and in their home communities."[90]

Lambasting the Curtis-Reed bill in a speech delivered at Randolph-Macon Woman's College in Virginia, anti-statist Republican William Borah of Idaho declared that the measure would ultimately lead to the federal control of education. Deny this as proponents might, their original intention and seminal legislation (the Smith bill) had sought the federal control and standardization of education. Their initial plan called for the establishment of a national system of schooling. When that scheme roused public outcry, proponents modified the bill. Let the present measure pass, warned Borah, and the original scheme would soon blossom to fruition. Broadening his attack beyond the bill, he asserted that Congress was glutted with legislation for the creation of new bureaus. "Anyone who will examine these bills," he concluded, "will find that the restless legislative mind does not propose to leave any activity, any business, free of governmental direction and surveillance."[91]

Politicians like Borah, Wadsworth, and Hill believed, as did most Catholics, that the Curtis-Reed measure was a watered-down version of the initial Smith bill and, if passed, would eventually lead to the original goal: a national system of education under federal control. Their opposition on such grounds was a boon to the church, for it sprang from constitutional and political beliefs, not religious ones. As such, their antagonism lent credence to Catholic opposition on the same grounds. When combined with the anti-bureaucratic attitude of Coolidge, it also showed that Catholics were, at least on the education issue, in the political mainstream.

Despite the rising chorus against further federalization, the tense situation in the House Committee compelled Burke to leave nothing to chance. Following the advice of Dowling's department, he thought the way to ensure the bill's demise was to offer a viable alternative. If, moreover, the NCWC advocated something positive, the testimony of Catholic witnesses would gain in sincerity, for most had urged upgrading the U.S. Bureau of Education. Burke asked Montavon to revise the Dallinger bill, which proposed the expansion of that office. A streamlined version of the parent measure, Montavon's rendition empowered the bureau to provide the same services as the proposed department in the Curtis-Reed bill but at one-sixth of the cost.[92] After Dowling approved the draft, Burke had Phipps introduce the legislation by guaranteeing the NCWC's support.[93] On 15 March Dowling issued a press release praising the Phipps bill as "a forward looking, constructive . . . statesmanlike measure," which would make the Bureau of Education "the most important educational research agency in the United States."[94] Enclosing a cover letter from Phipps and using his franked envelopes, Montavon sent a copy to every bishop, 800 college and university educators, and the entire Catholic press.[95]

Few papers gave it notice. The Cleveland and Cincinnati weeklies favored it as a safe alternative to the NEA bill; the Indianapolis and Saint Paul papers were matter-of-fact.[96] The *Pittsburgh Catholic* simply lumped the Phipps bill together with other education measures and warned broadly, "It seems . . . that all this legislation has a sinister background, and stealthily and insidiously is being laid the foundation stone of a legislative structure which will sweep away the cherished Catholic school system."[97] *America,* too, was cautious. Confident that Phipps was trying to avoid federal control, Blakely, writing under the pen name John Wiltbye, advised that the nature of bureaucracy was to expand. Opponents of federalized education, he said, did not have "to offer a substitute for every fantastic measure that issues from the Southern Masons or the National Education Association"; all they had to do was oppose it.[98]

After deliberating for two months, the House committee decided on 5 May to defer action until the second session—after elections. Perhaps feeling the sting of the NCWC's recent protest campaign in his home district, Reed informed Williams that although congressional sentiment was much more favorable to the bill in its present form, lacking as it did the subsidy, "the one thing that will have to be demonstrated in order to insure early and favorable consideration . . . is a sentiment 'back home' favorable to it." This advice made perfect sense because he was up for reelection. So too was Curtis, who gave Williams the same counsel: "Any organization is as strong in legislation in Washington as it is active in the congressional districts and no stronger."[99] The message was clear. The battle for a department of education would be won or lost in the constituencies.

The NEA received a second blow on 6 May when the Senate committee reported the Phipps bill as the only viable education legislation. The recent hearings had clearly shown that everyone believed the Bureau of Education was doing good research, but desperately needed more money and personnel. The Phipps bill would meet both needs while avoiding the controversial issue of whether or not to establish a department of education.[100]

The NEA kept these setbacks in perspective and gave them the best interpretation. Recognizing that the upcoming elections made it imperative to delay action on the Curtis-Reed bill, Williams told the annual convention that the Senate committee reported the Phipps measure as "a mere political gesture." The pressure on Congress to do something for education was so great that the committee had to take some action. Recommending the Phipps bill seemed the least dangerous political move. Noting that both Curtis and Reed advised pressuring congressmen at the grass roots, she argued that the success of the NEA bill rested with the association itself; it depended on the efforts of individual members. "Are we, the teachers of this nation, equal to the task?" asked Williams. "It may be a long fight. Have we the courage and the endurance to stand by our guns until the battle is won?" Answering her own questions, she declared that no one would doubt that American teachers "have just begun to fight."[101]

Strayer, chairman of the Legislative Commission, outlined the strategy. Each member had to impress upon her or his congressman and senators the reasons why the bill should pass. Nothing short of a personal interview would suffice. State and local educational associations must do the same, explaining the measure section by section to their representatives. At the same time the local press must be enlisted to disabuse the public of erroneous notions regarding the purposes of the legislation. Strayer was sure that the plan would work if the members got behind it, and he pledged that the NEA would continue its effort until the bill was signed into law.[102]

John McCracken, chairman of the ACE's Committee on Legislation, made a similar point to his organization, though he seemed resigned to the fact that success lay far in the future. In his view, the general public felt that the nation had too many, rather than too few, laws and wanted no more. Noting that the world was divided into two groups, "those who look for worldly salvation from the government and those who look for worldly salvation from education," he considered the recent setback for a department of education as wholesome, for it turned the thoughts of schoolmen "from government, as the source of all good, to our own broad job of education as the effective agent for our purposes, slow but sure and irresistible." Educators needed to educate the public on the need for such a department. He concluded by quoting Edmund Burke's maxim for the true lawgiver, " 'Time is required to produce that union of minds which alone can produce all the good we aim at.' "[103]

Rather than tutor legislators and the public, Catholics were making life miserable for Congressman Reed. Lally urged Warner Rexford, Chautauqua County surrogate, to run against Reed in the Republican primary. Eying a seat on the State Supreme Court, Rexford refused. Lally and his aids then bolted the Republican party to support Democrat John B. Leach, who focused his campaign on Prohibition and the Curtis-Reed bill. He opposed the latter on grounds that it was unconstitutional, would increase taxes, and federalize education. Lally sent a letter against the measure to every voter in Dunkirk, Reed's hometown. Although Leach lost the election, Reed lost Dunkirk, a city he had carried two years earlier by 1,500 votes.[104]

Meanwhile, the NCWC itself came under attack from certain Catholic quarters and even one of its own agents. When the record of the joint hearing was published in May, the *Fortnightly Review* began a year-and-a-half-long attack on the NCWC that quickly gained active support from within the Welfare Conference itself. The magazine enjoyed a potentially wide and powerful readership, for at the expense of Patrick H. Callahan it was sent free to every Catholic bishop, newspaper, and institution.[105] In the first of no fewer than thirty-one articles, Preuss started with Montavon's testimony. Irritated that a man virtually unknown to the Catholic population would claim to represent it, he angrily denied that Montavon, or the NCWC for that matter, had any right to speak for Catholics. Convinced that the Vatican had acted correctly in suppressing the conference in 1922, Preuss lamented, "Most of us will probably live to rue the fact that its original decree was not put into effect." Readers were to remember that in rescinding the suppression, Cardinal Gaetano De Lai had warned against identifying the NCWC with the hierarchy. "If it is not to be identified with the hierarchy, and is not elected by the laity, as Mr. Montavon admits," concluded Preuss, "just for whom or for what does it stand?"[106] A priest in Washington informed him that the article fell "like a bombshell" in the NCWC camp.[107]

Not long afterward, Preuss was contacted by James R. Ryan, an attorney who had resigned from the NCWC Legal Department in January 1926, shortly after Montavon took it over. Although Ryan ostensibly left the NCWC to enter private practice, he was probably piqued at being passed over for the directorship of the department. He was critical of Burke's appointment of Montavon because the man was neither a lawyer nor conversant, in Ryan's view, in domestic affairs. For more than a decade Montavon had resided in the Philippines and Peru, as a civil servant in the former and as American representative of the International Petroleum Company in the latter.[108] In fact, Bishop Edmund Gibbons, chairman of the Legal Department, had hired Montavon precisely because of his experience in public affairs. In Gibbons's view, the Legal Department already possessed a lawyer in Ryan and needed instead someone with more breadth.[109] Unaware that Gibbons had made the decision, Ryan held Burke responsible not only for Montavon's hiring but for the entire direction of the secretariat. He alleged that it was Burke's decision to remain neutral on the Curtis-Reed bill. "I strongly suspect, but cannot prove," he told Preuss, "that there was a 'deal' between certain officials of the N.C.W.C. and some of the proponents of this character of legislation." Driven to "a perfunctory opposition" of the bill by *America,* the NCWC laid plans for the subsequent endorsement of the Phipps bill. Ryan urged Preuss to assault that measure and requested copies of his article for distribution to key congressmen.[110]

In mid-June Preuss attacked Dolle's claim to represent all Catholic men in America. "His statement," said the editor, "will be the first intimation to millions of laymen . . . that they have given their adhesion or entrusted their religious rights and interests to Mr. Dolle's Council of Catholic Men." Of the more than 15,000 parishes in the country, fewer than 500 had local men's councils.[111] J. R. Ryan distributed copies of this new article to congressmen. In late June he visited the office of a New York representative, whose secretary informed him that a perturbed Burke had made the rounds on Capitol Hill trying to eradicate the bad impression made by Preuss's attack, which had caused something of a sensation and stimulated considerable demand for copies of the hearing.[112] Ironically, by subverting the NCWC in the eyes of congressmen, Ryan and Preuss were undermining the church's opposition to a bill that they themselves despised.

From July through September, Ryan fed Preuss unsigned articles, cowritten by Grattan Kerans, a rewrite editor in the NCWC press bureau. The two proposed to attack Burke, Montavon, and Dolle until the annual convention of the hierarchy in the hope that the bishops would sack them.[113] The most telling article appeared in August. Quoting De Lai's instructions reinstating the NCWC, it noted that agents of the conference were forbidden to trespass on the authority of a local bishop. For six years Burke and his agents had violated this injunction by advocating national legislation like the Sheppard-Towner Maternity Act and the Phipps bill. Because federal acts operated uniformly throughout the country, they did so in every diocese. Though some bishops opposed these measures, their antagonism was "nullified by the action of the N.C.W.C.," which supported them. If the NCWC was to avoid becoming a divisive or destructive influence in the American Catholic church, its agents must be chosen with the utmost care. "Their record thus far," concluded the article, "is one of arrogance and incompetence. Further intervention by the Holy See to correct these abuses need surprise no one."[114] The piece is significant because it was the ecclesiastical equivalent of the states' rights argument in politics. It portrayed the NCWC as a national Catholic bureaucracy undercutting the autonomy of the local bishop and diocese. The irony in the situation is noteworthy. While the NCWC opposed the Curtis-Reed bill and its predecessors as a bureaucratic violation of local self-government, Catholic critics of the NCWC now leveled the same charge against that organization with regard to diocesan autonomy.

The *Fortnightly Review* was not the only critic of the NCWC. As the Curtis-Reed battle was being fought, Curley found a new reason for concern with the NCWC: the church-state conflict in Mexico. In 1925 the Mexican government had begun a persecution of the church. Unknown to Curley, the bishops of that country had asked the NCWC not to agitate the matter for fear of intensifying the persecution. Curley helped Representative John Boylan of New York to draft a resolution urging Coolidge to sever relations with Mexico. Boylan went to Burke to solicit the NCWC's support. Needing the authorization of the Administrative Committee to take such action, Burke placed the matter before its members, who voted against public support for the resolution. Without mentioning the committee's role in the decision, Burke informed Boylan that the NCWC could not support his measure. Boylan relayed the information to Curley, who lost patience with the priest. Because this event occurred so closely in the wake of the Curtis-Reed contretemps, Curley now held Burke responsible for the NCWC's foot-dragging in every instance. He told the apostolic delegate, "At times like this, one does not know just who or what the N.C.W.C. is. In its ultimate analysis, it seems to be Father John Burke and nobody else." An early defender of the NCWC as the voice of the hierarchy, Curley was coming full circle to deny any identification of it with united episcopal action.[115]

Another prelate at odds with Burke and the NCWC was Cardinal O'Connell. Three days before the 1926 convention of the hierarchy, he summoned Burke to Oblate College in Washington for what turned out to be a rambling three-hour interview. O'Connell was concerned about the Phipps bill and wanted to know what the NCWC secretariat had done about it. He had heard that Burke had "sold out" on the education question by yielding to the federalizers. Moreover, the cardinal asserted that Phipps was anti-Catholic.[116] Although neither a member of the Klan nor sympathetic to its program, the senator had felt compelled to accept its help in his 1924 reelection bid, a campaign that he could not have won without Klan support.[117] This hardly qualified him as anti-Catholic, a fact corroborated by his willingness to aid the church by sponsoring its bill.

Without mentioning that Montavon had drafted the measure, Burke explained events leading to the NCWC's endorsement of it as an alternative to the Curtis-Reed bill. He noted that Dowling and his department had approved the Phipps bill, and the Administrative Committee had concurred. Again O'Connell asked what the secretariat had done about it. Burke sidestepped any admission that the secretariat had written it. "If your Em[inence].," he said, "means what have we done . . . in pushing the Phipps Bill, beyond what I have told you, I would answer 'nothing.'" He added that the Administrative Committee would press for its enactment only if the Curtis-Reed bill seemed likely to pass.[118]

O'Connell reminded Burke that the hierarchy had instructed the NCWC to make no public antagonism against the Curtis-Reed bill, but to stand squarely against further federalization (Muldoon's point back in January when he had counseled against taking any action). In O'Connell's view, the secretariat had violated this injunction in every respect. It had opposed the Curtis-Reed bill and supported the Phipps bill, which the cardinal considered to be yielding to federal control. Burke explained that the hierarchy's ambivalent instructions had put the secretariat in an awkward position. When it did not oppose the Curtis-Reed measure, the NEA and Congress interpreted its silence to mean that the bill was acceptable. The Administrative Committee, therefore, had no choice but to oppose it. The endorsement of the Phipps bill was a tactical maneuver to ensure the defeat of the Curtis-Reed measure. Because virtually every Catholic at the hearing favored a larger budget for the Bureau of Education with an increased investigative ability, the Administrative Committee backed the Phipps bill that contained both features, thereby lending sincerity to Catholic opposition to the Curtis-Reed bill.[119]

This set O'Connell off on a long dissertation about the two schools of American Catholics; this was actually a disquisition against Americanism in which the cardinal indiscriminately blended Dowling and the bishops of the Administrative Committee with the Americanists of the 1890s. In O'Connell's view, they lacked a solid understanding of Catholic principles and traditions and tried too hard to "remain at peace with America." America was a Protestant country, built on the principles of that religion. A Catholic must always distrust the government, said the cardinal. The state would always oppose the church on education. There were men in the 1890s who would "though they were Catholics have compromised our position. There were those who would do it today." Driving home the point that he considered Burke an Americanist compromiser, the cardinal concluded the interview with the advice, "Father Burke[,] no government has ever used a priest without squeezing him dry as a lemon and then throwing him out."[120] The interview produced no meeting of minds. It made clear to Burke, however, that O'Connell opposed the Phipps bill and considered the NCWC a needless, dangerous, Americanist experiment.

The next day Burke met with the Administrative Committee to discuss the Phipps bill. Reexamining the text, the committee concluded that it contained nothing unreasonable or objectionable.[121] The hurdle, however, would be securing the hierarchy's approval of Dowling's endorsement of the bill. Hanna explained to the hierarchy the secretariat's actions regarding the Curtis-Reed bill. When Dowling reported his endorsement of the Phipps measure, O'Connell asked if he had conferred with anyone before giving it the NCWC's blessing. Dowling replied that the entire Administrative Committee had been consulted. Curley expressed anger because Burke had tried to silence him in December on the Curtis-Reed bill and had failed to support the Boylan resolution on Mexico. He suggested that Burke was dictating church policy because in emergencies he acted without seeking the advice of the Administrative Committee. Gibbons responded that on those two issues Burke had contacted the committee and had carried out its instructions. Howard then announced his non-concurrence with Dowling's report. Praising the NCWC's opposition to the Curtis-Reed measure, he deplored its approval of the Phipps bill because that legislation was the entering wedge of federal control. "Once the principle of Federal supervision of Education is accepted," he warned, "and this measure is a near approach, there will be no end to the extension of this influence; and this will be prejudicial to the true interests of the nation and in time will be gravely injurious to religion." Dowling countered that the bill contained no trace of supervision and argued that the bishops had no warrant to read that into it. Interrupting further debate, Archbishop John Glennon moved acceptance of Dowling's report, which was carried.[122] This, however, did not heal the breach that was widening between the NCWC and prelates like Howard and Curley. Nor did it end criticism of the Phipps bill.

The Catholic press lined up against it, and leading the charge was the *Fortnightly Review.* An article ghostwritten by Ryan and Kerans hinted that Burke and Montavon had written the Phipps bill, "which bears the Senator's name but was drafted by men who should have known better." Readers were erroneously informed that Phipps headed the Klan wing of the Republican party in Colorado and therefore was not "the kind of official to whom the safety and interests of Catholic education could be entrusted." Yet his bill had recently been urged on the Catholic people by Burke, Dolle, and Montavon.[123] In other words, the magazine suggested that agents of the NCWC were at best stupid or at worst conscious traitors to the Catholic cause.

Responding to the *Fortnightly,* the Davenport *Catholic Messenger* rightly understood that the NCWC had backed the Phipps bill as an alternative to the Curtis-Reed bill. The editor even considered it a fairly good substitute because it gave education needed recognition. The NCWC had erred, however, in thinking that public opinion opposed the Curtis-Reed bill because it would establish a new federal department. In fact, people were antagonistic to any increased bureaucracy, and in this regard the Phipps bill was just as objectionable as the measure it was to replace. The NCWC, concluded the editor, suffered from bad advisors.[124]

America too lined up against the measure. The editor remarked that few proponents of the NEA bill believed that federal control of education could be achieved in one blow. Convinced that "the longest way around is sometimes the shortest way home," they would take what they could get at the moment: the glorified Bureau of Education in the Phipps bill. Later they would argue that such an agency was an anomaly unless it was "invigorated with plenary powers." In *America*'s view, the senator's measure was the seed that would blossom in a full-blown department of education.[125]

When Congress reconvened for the second session in December 1926, the Senate placed the Phipps bill on the calendar for action. Burke sought Hanna's advice. Counseling against further endorsement, the priest warned that if the NCWC Administrative Committee ordered positive action, it had better "be prepared to meet the criticisms that are sure to come, not from the non-Catholic but from the Catholic press."[126]

When the Executive Committee of the NCWC Department of Education met in January 1927, Dowling explained that at the meeting of the hierarchy, Howard had objected to the endorsement of the Phipps bill, while most bishops had sympathized with Dowling's support of it. Still, the hierarchy had not gone on record for or against the bill. James Burns, John Fenlon, and John A. O'Brien, three educational progressives on the committee, reaffirmed their conviction that the expansion of the Bureau of Education would be a step forward for schooling. Pace wanted to challenge the hierarchy to make a forthright declaration of its view of the proper relation of federal and state governments to education. In the present circumstance, he would settle for definitive advice on the endorsement of the Phipps bill. Reaffirming support for it, the committee asked Dowling to poll the hierarchy before making the decision public.[127]

Parsons obtained a copy of the minutes of this meeting and forwarded them to Curley. The Jesuit suggested that the NCWC was attempting to dictate church policy on education through this "unofficial" poll. Curley told Parsons to assume that none of the bishops had read *America*'s objections against the Phipps bill, so he should send each a letter explaining the dangers in it. Unless such action was taken, "Archbishop Dowling will probably give what will be more or less a rubber stamp approval to the Bill from the Hierarchy of the Country, and then 'America' will find itself in opposition to the Hierarchy."[128] Curly had developed a cynical attitude about his colleague in Saint Paul. To Dowling, he wrote that he had cast his lot in with *America,* a position tantamount to a vote of no confidence in the NCWC on education.[129]

While Dowling's canvass of the hierarchy on the Phipps bill was still in progress, the measure died a quiet death. On the day scheduled for its consideration, the Senate passed over it.[130] The NCWC itself finally abandoned the bill in April. The results of Dowling's poll showed that while the majority of the hierarchy favored it, several bishops condemned the measure outright and others feared that it was the entering wedge of federalization. Bowing to the dissenters, Dowling urged the Administrative Committee to discontinue support. The board agreed.[131]

Even after abandonment of the bill, the Ryan-Kerans axis pummeled the NCWC for supporting it and, like O'Connell, pictured the issue as Americanism redivivus. The two argued that Pope Pius XI would eventually condemn the NCWC as he had *Action Française,* a movement in France that identified nationalism with Catholicism, but in such a way that made the latter the servant of the former. The church's principal safeguard against nationalism was the autonomy of the local bishop. To call episcopal independence "deplorable provincialism" or "narrow parochialism" was to belittle a most advantageous feature of the church. The cowriters averred that it was easy to trace such language to those who wanted "to see the N.C.W.C. . . . recognized and operated as a 'great central agency' which should 'represent and protect' Catholic 'causes, rights and interests' in a 'national way.'" Although the NCWC lacked a mandate to act in such capacity, said the two, it had presumed to do so in no less a matter than "federal legislation affecting Catholic schools in the country at large," thereby invading diocesan autonomy in violation of "its conditional continuance under the decree of the Holy See."[132] Equating *Action Française* with Americanism, the *Fortnightly* warned: "Leo XIII checked a similar movement here with his famous brief 'Testem Benevolentiae,' but 'narrow earth-bound nationalism' is not by any means dead yet in America."[133] In other words, the NCWC was a promoter of Americanism. Try as the pair might for the remainder of the year, they were unable to convince the majority of the hierarchy or the majority of American Catholics of this charge.

The root of the ill will toward the NCWC on the part of Kerans, Ryan, Curley, O'Connell, Howard, and Parsons can be traced to the resurgence of educational progressivism in Catholic officialdom after the Oregon case, which forced a reconsideration of the federal education question, especially in light of the NEA's deletion of aid from the Curtis-Reed bill. Though suspicious of the NEA's motives in bringing forward a new measure, progressives, like Dowling, Carroll, Hayes, Pace, Burke, and Ryan were willing to consider it on its merits. While some were motivated in part by a concern that the church was gaining a reputation for obstructionism in national educational matters, all believed that the church should take some positive step for education. They were in the minority. If the years since 1919 had done anything, they had reinforced in the majority of bishops the correctness of opposition to federal control of, that is, increased federal participation in, education and the rightness of local hegemony. Conservative prelates also correctly sensed a shift in public opinion against increased federal bureaucracy. As this opinion would harden, so too would Catholic opposition. The hierarchy would be increasingly unwilling to accept alternatives to a department of education. As will be seen in the final chapter, the bishops would reach a point where they would consider absolutely no adjustment in the federal government's participation in education.

The differences between progressive and conservative prelates resulted in indecisiveness on the part of the hierarchy with regard to the new NEA bill, leading to tortured controversy over the NCWC secretariat. The bishops' decision to remain publicly neutral toward the measure while maintaining informal opposition backfired. When the secretariat took no formal stand, Congress and the NEA interpreted the church's silence as approval of the Curtis-Reed bill. The inaction of the secretariat caused Curley, O'Connell, and *America* to suspect that the NCWC had sold out the Catholic position. Their suspicions seemed confirmed when Burke followed the advice of the NCWC Department of Education and offered the Phipps bill as an alternative that the church could support. A mild, almost innocuous, substitute paying lip service to increasing the government's role in education, the measure was hardly the entering wedge for the creeping federalism feared by Parsons, Preuss, Howard, Curley, O'Connell, and others. That they saw it as such was testimony to the depth of their opposition to further federal bureaucratization. In their view, the church should favor no federal education legislation whatsoever, for sooner or later it would eventuate in the creation of a department of education. In short, the hierarchy's instructions had put the Washington secretariat in a no-win predicament. Burke and his staff were pilloried for carrying out orders that some mistakenly believed he himself had written. Worse, the NCWC had been forced to orchestrate virtually all the opposition at the 1926 hearing. The unprecedented high Catholic presence at that event served only to reinforce the religious issue and bring it to the fore.

6

The Religious Issue to the Fore

With the Curtis-Reed bill, the NEA had finally been able to unite the educational trust in support of legislation. The vast majority of administrative progressives favored increased federal recognition of and participation in education. Yet segments of the trust endorsed the measure for different reasons. Progressives like Charles Judd, Samuel Capen, and members of the ACE viewed the proposed department primarily as an agency for educational research. The NEA had removed aid from the bill precisely to gain the backing of such educators. George Strayer, Charl Williams, and other leaders of the association, however, remained committed to the NEA's original program, namely, a department and aid, now sought in two steps. The beauty of the Curtis-Reed bill was that it allowed each element of the trust to read its own agenda into it.

The deletion of aid from the bill cut two ways. While it won over an alienated segment of the trust, it weakened support among teachers. A considerable portion of the aid package was to have been used to increase their salaries. With such incentive removed—or at least pushed into the indefinite future—grassroots support for the legislation began to wither. It eroded even further as more and more newspapers lined up against the Curtis-Reed bill. Thus, the NEA was faced with the new task of building morale within its own ranks.

The NCWC had quickly realized that the NEA was simply seeking to enact its program in parts. It was also aware that both the Southern Jurisdiction of the Scottish Rite and the Ku Klux Klan viewed the Curtis-Reed bill as the means for compulsory public schooling. Blocked by the Supreme Court from using state legislation to achieve that goal, the sole means left to accomplish it was the establishment of a department of education with massive federal aid at its disposal. Only the improvement of public schools to the extent that private ones could not compete with them would suffice to drive parochial schools out of existence. Yet, shorn of federal aid as the bill was, it appeared to be harmless to parochial schools, a point that representatives of the NEA had argued to the NCWC. Thus, the church's objection to the measure in 1926 seemed to be proof of its opposition to public education and by extension to the nation itself, at least that would be the interpretation given it by Protestant observers and the Klan wing of Masonry. With the Catholic Alfred E. Smith destined to become the Democratic candidate for president in the upcoming election, the religious issue was sure to figure in the next round in the battle for a department of education.

If the NEA's legislative program was to succeed, the first task was to reinvigorate support among teachers. "One of the things which all of the legislative workers in Washington, representing great national organizations like ours have learned," Williams told the 1927 convention of the NEA, "is that resolutions and statements have lost their force and effect." Congressmen wanted to know how many people supported a bill and were "willing to fight for it in their own respective congressional districts." This was the lesson the NCWC had recently taught Republican Daniel Reed, chairman of the House Committee on Education and sponsor of the Curtis-Reed bill. The NEA was slowly awaking to that fact. Many legislators had told Williams that they personally favored the bill but that educators back home opposed it. The NEA's enemy was within, as Williams made clear. First, there was a lethargic and sometimes timid membership, cowed by the opposition of the press. "I do not know what is so deadly about a half column of print," she told conventioneers, while reminding them that a number of papers and magazines supported the measure. Second was a division in the educational community: some local chapters of the NEA wanted to boom the bill, but were blocked by a principal or superintendent. Finally, other chapters did nothing, hoping that the trick of passing it would somehow be turned in Washington by sensational methods. "I am the last person in the world to censure a congressman who hesitates to do anything for us before we ourselves decide what we want," concluded Williams. The battle would be won or lost at the local level. There were thirty-four state educational associations affiliated with the NEA, each organized to promote the bill. Moreover, thirteen of the remaining fourteen states had educational committees to push it. She urged listeners to show the same pluck that inspired the women's suffrage campaign.[1]

Once schools were in session, Williams encouraged members of the NEA to enlist the support of the local Parent-Teachers' Association (PTA). Each branch of the PTA was to be encouraged to appoint a committee to visit congressmen and senators before they returned to Washington in December. If this could not be done in person, the PTA should do so by letter. In any case, each unit should report the response of the legislator so that the NEA knew where the congressman stood.[2]

Two weeks later, Williams circulated a broadside of an editorial from the Hearst papers decrying the rate of illiteracy. According to the 1920 census, six out of every one hundred Americans was unable read. Using comparison for impact, the editor claimed that nine European nations had a lower rate than wealthy America, whose five million illiterates equaled in number the population of Australia. In the face of a lamentable problem, "Uncle Sam's entire interest in education is represented by a little Bureau of Education, an insignificant branch of the Department of the Interior," which in 1926 had a budget of $220,000. The conclusion was clear. The nation needed a department of education.[3] This was alarmism and a half-truth. Indeed, in 1920 the rate was what the editor claimed, but it had been steadily declining since 1870 and by 1930 would be four per hundred—and that without a department of education.[4] No matter, the editorial made good advertising, and Williams spread it abroad to "convey to our members some of the spirit which led temperance forces who fought for half a century before their aim was achieved." She urged members to build a favorable press by asking two or three influential citizens in their communities to give local papers statements favoring the NEA bill.[5]

Williams's words and actions indicated two things. First, support among teachers for a department of education had withered as political opposition to the measure increased. No doubt, the NEA's abandonment of federal aid—with the attendant lure of higher salaries—had contributed to the malaise. Teachers now seemed content to let their delegates to the annual convention pass resolutions in favor of the bill and let it go at that. The campaign for "democratic," top-down reform of education was fizzling. Second, a groundswell of support from the press was missing. Rather, major dailies had lined up against the bill. This is not to say that the NEA enjoyed no support, for some journals did promote the measure, most notably the Hearst papers, which would become more sensational in their endorsement. The fact remains that whether the press reflected or drove public opinion, the tide against the bill had swelled through the middle years of the decade.

Alert to the NEA's propaganda crusade, the NCWC rallied the National Council of Catholic Women. Agnes Regan, the executive secretary, told her sisters that the NEA was trying to sell its bill "through an educational campaign which is to be conducted with consummate patience and unrelenting vigor." Warning that such a drive might be effective, she advised members to ensure that community leaders were conversant with the pros and cons of the NEA bill. Catholic mothers should make their opposition known to journals devoted to the interests of women and children. Moreover, they should applaud newspaper editors who opposed the measure and send protests to those who favored it. Regan suggested taking to the airwaves through broadcast debates on the bill. Of course, Catholic women were also to voice discontent to their senators and congressmen.[6] In the fall, James H. Ryan, executive secretary of the NCWC Department of Education, encouraged delegates at the annual convention of the NCCW to campaign against the measure by placing two NCWC pamphlets in the hands of teachers, students, and women who belonged to organizations supporting the legislation. The first was his *Curtis-Reed Bill—A Criticism;* the second, *Editorial Opinion and the Curtis-Reed Bill.*[7]

Meanwhile came an important announcement that virtually escaped notice. In a brief back-page item running scarcely a few lines, the *New York Times* reported in April 1927 that Secretary of Commerce Herbert Hoover, who had earlier favored the Sterling-Towner bill, did an about-face, announcing his opposition to the establishment of a department of education. "I can see no critical condition or pressing need in education to warrant such a step," declared Hoover. "Under its own sails the greatest industry in the country [education] is now accomplishing results which rival those of any people in the world."[8] His reversal was probably due to the Curtis-Reed bill's abandonment of federal aid. As Hoover explained in his memoirs, he had favored "Federal aid to backward areas where there was genuine need" because the deficiencies in those regions affected the whole nation. While "certain very vocal educational associations" had advocated such aid, federal funding was not, he said, their true end; rather, they "wanted the Federal head under every state and county tent"—the camel's nose to which Paul Blakely had alluded a year earlier. "Free and effective public schooling," said Hoover, was "the most successful accomplishment of our system of local government."[9] Clearly, he believed the aim of the NEA was federal control of education. His shift in attitude on the education issue was significant because within a year and a half he would win both the Republican nomination for the presidency and the office itself. His volte-face did not augur well for the NEA and its legislative program.

With greater notoriety and more significance in the short run, Calvin Coolidge broke a several-year silence on the issue in November 1927. Having abandoned the Reorganization bill, he announced that he still favored the creation of a department of education and relief. A month later in his state of the union address, he noted that the federal government had always fostered and encouraged education. "The general subject is under the immediate direction of the Commissioner of Education," he told Congress. "While this subject is strictly a State and local function, it should continue to have the encouragement of the National Government. I am still of the opinion that much good could be accomplished through the establishment of a Department of Education and Relief, into which would be gathered all of these functions under one directing member of the Cabinet."[10] In short, Coolidge reiterated his stance of 1924–25: education was a local matter. The federal interest in it should remain under a commissioner, though within a mixed department under an able administrative secretary. By implication, there would be no separate department of education. Supporters of the NEA's program, however, would interpret this endorsement of a mixed department, along with the Republican party's 1924 plank in favor of such an agency, as a commitment in principle to the establishment of a separate department, something that was patently untrue.

On the opening day of Congress in December 1927, Reed introduced the NEA bill in the House; a week later, Republican majority leader Charles Curtis did so in the Senate. Virtually identical to its predecessor, the measure added a National Council on Education, composed of state superintendents, to serve as an advisory committee to the secretary of education. The idea of such a council had originated in the Sterling-Towner bill and was kept alive in one form or another in the Sterling-Reed, the Dallinger, and the Phipps bills.[11]

The next day, Reed pleaded for the measure on the floor of the House. While the Scottish Rite and the Ku Klux Klan might vilify Catholic objectors as enemies of public schools, congressional advocates of the legislation knew that most who opposed it within the Capitol were states' righters, who did not belong to that denomination. Such men had to be convinced that Congress had the power to enact such legislation. Noting that the creation of new secretariats had never been undertaken hastily, Reed argued that the American situation was so altered that the national interest now warranted a department of education. In his view, the general welfare clause in the Constitution empowered Congress to create one. In support, Reed marshaled the opinions of Justice Joseph Story, Alexander Hamilton, and James Monroe, and cited two precedents: the Smith-Lever Act and the Smith-Hughes Act. Furthermore, the distribution of information—in this case educational information—was an accepted function of the federal government, already being done by the Departments of Agriculture, Commerce, and Labor. The only adequate channel for the dissemination of school studies was a department of education.[12]

The NEA decided against pressing for consideration of the bill until after the presidential election of 1928. During the first weeks of Congress, William Davidson, who replaced George Strayer as chairman of the Legislative Commission, spent several days in Washington conferring with Coolidge, Curtis, Reed, and other congressional leaders. After also studying the complexion of the Education Committees of both houses, he decided that "a vigorous attempt to force the bill out of committee would be unwise before the party conventions and the November elections."[13] Although this may have been the only reasonable decision, it probably ensured the bill's demise because the second, short session of Congress was always choked with necessary legislation.

Still, proponents remained active, especially the Ku Klux Klan and the Southern Jurisdiction of the Scottish Rite. With Al Smith, a Catholic, running for president, religious tensions were high, spilling over to the school issue. Catholic opposition to the Curtis-Reed bill the previous year was viewed by some as evidence of Rome's attempted political domination of America. Indeed, it seemed that Catholic success in the Oregon school case and in blocking the apparently innocuous Curtis-Reed bill served only to make antagonism to the church more overt on the part of such groups. In November 1927 the *Fellowship Forum* (Washington, D.C.), mouthpiece of the Klan wing of Southern Scottish Rite Masonry, published a book entitled *Proof of Rome's Political Meddling in America.* Its purpose was to demonstrate that the church intended "to make America Catholic," that is, to seek control of the United States through political means. According to James S. Vance, a former grand commander of the Rite who edited the *Forum,* the plot called for Smith either to become the Democratic nominee for president or split the party in trying, thus ensuring the defeat of its candidate. The headquarters of this campaign was the secretariat of the NCWC, which, as the book pointed out with foreboding, was "located in Washington, the National Capital, almost within a stone's throw of the White House, and on a broad avenue leading almost directly to the Capitol building itself." The volume contained a map of the District of Columbia "showing the strategic location of important Roman Catholic institutions with reference to their accessibility to the Capitol, the White House and the Government Departments." The foot soldier of this Catholic offensive against America was the immigrant, a fact that accounted for the church's opposition to the National Origins Act (1924) and the Catholic attempt to have it repealed.[14]

Evidence in proof of this plot was to be found in the reports made to the hierarchy by NCWC department heads. Somehow, Vance had come into possession of the reports for the years 1920 through 1922. The book simply reprinted a number of them with emphasis added to call attention to pertinent passages. Particular stress was given to the Administrative Committee's report of 1921 recounting the effort to kill the Smith-Towner bill. Archbishop Edward Hanna noted that the NCWC had organized a national protest against the Smith-Towner bill and had secured Democrat William King of Utah to defend the Catholic position on the floor of the Senate. With sinister implication, the book also directed attention to Archbishop Austin Dowling's remark that the NCWC's Department of Education was to serve as "A CONNECTING AGENCY BETWEEN CATHOLIC EDUCATION ACTIVITIES AND GOVERNMENT EDUCATION AGENCIES," as well as to be an organization for safeguarding Catholic education. "In nothing have the political and the lobbying activities of the Roman Catholics been more pronounced than in their fight to defeat every effort for Government aid for public education and the movement to create a separate Department of Education," declared Vance. "It is in their fight against the plan to create a Federal Department of Education that they have been most bitter, have organized their most powerful lobby and have shown more clearly than probably with reference to any other matter that the Roman Catholic church in the United States is 'in politics.'" To demonstrate the church's current effort in this regard, he reproduced Ryan's speech to the NCCW, delivered in October 1927, encouraging the ladies to distribute NCWC pamphlets against the Curtis-Reed bill to teachers, students, and women who belonged to organizations promoting the measure.[15]

Mainline Scottish Rite Masonry picked up the tune. In March 1928 the *New Age* reprinted an article from the *Central Christian Advocate* written by H. E. Woolever, editor of the *National Methodist Press.* Woolever accused the Catholic church of using machine politics to keep the education bill from reaching the floors of the House and the Senate. Like Vance, he noted that for years the NCWC Administrative Committee had boasted in reports to the hierarchy that it had manipulated Congress to keep the bill in committee. Worse, the church had packed the present House Committee on Education. The new Congress expanded the latter's membership from fifteen to twenty-one. Although only 8 percent of congressmen were Catholic, five committee members (almost 25 percent) belonged to that denomination and three others were sympathizers easily controlled by the church. Woolever singled out Democrat Loring Black of New York as the principal spokesman of the Catholic hierarchy at the 1926 hearing. "It was pathetic," said the Methodist, "to see him manipulated like a Punch and Judy marionette by a group of Roman Catholic sisters who were present." Sad was the fate that placed such an important bill in the hands of a man of this caliber. Commenting on Woolever's article, the *New Age* remarked that although an unidentified congressman admitted that the Curtis-Reed bill should be enacted, he intended to vote against it because there were enough Catholics in his district to ensure his defeat.[16] All of this convinced the Masonic editor that the church was bent on destroying the public school system, a plan that included blocking the NEA measure.[17]

Here continued the fundamental misconception of 100-percenters, like those of the Klan and the Southern Jurisdiction. Catholics opposed the Curtis-Reed bill and its predecessors because many believed that it would lead to federal control of education, and some, perhaps most, because they feared that it would spell the end of parochial schools. Klansmen and Southern Masons, however, interpreted this opposition, willfully or mistakenly, as proof that the Catholic church opposed public education and intended to subvert the American form of government. To be sure, conservative Catholic clergymen like Paul Blakely had made inflammatory statements about public schools. Some Catholic ethnic groups, like the Slovaks, denounced them as anti-religious. Such remarks, however, were aimed not at the destruction of public education, but at convincing Catholic parents of the need to send their children to parochial schools. Yet there were too few of the latter in the nation to meet the need. In 1920 only about half the parishes had schools, and if mission stations and chapels were counted, the proportion shrank to 35 percent.[18] In other words, it ill served Catholic parents or their church to denounce, let alone seek the destruction of, public schools because so many Catholic children depended on them. As for the charge that the church was attempting to subvert the American form of government, nothing could have been further from the truth. Far from undermining the government, Catholics were becoming adept at using the political system to protect the interests of the church and further its point of view.

Actively promoting his bill, Reed had his December speech printed in pamphlet form and sent a copy to each of his colleagues in Congress. With that, the Capitol and House Office Building buzzed with rumors of a forthcoming hearing. The notion remained hearsay until early April when word came that although the Senate committee had decided against holding a hearing, Reed's House committee would do so at the end of the month. Opponents were assigned the final two days.[19]

As in 1926, the NCWC lined up speakers against the bill, but this time the mood was different. There was no urgency. Perhaps the Senate committee's unwillingness to participate reinforced a conviction that nothing would come of the bill. Whatever the reason, the NCWC secured about twenty witnesses to appear before the committee and got another fourteen to mail in statements. This contrasted sharply with the 129 proponents mustered by the NEA. The NCWC forgot to make use of Mark Lally, who had mounted the protest campaign in Reed's district during the 1926 hearing. At the eleventh hour, however, Lally himself contacted William Montavon to ask about a repeat performance. Although a protest was launched, Lally confessed that it was not what it might have been because of the short notice.[20]

The religious issue was more in evidence at the 1928 House hearing than at any previous congressional inquiry. While proponents had the floor, the issue made its presence felt through the questioning of two Catholic congressmen, Democrats Black and John Douglass of Massachusetts. Both tried, with more or less success, to force proponents into damaging admissions. The educational trust argued in the main that the Bureau of Education was inadequate to the task of conducting the studies and gathering the information necessary for the advancement of the nation's schools. Chronically underfunded, it would never receive adequate appropriations until elevated to departmental status. When Black asked Davidson how "attendance at Cabinet sessions [would] help the department be a fact-gathering agency," he responded that a secretary would lend greater importance and enhanced dignity to education. "Facts are facts," retorted Black, "irrespective of the dignity of the explorer." The congressman later asked point-blank, "Can you put your right hand up before God and say you have abandoned the theory of Federal aid?" While admitting that some members of the NEA had not abandoned that hope, Davidson answered that the quest for aid had "nearly reached the vanishing point." Douglass asked Davidson if the NEA believed that the public school system was so retarded as to warrant federal intervention. Like Hoover, Davidson admitted that the system had made "untold progress"; unlike Hoover, he contended that it could have made more progress had it proceeded "on a sound scientific basis." Douglass then cornered him into admitting that the results of federal fact-finding would constitute standards for the educational community. Davidson noted, however, that the department would have no authority to enforce those standards—facts—it would simply disseminate them. "Then, what is the value of the act?" wondered Douglass. A department of education, said Davidson, would prove as valuable as the Departments of Agriculture, Commerce, and Labor, a point long argued by proponents.[21]

During the testimony of John A. H. Keith representing the NEA and Charles Judd representing the North Central Association of Colleges and Secondary Schools, Douglass tried to box them into acknowledging that scientific studies were already being conducted by the NEA and private agencies like the Carnegie Foundation. Keith admitted that the NEA did make investigations into various aspects of education, but argued that its financial and personnel resources were inadequate to the task. Douglass then maneuvered him into the admission that various states had done studies on matters of significant educational interest and that those were made available through the NEA. Keith lamely contended that the NEA was not an appropriate clearinghouse and could not properly do the work. More convincingly, Judd noted that private foundations made grants for educational studies only for matters of interest to them. The educational community needed investigations of numerous topics beyond the scope of such interests. The federal government alone was capable of funding and conducting the many necessary studies. These were valid points that reduced Douglass to silence on the issue.[22]

Judd willingly conceded a point that Douglass had been unable to get Davidson or Keith to admit, namely, that the educational trust sought the standardization of education. Convinced of its desirability since early in the decade, Judd had altered his thinking on the means of attainment. No longer did he believe that standardization should be accomplished by Congress on the basis of studies conducted by the proposed department. Rather, it should be carried out by regional accrediting agencies like the North Central Association, which he chaired. Thus, the process of standardization would rest not with the people of the nation expressing their will through the federal government, but with the educational community itself, a point with which Samuel Capen concurred.[23]

John McCracken, chairman of the ACE's Legislative Commission and a friend of Black, tried to mollify Catholic opposition with a personal touch.

During the war, he and Father Peter Guilday, a professor at Catholic University, had collaborated with the government on the classification of all public and private colleges and universities in the mid-Atlantic states. "We never had any difficulty working together in the common cause," remarked McCracken. "One thing I learned in that service was that there could be quite as much suspicion, jealousy, and misrepresentation between institutions conducted by different orders of the same church as between State and church institutions or between Catholic and Protestant, and that a neutral Federal officer with no ax to grind might be very helpful to all concerned." If the investigative work of the proposed department were done in a scientific way, said McCracken, it would prove beneficial to public and parochial schools alike. "We may interpret the facts differently," he observed, "but if we are interested in knowledge at all, we all want to know the facts." He reinforced his argument with the story of how a "loyal son" of the Catholic church had "strayed" into Lafayette College, a Presbyterian school, where he was required to take a course in biblical geography. The young man began to have scruples because the class was taught by a Protestant. On making this known to the college's president (McCracken himself), the latter accompanied the lad to his parish priest to lay the matter before him. The wise old clergyman replied that it probably made little difference in the location of Jerusalem if it were taught by a Protestant or a Catholic. "The knowledge that is common to us all forms so large a part of the stuff of education," concluded McCracken, "that if we are granted tolerance in other fields, we can without risk to cherished principles cooperate on the ground that is common to us all." For Black, the issue was not common knowledge or tolerance, but the role of the federal government in education. He wanted Washington to leave both public and private schools alone.[24]

Then Black, Democrat Brooks Fletcher of Ohio, and Republican E. Hart Fenn of Connecticut drove home the point that popular support for the measure had dried up. In their view, demand for passage of the bill should come from the people, not educators. Constituents had stopped sending letters advocating it. When McCracken suggested that perhaps people had wearied of writing after ten years of fruitless struggle, Black countered that support had disappeared when federal aid was dropped from the bill. With that "all-important feature" gone, popular interest had dried up. Neither Black nor Fenn could understand why it was necessary to create a department of education to conduct scientific investigations when an adequately funded Bureau of Education could perform the function.[25]

Previously an advocate of a federal commission on education, Charles R. Mann, director of the ACE, had since become convinced of the inevitability of a department, but was willing to let the system "evolve and be sure of each step as we go along." To further the evolutionary process, he proposed a modest bill that appropriated $500,000 to the Bureau of Education for a two-year study of the organization, administration, and financing of secondary education and its articulation with higher education. In essence, he was suggesting that the educational trust postpone enactment of the Curtis-Reed bill for two years so that Congress could see the value of the scientific investigation of education.[26]

While the questioning of Douglass and Black was at times testy, it was always confined to the political dimension of the matter; the religious issue remained for the most part politely in the background. Proponents knew the two congressmen were Catholic but, with the exception of McCracken, chose not to address the fact directly. The atmosphere of the hearing changed when opponents had the floor. During that time, Elmer Rogers, a Southern Scottish Rite Mason, who for two years had published an article on some aspect of the Curtis-Reed bill in each issue of the *New Age*, was constantly at the side of Republican John Robsion of Kentucky feeding him information to use against Catholic witnesses. During the testimony of two NCWC representatives, Charles Dolle, executive secretary of the NCCM, and his counterpart Agnes Regan of the NCCW, the atmosphere became electric, ultimately leading to an angry exchange.[27]

At the heart of Dolle's statement was the suspicion that the Curtis-Reed bill laid the foundation for the eventual standardization of education. His argument ran thus. The proposed department aimed at improving teaching methods, curricula, school organization, and the finances of education, to be accomplished through research and dissemination of information. The only thing that would keep the proposed department from imposing its findings on states was the lack of federal aid; if federal aid materialized, however, local governments might be induced to accept the results of the research. After passage of the present bill, a subsidy would be demanded because the NEA had drafted the predecessors of the present bill, which included aid. Moreover, Williams had declared that it was inconceivable that the association would ever quit seeking federal aid. Once a department was established, its advocates would press for money to put teeth in the agency.[28]

Democrat Ole J. Kvale, a Minnesota Lutheran, agreed with Dolle, but was disappointed that he failed to mention the real motivation behind Catholic opposition. In view of the splendid work that Catholics were doing for education, Kvale said that Dolle had "a perfect right" to express "the one main reason for the objection to this bill," namely, that "many people in the Catholic Church as well as other churches are afraid of attempts to standardize education which they see coming, a standardization which will conflict with their system of education." Dolle responded that he came to represent not the religious interests of Catholics, but their civic interests. The relentless Kvale answered that he, too, opposed the bill as a citizen, but he also objected to it as a Lutheran and offered "no apologies for my opposition on that score." This admission was quite fine for Kvale because Lutherans were Protestants, as much opposed to Rome as other Protestant Americans, no matter how much some of their fellow citizens might view their support of Lutheran parochial schools as misbegotten. Reluctantly admitting that religious affiliation did play a part in Catholic antagonism, Dolle emphasized that Catholics offered their objection as citizens.[29]

Robsion attacked, asserting that Catholics had always opposed public schools, had objected to the creation of the Bureau of Education, and were now fighting the establishment of a department of education. In his view, Kvale had hit the mark. For ten years the real reason behind Catholic antagonism was fear for parochial schools. Robsion declared that if he thought the bill threatened harm to them, he would oppose the measure himself. Kvale assured him that Catholics objected to it because they believed that it definitely would lead to the injury of their schools. Douglass denied Robsion's charge that the church had always opposed public education. Qualifying his remark, Robsion admitted ignorance of the church's present attitude toward public schools, but contended that it had opposed them in the past. Douglass again denied this, asserting that Catholics were simply attempting to protect their own educational system. In an effort to defuse the situation, Dolle observed that the church considered the bill just as dangerous for public schools as for Catholic schools. He wanted it clearly understood that the measure would open the way for federal control of *all* education. Arguing that Congress knew such control was unconstitutional, Reed held that "Uncle Sam" would protect Catholics against it just as he had against the Oregon school law. Dolle remained unconvinced.[30]

His time before the committee was tame compared with Regan's, whose statement covered the same points. After thirty years in public education, including a post on the San Francisco School Board, she had left that career to become executive secretary of the NCCW. At the opening of her remarks,

Rogers entered the room with five books for Robsion. Throughout her testimony the congressman pored over the material, noting items brought to his attention by the Mason.[31] When Regan finished, Robsion observed that practically every Catholic witness had opposed the bill. Without wishing to be offensive, he asked if it had not been the policy of the church "to look with somewhat of distrust upon the public schools of the country." Emphatically denying it, Regan argued that Catholics were as interested in public education as other Americans. They paid taxes like everyone else, and only a quarter of their children attended parochial schools. They objected to the bill as American citizens. Robsion then read a passage from *A Manual of Moral Theology* by Reverend Thomas Slater, S.J.: "The church condemns all non-Catholic schools, whether they be heretical and schismatical, or secularist, and she declares that as a general rule no Catholic parent can send his young children to such schools for educational purposes without exposing their faith and morals to serious risk, and therefore committing a grave sin."[32]

Before Robsion could cite other authorities, Regan protested to Chairman Reed against this line of discussion. She was objecting to the bill on civic grounds and Robsion turned the matter into an issue of religion. When Reed made no move to intervene, Regan accused Robsion of using the same tactic that had been employed in previous hearings, namely, when a Catholic came before the committee, make it appear as if the person opposed the bill because it involved a religious question. Robsion countered that church leaders "strenuously opposed" the establishment of both public schools and the Bureau of Education. "No; they did not oppose public schools," interjected Douglass. "You are entirely wrong. The Catholic Church in this country has never taken any stand on public schools as such." Robsion explained his point: although opposed to public schools at their inception, the church had discovered that its fear was groundless; so it was now. Regan again objected "to the injection of this religious issue into the hearing coming from a person [Rogers] who is not a member of this committee. I resent it as an American citizen, and I think it is dragging in religious prejudice where it does not belong."[33]

At this point, another Catholic congressman tried his hand at countering his colleague. Louis Monast, a Rhode Island Republican, told Robsion that he misunderstood the church's position. Whenever possible, Catholics favored sending their children to parochial schools. "They do not oppose the public school system. They are entirely in favor of it and do really contribute through their taxes in the same way as other citizens." Robsion continued to explain his view while Regan continued to protest. It was a frustrating, irritating experience.[34]

In fact, there was truth on both sides. Regan, Douglass, and Monast were correct in asserting that the church had not opposed the establishment of public schools. It had tried unsuccessfully to accommodate itself to them and was forced to establish parochial schools to protect the faith of Catholic children. The passage Robsion quoted from Slater's book did indeed articulate the current, official position of the Catholic church in places where there existed a union of church and state. The church viewed religion not as a separate academic subject alongside all others, but as the foundation upon which the correct understanding of all other disciplines rested. Religion served not only as a bedrock for them, it was also to suffuse and permeate them. Because a purely secular education tended to undermine or at least relativize the importance of religion, it jeopardized the faith and morals of young Catholics so educated. For that reason, Catholic parents had a moral obligation to place their children in parochial schools. To be sure, there were Catholic clergymen, like Blakely, who made intemperate remarks about public education, but most Catholic lay people, like Regan, Dolle, Douglass, and Monast, felt otherwise. There were simply not enough Catholic schools in the United States to accommodate all Catholic children. Thus, many, if not most, Catholic parents had to send their children to public schools. Rather than attack such schools, they depended on and supported them.

In some respects the hearing was meaningless. First, only the House participated. That the Senate refused to join or hold one of its own gave mute testimony of its intention to take no action on the bill. Although the hearing gave the measure a high profile, and Reed claimed that it demonstrated that a department of education must be established, it seems to have swayed only one committee member. Whereas Monast had gone into the hearing definitely opposed to the bill, he came away with a sense that antagonists objected less to the measure itself than to what might eventuate from it. Still, the NCWC thought that he would reject the legislation because 1,800 of his constituents had protested to him against it. If he could be counted as opposed, then the committee was evenly divided: seven members opposed the bill, namely, Democrats Kvale, Black, Douglass, René De Rouen (Catholic) of Louisiana, Vincent Palmisano (Catholic) of Maryland, and Republicans Fenn and Monast; seven favored it, namely, Republicans Reed, Robsion, Willis Sears of Nebraska, Elmer Leatherwood of Utah, and Democrats Fletcher, Bill Lowry of Mississippi, and Malcomb Tarver of Georgia; the remaining seven were undecided or noncommittal.[35] Those who opposed the measure generally represented the urban eastern wing of each party, while those who favored it represented the western and southern rural wings of their parties.

A by-product of the hearing was the passage of a modified version of Mann's bill for a study of high schools. At the urging of the North Central Association, supported by the ACE and the NEA, Congress appropriated $225,000 for the Bureau of Education to conduct a three-year study of the organization, administration, financing, and work of high schools as well as of their articulation with elementary and higher education.[36] While it was Mann's hope that such a study would convince Congress of the need for a department of education to carry out such scientific investigations, the venture would prove fruitless in that regard.

The hearing was a keen disappointment to McCracken. At the opening of the Seventieth Congress, he had predicted that the Curtis-Reed bill would receive more attention than it had in the previous Congress.[37] Now viewing his prediction as "a delusion," he reported to members of the ACE that Congress was like a balloon drifting in the wind of public opinion. Its response to the Curtis-Reed bill was: "There is no popular demand. We do only those things we are compelled to do by pressure from without." In McCracken's view, both political parties were working to ensure that the issue would remain "Hush! Hush!" Given that Al Smith was likely to receive the Democratic nomination for the presidency, neither party wanted to "wake the baby." Any issue on which Catholics or their enemies had taken a stand was "taboo." Like Woolever, McCracken remarked that the membership of the House Education Committee had been expanded, with Catholics holding five seats. Admitting that their membership was about proportional to their number in the general population, he noted that it indicated "an interest in public education among Catholic members of the House somewhat out of proportion to the total number in the House." Friends of a department of education had, he reported, assiduously avoided the question of church and state in education. They believed that the public and private systems were mutually beneficial, spurring one another to improvement. They hoped that when a department of education was finally established, it would "come as the department of all educational enterprise." Urging Catholics to see the wisdom of this position, McCracken warned, "For as surely as they carry their opposition to the point of attempting to frustrate the legitimate aspirations of American education as a whole, they will antagonize state education to a point that may endanger the very existence of parochial and voluntary education."[38] To be sure, this was an ominous warning that bespoke the frustration of the educational trust, but it was one that ultimately proved unfounded.

If the NCWC took the hearing in its stride, it prepared with purpose for the national party conventions. The NEA was sure to seek an endorsement of the Curtis-Reed bill, and Montavon considered the wording of the 1924 Democratic education plank, written by Ryan and Edward Pace, as too elastic. By allowing the federal government to offer "counsel, advice or aid" for schooling, warned Montavon, the plank left "open the door for practically everything that the advocates of a Federal Department of Education have been demanding." The task of changing the wording, while at the same time blocking the NEA, would be exceedingly delicate, especially because Smith seemed certain to win the Democratic nomination. His candidacy compounded problems all the way around, making it practically impossible for the church to apply organized pressure on either party. So Montavon recommended that Burke send politically adroit individuals to work both conventions.[39]

Burke sent Montavon himself to handle the Republican convention in Kansas City and asked the NCCW to assign him a companion. It appointed Mrs. August Kech, a woman familiar with the political process. The two were to ensure that the party adopted no plank for a department. Arguments were to be based solely on the best interests of the country, without anger or threats of any kind. If the two agents were questioned about the Catholic attitude, their response was to be that, personally speaking, they thought the majority of Catholics opposed the Curtis-Reed bill. Burke suggested to Montavon that it might be opportune for the Republicans to adopt a plank against increased bureaucracy along the lines of President Coolidge's recent address to the Daughters of the American Revolution. Warning that tyrants had always held people in bondage with the plea that it was for their own good, and noting that the populace usually submitted because it was easier than self-rule, Coolidge had urged a return to local self-government. So much did an anti-bureaucracy plank appeal to Burke that he drafted a model and sent it to William Joseph Donavon, assistant U.S. attorney general, for transmittal to the convention.[40]

En route to Kansas City, Montavon and Kech devised a strategy for carrying out their task without implicating the NCWC or the church. Accordingly, she was to identify herself with the Pennsylvania delegation and work in close contact with its representative on the platform committee. Donning the mantle of a foe of increased bureaucracy, including the scheme for a department of education, Montavon was to make representations to the party leadership, which was trying to calm a restive farm bloc that sought a bureaucratic solution to its economic woes.[41]

The prosperity of the 1920s had bypassed farmers, who were experiencing an economic depression due to low prices for agricultural products and severely weakened purchasing power. In 1927 Congress had passed the McNary-Haugen bill that established a Federal Farm Board to work with cooperatives in disposing of surplus goods. The measure created a stabilization fund to help farmers absorb losses, and authorized the Farm Board to raise domestic prices to tariff level. The bill was one of the few progressive measures the farm bloc was able to pass over conservative Republican opposition. Coolidge vetoed it with a stinging message. Congress reworked the bill to meet his objections and passed it again in 1928, only to have it returned with an even more scathing veto. Thus, American farmers decided to carry their fight to the national party conventions, something that Montavon was eventually able to turn to the advantage of the NCWC.[42]

In a brief meeting with James C. White, secretary of the Republican National Committee, Montavon explained that he was in Kansas City to keep the party from endorsing the Curtis-Reed bill. Assuring him that there was little need for worry, White declared his opposition even to the party's repeating its 1924 endorsement of a department of education and relief, a plank he himself had written.[43]

On the convention's opening day, Montavon attended the first meeting of the platform committee. Speaking on behalf of the NEA, Mrs. Frederick Bagley, president of the National Committee for a Department of Education, recalled Coolidge's frequent references to the need for a department of education and relief. Arguing that the party's 1924 plank in favor of such a secretariat constituted a commitment in principle to a separate department of education, she exhorted the committee to adopt a plank drafted by the NEA in support of the creation of one. As she finished her statement and distributed copies of the plank, the session was interrupted by more than 300 banner-waving farmers demanding relief. With adjournment, Montavon sounded out the vice chairman, Senator Hiram Bingham of Connecticut, to gauge reaction to Bagley's request. The committee was so preoccupied with other matters, said Bingham, that to avoid controversy, it might simply readopt the 1924 plank. Montavon urged the senator to remember his opposition to increased bureaucracy. Bingham agreed to do his best to keep the 1924 endorsement out of the platform.[44]

Throughout the night the drafting committee met in executive session. Maintaining contact with it, especially via White, Montavon suggested that the best way to derail the farm bloc would be to take a firm stand against further bureaucracy by deleting the 1924 endorsement of a department of education and relief and by adopting a home-rule plank in accord with Coolidge's address to the Daughters of the American Revolution. By sunrise the committee had scrapped the educational endorsement, but had made no move toward home rule, though Bingham and others were working toward this end.[45]

In the morning, Montavon met with Kech, who by this time had thoroughly identified herself with the Pennsylvania delegation. He urged her to contact that state's powerful political boss, Senator-elect William Vare, a member of the platform committee, to encourage him to vote for a home-rule plank. Together with colleagues in the delegation, Kech made the representation, and Vare promised to do what he could. Apparently, he and Bingham did quite a bit. In its final form the platform not only omitted the 1924 plank for a department of education and relief, but also concluded with a home-rule plank that deplored further federal encroachment on regional affairs and encouraged local communities to become self-reliant and solve their own problems.[46] While it would be too much to claim that Catholic lobbyists were solely responsible for the inclusion of a home-rule plank in the platform, they certainly played a key role in its adoption.

With regard to the presidency, the farm bloc opposed the nomination of Herbert Hoover and had intended to support George Norris, until he announced that if the party nominated a regular, he would back the Democrats, which he eventually did. The bloc then turned to Frank Lowden, governor of Illinois, who demanded endorsement of the McNary-Haugen bill in the party platform as the price of his candidacy. Coolidge had let it be known that he would consider such an endorsement as a repudiation of his administration. So, the bloc was reduced to running favorite sons, none of whom had a chance.[47] When the Old Guard threw its support to Hoover, he was assured of victory. Like Harding and Coolidge before him, he abhorred the thought of government competing with business. He promised that if the lame-duck Congress did not resolve the agricultural issue, he would call a special session for the creation of a Federal Farm Loan Board to supply money for the establishment of a farmer-owned stabilization agency.[48]

The Democratic convention presented a far more risky situation for the NCWC, with much less at stake. Burke's sole objective was the deletion of the word "aid" from the 1924 education plank, yet political realities would render this impossible. By 1928 the party was sharply divided with urban, Catholic, immigrant and hyphenate wets joining urban, middle-class, Protestant wets against rural, Protestant, drys. Indeed, Prohibition had emerged as an issue symbolic of the hegemony of old-stock, rural Americans.[49] As the convention approached, tension mounted because Smith, a thoroughgoing wet who drank in defiance of the Constitution, was a shoo-in for the nomination. His Catholicism raised questions in Protestant minds about dual allegiance to Washington and Rome. Though the editor of *The New Republic* advised that the Democrats could win the election only by espousing the New Nationalism of the late Theodore Roosevelt, Smith wanted a Jeffersonian states' rights platform opposing further federalization, something that the Republicans were already doing quite well.[50]

Burke sent Montavon to Houston to oversee Catholic interests. Once there, he quickly realized that there would be a fight to the finish between the party's urban wing led by Smith and its rural wing, which opposed him on account of his wetness and religion. For several days prior to the convention, a dry headquarters operated out of the Rice Hotel, and local Baptist and Methodist churches held revival meetings and sent word to coreligionists around the country, urging them to pray against the anti-Prohibition movement that would overturn "Protestant Civilization." In Montavon's opinion, these churches intended to secure a bone-dry plank in the party platform so that if nominated, Smith would repudiate it, thus either ensuring his defeat or at least giving Democrats who opposed him on account of religion a safe pretext for rejecting him. Thus, the membership of the platform committee assumed great importance. An attempt to secure the appointment of only bone-dry delegates met with partial success, resulting in a split committee. Leading the drys were Senator Carter Glass of Virginia and former Secretary of the Navy Josephus Daniels of North Carolina. At the head of the wets was Senator Millard Tydings of Maryland.[51]

Taking quick stock of the situation, Montavon saw that it would be dangerous to make any avowedly Catholic representation to the committee. Conferring instead with Lillian Westrop, an alternate from Ohio, he explained the NCWC's desire to have the word "aid" removed from the 1924 plank. She agreed to lay the matter before former Secretary of War Newton D. Baker, Ohio's delegate on the platform committee and a states' rights advocate. He refused to do anything because Smith's attitude toward public education had already been questioned. In fact Baker claimed that charges were made that Smith, if elected, would destroy public schools, therefore any tampering with the education plank would involve the committee in an uncontrollable fight. If Montavon could not get the word "aid" out of the plank, he would try to forestall any endorsement of a department of education. He urged Westrop to impress on Baker how contradictory it would be for Democrats who were promoting states' rights to endorse the creation of such a secretariat.[52]

The educational aspect of the hearing held by the Platform Committee was reassuring. Three people spoke about schooling and none seemed very interested in a department of education. Strange to say, the NEA sent no official representative. More remarkable, Dr. Starlin Marion Marrs, Texas superintendent of schools and a member of the NEA's National Council on Education, denied desiring a department. The sole voice in favor of one was Mary Simkhovich, representative of the Women's Democratic Union of New York City. The Union's platform, drafted by, among others, Ida Tarbell and Eleanor Roosevelt, contained thirteen planks, the last calling for a department of education. Recognizing that the Union's endorsement of a department was simply a gesture to make it perfectly clear that Smith was not opposed to public education, the committee ignored the idea and readopted the 1924 plank, including the word "aid." Given the situation, reported Montavon, this "was the best that it was possible to obtain from this Resolutions Committee."[53]

Indeed, the religious question was volatile, as events surrounding the Prohibition plank showed. On the evening of 27 June, some 2,000 bone-dry advocates met at the Second Baptist Church where speakers whipped them to fever pitch. They then marched on the convention and threatened to bolt the party unless it adopted an appropriate plank. While the protesters disrupted the general session, the platform committee was holding a public hearing that all but degenerated into a donnybrook. Using the occasion as a forum, Tydings delivered a brilliant speech against the Volstead Act, which he argued was unenforceable and violated state sovereignty. He proposed a plank calling for repeal of the law, thus leaving enforcement of the Eighteenth Amendment to the states. Glass, Bishop James Cannon of the Methodist Church of Virginia, and others objected so furiously that physical violence was avoided only with difficulty. In the end, Glass convinced the committee's drys to accept an innocuous blanket plank condemning Republican law-breaking while pledging Democrats to make "an honest effort to enforce the Eighteenth Amendment and all other provisions of the Federal Constitution and all laws pursuant thereto." Only Governor

Dan Moody of Texas held out for a bone-dry resolution, even threatening to bring in a minority report. When the convention took up the platform the next evening, Moody found himself a minority of one and reluctantly conceded in favor of the law-enforcement plank. Reported Montavon, "In this way ended what, by a great majority of observers in Houston, was declared to be the fiercest fight for religious liberty ever waged in the United States."[54]

If that was true, it went for naught. Smith received the nomination, and fulfilling the expectations of anti-Catholic dry Democrats, accepted the party's nomination with a telegram declaring that the solution to bootlegging and lawlessness was "real temperance" rather than Prohibition. This repudiation of the mild compromise plank opened the way for insurgency. To be sure, Smith's wetness and Catholicism (the two were historically interchangeable: Rum and Romanism) played a part in his loss of the election, as recent historical studies have demonstrated. Yet even if dry Democrats had not bolted the party, it is unlikely that Smith could have won the election, for the nation was riding a tide of prosperity associated with the Republicans. During the campaign, he portrayed himself as a pro-business candidate in favor of limited federal government, both Republican strong points on which it was difficult to compete. A constellation of personal characteristics contributed to his defeat. He was an open wet, Catholic, urbanite who reveled in his city roots, spoke with a Bowery accent, and chomped on an ever present cigar. If his religion and wet proclivities offended Protestant agrarians, his demeanor offended genteel Protestant city dwellers. He lost in a landslide.[55]

When the NEA met in convention in July, Davidson interpreted recent events as favorably as possible. Noting that Coolidge, in his December 1927 state of the union message, had called again for a department of education and relief, Davidson interpreted the president's words as an endorsement of the principle behind the Curtis-Reed bill. As further proof, he cited the Republican party's 1924 plank in favor of a department of education and relief—without mentioning its failure to be readopted. Even better, because the Republicans had nominated Curtis as running mate for Hoover, the NEA bill now enjoyed the sponsorship of both the chairman of the House Committee on Education (Reed) and the Republican vice presidential candidate. As the NCWC feared, Davidson interpreted the Democratic plank as a commitment "to the general principle for which this association has stood since 1918 in its battle for a Cabinet post for education."[56]

Davidson certainly put the best face on things, a face not altogether in accord with reality. Coolidge favored a department of education and relief, not a separate department of education, a point he had made patently clear to the NEA as early as 1924. As for Curtis's nomination to the vice presidency, he received it, not for his feckless sponsorship of the education bill, but as a sop to the farm bloc because he hailed from Kansas and supported the McNary-Haugen bill. As historian George Mayer has pointed out, however, Curtis had little in common with farm-bloc dissidents. A politician who made "a religion out of regularity," he broke with the administration on the McNary-Haugen bill only because his constituents supported it, and he knew Coolidge would veto it.[57] Davidson's remark about the Republicans' 1924 education plank was disingenuous in light of his omission of its deletion from the current platform. With regard to the Democrats, their plank was intended in no way to endorse a department either directly or in principle.

Having primed convention delegates with a hopeful report, Davidson outlined the NEA's strategy. It called for pressure on the House and Senate Education Committees to report the Curtis-Reed bill during the final session of Congress. Like McCracken, he admitted that the only thing that impressed Congress was grassroots support in the constituencies, a support that had been noticeably lacking even among teachers in his home state of Pennsylvania. The 3,100 members—regular and adjunct—of the NEA's Legislative Commission were powerless to force the bill out of committee, he told conventioneers; "the whole thing depends on you and the 1,000,000 teachers of the Republic." The nation's educators had to be mobilized so that "they will insistently and persistently fusillade the Committee on Education in the House . . . and the Committee on Education and Labor in the Senate."[58]

The desired pressure was not forthcoming. The association's "democratic," top-down reform had lost the ear of teachers at the grassroots. The NEA failed to get the Curtis-Reed bill out of committee during the lame-duck session. By February 1929 the association abandoned hope and began formulating its strategy for the new Congress, which President-elect Hoover was to call into special session to provide relief to the farm bloc. From sources close to the NEA, the NCWC learned that the association planned to mount a concerted drive for the bill as soon as the next Congress convened. The program included "a method of 'hand-picking' the members of the [Senate] Committee on Education and Labor so that the measure will meet with little opposition." Alarmed by this news, the Executive Committee of the NCWC

Department of Education commissioned its new executive secretary, Father George Johnson, to take countermeasures. He was to interview congressmen personally to secure the appointment of antagonists of the bill to the Education Committee.[59]

During the lame-duck session, representatives who would be returning in spring for the new Congress got a preview of coming attractions. In mid-February 1929 Robsion introduced a bill to create a department of public education. The legislation differed from the Curtis-Reed measure in minor respects: it neither transferred the Board of Vocational Education to the proposed department nor did it establish a Federal Council on Education. In outlining the purposes of the proposed secretariat, the bill appealed to the general welfare clause as the basis for congressional action. The department was to "aid and encourage the public schools and promote the public educational facilities of the Nation so that all the people . . . shall have larger education opportunities and thereby . . . make more general the diffusion of knowledge, and provide for the general welfare. . . ." Because the department was to deal exclusively with public education, it would constitute no threat to parochial schools. Moreover, each state was to maintain complete control over all phases of education within its borders.[60] The bill embodied the very split about which McCracken had warned. He had argued that Catholic obstructionism might inject the church-state conflict into the education issue. Robsion's bill clearly indicated that it would promote only public institutions. Moreover, some Catholics interpreted the title of the proposed department as intended to make the legislation's antagonists appear as enemies of public education.

Six days before Congress adjourned, Robsion addressed the House on his bill. Thrice, his words drew applause as he told colleagues that the measure simply aimed at helping pupils and teachers in public schools, that public education ought to be dignified with a secretary in the president's cabinet, and that the founding fathers had drafted the Constitution to ensure domestic tranquillity and to promote the general welfare, the very purposes of his bill. When the applause died away, he launched into a lengthy historical survey of both the general welfare clause and federal participation in education. With regard to the former, Justice Joseph Story, Alexander Hamilton, and James Monroe had contended that Congress possessed the power to appropriate funds for just about anything so long as the money was used for a general and not a particular, local need. To show that Congress had exercised this power, Robsion cited the Land Ordinance of 1785 that set aside a section of land in each township in the Northwest Territory for the support of education, the Smith-Lever Act that appropriated federal funds for education in agriculture and home economics, the Smith-Hughes Act that authorized money for vocational education, the Smith-Bankhead Act that did the same for the educational rehabilitation of war veterans, and the establishment of the Bureau of Education. If Congress had the power to do all the foregoing, concluded Robsion, it could also establish a department of education.[61]

Having argued the constitutionality of enacting the bill, he addressed the question of why passage had yet to occur. It was not for lack of support. Although opponents of a department liked to claim that the measure was backed by only a self-serving organization (meaning the NEA), in fact some thirty-five to forty national bodies favored the legislation. The bill, moreover, allegedly had significant support on Capitol Hill itself. In the years since 1925, asserted Robsion, Congress had been polled three times on the Curtis-Reed bill, each canvass revealing a comfortable majority in favor. Yet the legislation never came to the floor for action; it had "been smothered in committee by men being placed or getting on the Committees on Education of the House and Senate who are opposed to this measure."[62] Clearly, the membership of the two panels had become an issue.

Appealing to the emotions, Robsion noted that in 1928 Congress appropriated nearly $1 million for the Bureau of Education but mandated that the commissioner devote almost three-quarters of the money to "looking after reindeer, health, education, and so forth . . . of people in Alaska." Pleaded Robsion, "Are not the 30,000,000 'dears' of the 48 States of much more importance to the welfare of this country than the 'deers' of Alaska?" He demanded that the entire appropriation be spent on educational research, something that would really benefit the nation. Returning to animal husbandry, he pointed out that the Department of Agriculture engaged in extensive research on the handling, production, and marketing of crops, cattle, and poultry, thereby adding to the progress and prosperity of the country. "Can it be," he concluded, "that we regard the welfare of chickens, pigs, cattle, sheep, roads and instruments of war of more importance than the 30,000,000 children of the land seeking education?" The answer was obvious: the general welfare demanded a department to promote educational research.[63]

His words drew an immediate response from Black, who commended Robsion's frankness in admitting that the nation's educational problems could be solved with federal money. That confession was more than the NEA had been willing to make during the past several years. Originally, said Black, the association had been open about seeking aid. When it became apparent that funds would not be forthcoming, the appropriation was dropped from the education bill, and the NEA began claiming that it sought only the creation of a department. Neither were the opponents fooled nor were Robsion's figures correct. Wrong in part himself, Black asserted that the federal government had spent $60 million on education in 1928 (in fact, the sum was $6 million). Denying that states needed federal funds, he accurately pointed out that in the same year they outlaid $2 billion on schooling the nation's 25 million students at a cost of about $80 per child. States and local communities were vying to outdo one another in the support of education, a competition that federal money would probably kill. Challenging Robsion's barnyard rhetoric, Black made an emotional appeal of his own: "What are our children that they are to be taken care of like we take care of chickens, hogs, and pigs? You could take care of chickens or hogs . . . under the guidance of the Federal Government without doing them any harm, but you can not interfere with the child. The child belongs to the parents . . . and the further the Federal Government keeps away from the real guidance of the child, at the mother's knee, the better it is for the child. . . . [T]hese research operators . . . want to operate on the human soul and the human mind and the human heart as though they were operating on guinea pigs." Like Robsion's speech, Black's drew applause.[64]

Similar to Reed's speech earlier in the Congress, Robsion's was a sustained argument for a broad construction of the Constitution, a point that he tried to emphasize in the bill itself by referring to the general welfare clause. He knew that congressional opponents were states' righters, so the burden of proof rested on advocates to show that Congress had the power to act. In his view, there were ample legal precedents for the federal government to enter the field of education. Like Reed and other advocates, Robsion considered the Department of Agriculture to be one of these precedents. If that agency promoted the general welfare through research and dissemination of information, so too would a department of education. Black, an ardent states' righter, adroitly—and unfairly—turned the argument into one of federal aid. He made it appear as if Robsion's appeal for the expenditure of the bureau's entire appropriation on research was a plea for aid to schools. Black then argued that states and local communities were adequately funding their schools and that a federal subsidy would only kill such an initiative. It was a clever ploy that won a favorable response.

The reaction of the House to the two speakers indicated that the issue of the federal role in education stirred feelings on both sides of the question.

Although there is no way of substantiating Robsion's claim that the majority of legislators supported the Curtis-Reed bill, it certainly enjoyed a considerable following in Congress. Equally certain, proponents in and out of Congress believed that the measure had died unfairly at the hands of stacked committees. Robsion's speech served as a public challenge to rectify this situation in the next Congress. Although it is hardly plausible that the House Committee had been stacked against the bill—in fact, the membership was evenly divided—the matter was otherwise in the Senate. Eight of the eleven members of the Committee on Education and Labor are known to have opposed the Curtis-Reed bill, namely, Republicans James Couzens of Michigan, William Borah of Idaho, Lawrence Phipps of Colorado, Jesse Metcalf of Rhode Island, Hiram Bingham of Connecticut, Frederick Gillett of Massachusetts, and Democrats Andrieus Jones of New Mexico and David Walsh of Massachusetts. For the most part, this was a combination of urban eastern Republicans and an urban eastern Democrats. The NEA hoped to rectify this situation by securing members on both committees who were favorable to the bill. As fate would have it, however, Congress would have less to say about the education issue than would the new Hoover administration.

Several things became apparent from this phase of the department of education battle. First, with federal aid deleted from the Curtis-Reed bill, popular support had dried up. Major papers had lined up against the bill, and even the NEA admitted that many local journals opposed it. Advocates were limited to the educational trust, some local associations of teachers, PTAs, women's clubs, Methodist associations, the Southern Jurisdiction of the Scottish Rite, and the Ku Klux Klan. Although there may have been considerable support for it in Congress, representatives were unwilling to press for the measure absent a desire at the grassroots. Second, the House of Representatives expanded membership on the Education Committee and increased the number of Catholics who held seats so that their number accounted for nearly a quarter. Why the House leadership did this is uncertain. No evidence has surfaced to indicate Catholic connivance. More likely, Republican leaders sensed that Catholics were allies in opposition to increased federal bureaucracy and sought to exploit this sentiment to advantage. Third, the religious issue had moved to the fore. Catholics increasingly became identified, rightly or wrongly, as the principal opponents of the drive for a department of education. To be sure, they opposed the idea, but they were far from alone. Still, Southern Scottish Rite Masons and Klansmen viewed them as enemies of public schools, if not of American civilization. Even the irenic and friendly McCracken was beginning to consider Catholics as obstructionists, to say nothing of how the adversarial Robsion viewed them. It was almost as though proponents became more stridently anti-Catholic the more that success receded from grasp. Finally, the apparent Catholic complacency toward the battle in Congress indicated that the fight was fast coming to a close in their favor, though its legacy would result in a certain amount of self-righteousness and a deepened sense of intransigence.

7

The Triumph of Home Rule

The events surrounding the drive for the second Curtis-Reed bill revealed the depth of antagonism toward the Catholic church. Political observers were convinced that the congressional committees on education had been stacked with opponents to ensure the measure's defeat. In the coming Congress, representatives themselves would echo the charge. Never in the period under study had the religious issue been so frankly addressed in such a confrontational manner in committee hearings. The exchanges indicated the depth of feeling about Catholic objections to legislation, apparently harmless to their schools. Yet in Catholic eyes, the harmlessness was just that: apparent. The same groups that had backed compulsory public schooling had thrown themselves with renewed vigor behind a department of education. In the waning hours of the lame-duck Congress, Republican John Robsion of Kentucky had introduced a new bill for the creation of a department of public education, thus framing the issue so that Catholic opposition would be construed as opposition to public schools. The bill was to be reintroduced in the next Congress, and for the first time, politicians would blatantly court the Ku Klux Klan and the Klan wing of the Scottish Rite Masonry.

While the struggle over the Curtis-Reed bill sharpened the focus on the religious issue, the NEA found that measure had dissipated interest among teachers. With the absence of federal aid, the measure offered them no tangible gain. Though the leaders of the NEA had tried valiantly to rally the rank and file to the cause, their efforts at top-down, democratic reform of schooling had collapsed. Grassroots support was dead. In the next round, the officers of the NEA and the educational trust would themselves carry the banner forward for a department of education and a renewed drive for federal aid.

The final episode would take place within a starkly drawn context. The 1928 election had brought into sharp relief an urban-rural rift in the nation. At the Republican national convention, the farm bloc had attempted to compel the eastern leadership to come to grips with the economic plight of agrarians. At the Democratic counterpart, rural advocates had portrayed the issue as a conflict of civilizations: urban, wet, immigrant, Catholics versus rural, old-stock, dry, Anglo-Saxon, Protestants. One of the concessions that Republicans had made for victory at the polls was the promise to call a special session of Congress to address agricultural issues. For its part, the farm bloc would endorse a department of public education and federal aid for rural schools.

A question mark in the next phase of the battle for a department of education was the new president. Although Herbert Hoover's interest in schooling was well attested, no one knew where he stood on a department of education.[1] His 1927 reversal on the issue had drawn only the briefest of back-page coverage in the press, something that had escaped the notice of virtually everyone. During the presidential campaign, the educational issue never surfaced because the public focused on Al Smith's urban background, his religion, and his stand on prohibition.[2] Preliminary indications regarding Hoover's position raised more questions than they answered. In early February 1929 Hoover nominated Dr. Ray Lyman Wilbur for secretary of the Department of the Interior, which housed the Bureau of Education. A Hoover classmate at Stanford University, Wilbur was president of their alma mater. The selection of an educator prompted the San Francisco *Monitor* to urge Catholics to vigilance in case the appointment marked the first step toward a department of education. In contrast, the Saint Paul *Catholic Bulletin* had no fear that Wilbur's nomination portended a new secretariat because for Hoover to support the creation of one would constitute a reversal of his previous stand against increased bureaucracy.[3]

The inauguration did little to clear up matters. Giving education prominent notice, Hoover admitted that although schooling was "primarily a responsibility of the States and local communities," the nation, too, was "vitally concerned in its development everywhere to the highest standards and to complete universality." As the population expanded and both science and technology boomed, national life became more complex, demanding more leaders from every occupation and strata. "If we would . . . constantly refresh our leadership with the ideals of our people," warned Hoover, "we must draw constantly from the general mass. The full opportunity for every boy and girl to rise through the selective processes of education can alone secure us this leadership."[4]

Clearly, for the new president education was a national concern, but because he left unsaid what he intended to do about it, speculation continued. Archbishop Austin Dowling confided to George Johnson that Hoover's words foreshadowed a federal department of education. When Johnson shared Dowling's opinion with NCWC department heads, Justin McGrath of the press bureau said that many Washington correspondents agreed. Noting that just as many held otherwise, Johnson advised that since there was no likelihood that a bill for a department of education would be considered during the short special session of Congress, "it would be unwise for the Catholic press to discuss the question at this time or for the N.C.W.C. to send out any alarm based on the President's statement."[5]

Following his advice, the NCWC news service interpreted positively both the inaugural address and Wilbur's appointment. Rather than signaling a desire for a department of education, Hoover's words and actions indicated his intention to gather the government's scattered educational agencies in Wilbur's Department of the Interior. The NCCW agreed that the government's educational activities would "be concentrated and expanded in the Department of the Interior which, from a practical point of view, will virtually become the department of education." Noting that the NEA and others had yet to abandon their advocacy of a separate department, the news service expected them to "accept the compromise plan of giving more attention to educational matters in the Department of the Interior."[6] This was wishful thinking.

The NCWC's assessment of the situation received corroboration from a source close to the president. In mid-March, Richard Washburn Child, a Hoover intimate, argued in the *Saturday Evening Post* against the evils of bureaucracy. Concerned that increasing centralization meant long-distance government instead of local self-rule, he singled out schooling as a case in point. "The menace of a cut-to-pattern national educational system, with the tyranny of bureaucracy, is upon us," warned Child. "Proposals leading to colorless uniformity and sheep herding would tend to reduce all local, traditional manners, customs and convenience to a common level. Upon this issue may depend the breakdown of a centralized government in Washington which has undertaken too much. And upon it may depend also the decay of community responsibility which undertakes too little." Child considered the federal education issue critical: there the battle against bureaucratization would be won or lost. The stakes were high because communism found it easier to take over a centralized country than one whose government was highly localized. Child's point was that "Hoover—Or Some Other" would have to make the turn away from bureaucracy, and education was the place to start.[7]

The NCWC Press Department publicized the article in its editorial service. Quoting essential portions, the editor stressed Child's relationship with the president and commented that many Washington observers believed that his "treatment of the problem of centralization had White House inspiration." If that were true, all rumors that the administration favored a separate department of education were false.[8] Taking their cue from the NCWC, several Catholic papers around the country played up the implications of Child's view.[9]

Undaunted by Child's words, Robsion introduced his bill for a department of public education on the opening day of the special session of the Seventy-first Congress.[10] Soon, two Republican members of the farm bloc, Representative Charles Brand of Ohio and Senator Gerald Nye of North Dakota, introduced a bill for the promotion of rural education through an annual appropriation of $100 million for two years to be spent on teachers' salaries and elementary schools in non-urban areas. Funds were to be apportioned to a state in the proportion that its rural population bore to the rural population of the United States. Like the Smith-Towner and Sterling-Towner bills, the funds were to be offered on a matching basis.[11] Whereas earlier proposals had earmarked only $25 million for rural education, the entire appropriation of the Brand-Nye bill was destined for that purpose, amounting to a significant redistribution of wealth because populous urban areas would be taxed to pay for the schooling of farm children.

Brand argued that wealth had become concentrated in cities, while the vast majority of children remained in rural areas. Once children were educated at rural expense, many left the farm for the factory, draining agricultural areas of future producers as well as the money that had educated them. " 'Home Rule' is a fine phrase that became popular when wealth was about equally distributed," said Brand. " 'States Rights' still has the ring of bygone days but how are 40 States to collect taxes from wealth that has escaped into 8 States?"[12] In short, Brand contended that the federal government had an obligation to redistribute wealth in order to equalize educational opportunity. In the Brand-Nye and the Robsion bills, Congress had before it in two separate measures what amounted to the old Smith-Towner and Sterling-Towner bills, this time aimed solely at public education and rural schools.

If Child's article offered unofficial insight into Hoover's thinking about the federal role in education, a more authoritative view was soon available. Addressing the ACE on 3 May, Wilbur warned that there was "a distinct menace in the centralization in the national government of any large educational scheme with extensive financial resources available." Echoing cautions voiced by Catholics, he continued, "Abnormal power to mould and standardize and crystallize education which would go with the dollars, would be more damaging to local government, local aspiration and self-respect, and to state government and state self-respect, than any assistance that might come from the funds." True citizenship was learned in the family and the local community. Although the federal government had a role in education, Wilbur limited that function to the development of "methods, ideals and procedures . . . to be taken on their merits." "A Department of Education similar to other departments of the Government is not required," he concluded. "An adequate position for education within a department, and with sufficient financial support for its research, survey and other work, is all that is needed."[13]

His words delighted the Catholic press. The editor of the Saint Paul *Catholic Bulletin* considered them a "refreshingly clear and outspoken" condemnation of a department of education. The secretary's bold conclusion led the Boston *Pilot* to proclaim that the federal education question was "a vanishing issue." "If Secretary Wilbur speaks for the President on this matter, there will be great disappointment in some circles," commented the editor of the Omaha *True Voice*. His colleague at the Cleveland *Catholic Universe Bulletin* was more specific, remarking that although Catholic opposition had postponed the realization of the NEA's dream of federal control of education, the association remained undeterred. Said the Cleveland paper, "The declaration of Secretary Wilbur, himself a schoolmaster, may be expected to discourage further demands in that direction."[14] The editors were too sanguine.

Secular newspapers also interpreted Wilbur's words as meaning that there would be no department of education established on Hoover's watch. The *New York Times* backed away from the half-hearted endorsement that it had given for a department in 1925 to now give a lukewarm reception to Wilbur's plan. Readers were informed that Wilbur intended to use the various existing federal educational agencies to serve as clearinghouses for information, leaving states and local communities free to use them as they saw fit. While remarking that this method "concededly has its limitations and difficulties," the editor thought that it should be "commended as at least a valuable expedient at our present stage of political development." After this backhanded compliment, he affirmed that Wilbur's method might be a valuable means of curbing further federal intervention in general. "It may," said the editor, "tend to check the impulse of men to rush to Washington to get something done, if they are more frequently told when they get there merely what ought to be done, and then are sent away to do it themselves. It would not be a bad plank to put in a party platform: 'Centralize information but localize execution.'"[15]

Other papers came out squarely behind Wilbur. The New York *World* welcomed his message as "a decisive statement . . . needed to clear the air," one to be "applauded" by all who opposed the federalization of education. The Louisville *Courier Journal* in Robsion's home state rejoiced that "a separate Department of Education is not favored by the Hoover Administration . . . a distinct victory for the cause of local self-government, as opposed to centralized bureaucracy." Like Representative Loring Black, the editor considered Robsion's "unique argument of placing school children on a plane of equality with swine . . . odious." The Philadelphia *Evening Bulletin* believed that the Bureau of Education "such as now exists can adequately handle all that Washington needs and ought to do," a point with which the *Cleveland Plain Dealer* agreed. The most prescient remark came from the Atlanta *Constitution*, which noted that the NEA would probably "hotly urge" the establishment of a department of education, "but with the administration dead set against the project, the discussions must necessarily be mostly academic." And so they would be. Such sentiments as these were echoed by at least thirty-six other papers. The NCWC gathered the opinions together and published them in pamphlet form.[16]

While the NEA must have felt the blow of Wilbur's words, it interpreted them in a way that enabled it to continue the drive for a department of education. In an address at the founders' dinner of the annual convention of the National Congress of Parents and Teachers, James Crabtree considered Wilbur's desire to enlarge the Bureau of Education as something of a triumph. No other secretary of the interior, declared Crabtree, had shown much interest in national education or the bureau, something that was untrue. Franklin K. Lane, who held that post under Woodrow Wilson, was an ardent educational nationalist more sincerely committed to a federal presence in that field than was Wilbur. In Crabtree's view, the bureau's sudden significance resulted from the united efforts of the NEA, the PTA, and others in agitating for a separate department of education. If proponents of the latter slackened their efforts now, this newly awakened concern might dry up. "The most effective help that can be given Secretary Wilbur for the enlarged bureau," declared Crabtree, "is to increase the efforts for a Department of Education."[17] This speech was a willful misrepresentation of Wilbur's intentions.

In mid-May, Wilbur took a surprising step by requesting that Hoover appoint a National Advisory Committee on Education (NACE) to study the federal government's role in that field. John Burke visited Hoover to ask for Catholic representation on the committee. The president informed him that the committee would recommend a plan for coordinating the various federal educational agencies. During the conversation, Hoover intimated that he was toying with the idea of placing all educational agencies under an assistant secretary as a way of avoiding the establishment of a department of education. Three days after this interview, Wilbur telephoned Burke to inquire if Edward Pace might serve on the NACE.[18]

Accompanied by Johnson, Burke went to discuss the matter with Wilbur. The conference offered insight on his thinking, with Wilbur expressing dissatisfaction with the present state of affairs. The Board of Vocational Education, for example, possessed increasingly larger appropriations, which he was convinced would lead to increasing federal control. In his view, Congress ought to halt federal subsidies for all educational purposes. Moreover, the Bureau of Education should be confined to doing research and disseminating statistical information. Wilbur hoped that the NACE would reach the same conclusions and make the appropriate recommendations. This was welcome news to Catholics, confirming that the administration and the church held identical views on the federal role in education. Besides asking for Pace, Wilbur wanted someone from the National Catholic Educational Association (NCEA) on the committee. Burke recommended Johnson, who was both secretary general of the NCEA and executive secretary of the NCWC Department of Education.[19]

Burke considered the appointments of both Johnson and Pace to the NACE an event whose significance was best summed up by Bishop Samuel Stritch of Toledo. On learning of Johnson's appointment, Stritch remarked that it represented "a great opportunity for the cause of Catholic education and education in general for the United States" because for the first time the federal government recognized the church "as one of the great educating agencies of the Country." "So far educators, while willing here and there to drop a faded rose on our hands, have ignored us and considered us as something apart," commented Stritch. "Our philosophy and our objectives were given scant consideration." Regional accrediting bodies had failed to view Catholics "as a people with definite and worthy policies in the field of education." Rather, they saw them as "pitifully lagging behind in the field of education," to be tolerated only for social and political reasons. Stritch attributed the change in attitude to the testimony of Catholics before congressional hearings on the various educational bills throughout the decade. The government now recognized that Catholics were endowed with ideas that might be worthwhile, and that they had a part to play in the education of the nation.[20]

Expressing the separate-but-equal sentiment that characterized Catholicism during the early decades of the century, Stritch correctly understood that the government—at least the administration—recognized that Catholics had something to offer education. That the educational trust shared this view is unlikely. It is also quite possible that what the administration valued in the Catholic position was the church's opposition to any increased federal participation in education, something that Hoover and Wilbur also shunned. Both the administration and the church viewed education as a local matter. At least on the federal education question, Catholics were more in tune with the Republican leadership and mainstream editorial opinion than were progressive educational administrators.

The NACE met for the first time on 7 June 1929. Comprising fifty-two members, it was a who's who of the educational trust, including Charles Mann, director of the ACE; James Crabtree, secretary of the NEA; Uel Lamkin, president of the NEA; Frank Cody, superintendent of schools in Detroit; Samuel Capen, chancellor of the University of Buffalo; Henry Suzzallo of the Carnegie Foundation; William Davidson, chairman of the NEA's legislative commission; William Bagley of Columbia University; Charles Judd of the University of Chicago; Lotus Coffman, president of the University of Minnesota; James R. Angell, president of Yale; Edward Elliott, president of Purdue University; Ellwood Cubberley of Stanford University; and George Strayer of Columbia University—the last six were founding members of the Cleveland Conference. After appointing Mann as general chairman and Crabtree as general secretary, Wilbur made his position clear. Mindful of the history of local development in public education, the NACE was to determine the federal government's future role in that area. Wilbur warned that people increasingly looked to the national government for leadership and to Congress as if it were "a glorified Red Cross agency," a tendency that must be curbed in education. The NACE was to consider a list of questions drafted by Wilbur and slanted in favor of home rule: What kind of federal activity would strengthen local autonomy? What kind of coordination of federal agencies would best limit federal activity? What justified the continuation of existing federal subsidies? What evidence showed that these monies weakened local responsibility? The questions clearly indicated the direction in which Wilbur wanted the committee to go.[21]

Taking charge of the meeting, Mann outlined the present scope of federal involvement in education. Two speakers from the Bureau of Education addressed the issue of federal subsidies. The NACE then divided into three subcommittees: one to study the best organization of federal educational agencies, a second to study federal aid to colleges and universities, and a third to study federal aid to common schools and high schools. When the groups reported their inability to formulate any definite plans before the fall, Mann adjourned the NACE sine die.[22]

Despite Hoover and Wilbur's intention to limit the federal role in education, the NEA went ahead with its program. Davidson reported at the annual convention that Hoover's inaugural address was "widely interpreted in the press as indicating a desire to create a Department of Education." He neglected to mention Child's article in the *Saturday Evening Post* and Wilbur's address to the ACE, both indicating that the president opposed such a secretariat. Moreover, Davidson called the establishment of the NACE "unquestionably the most far reaching step" in the movement for a department—the furthest thing from Wilbur's mind. "The best way to help the Secretary of the Interior achieve his goal, whatever it may be," concluded Davidson, "is to continue our activities for a Department of Education."[23] The NEA "emphatically" reendorsed its commitment to the establishment of such a department. More circumspectly, the convention called on Congress to fund a study aimed at remedying the inequality of educational opportunity suffered by children in rural areas, in imitation of its recent appropriation for a study of secondary education.[24]

The NCWC had an inside observer on the NEA's Board of Directors at the convention. Agnes Bacon, Catholic state director of the NEA in Rhode Island, reported on the meeting. Her memorandum was revealing. In a farewell address to the board, outgoing President Lamkin, a Mason and a member of the NACE, told colleagues that they must consider the NEA supreme in matters of schooling. They must forget the U.S. Bureau of Education and work to make the NEA the only authority on education in America.[25] Even allowing for rhetoric on Lamkin's part and bias on Bacon's, this attitude raises questions about the motive behind the NEA's strong desire for a department of education. Catholics had long suspected that the NEA sought supremacy in education and aimed at using a department to achieve it.

The association's resolution on rural education was the prelude to a renewed drive for federal aid. Although at the 1928 House hearing, Davidson had told the Committee on Education that the NEA's commitment to federal aid had all but reached the "vanishing point," it quickly reappeared. Prior to introducing the rural education bill, Brand had consulted Crabtree, who told him, "Your idea is sound whether you can ever force its acceptance on the part of Congress or not!"[26] After the NEA convention, Crabtree stumped for the bill at the fiftieth anniversary of the Fairview School District near Elmwood, Nebraska, where he had begun his career. Repeating Brand's argument about the concentration of wealth in the cities and the flight of rural children from farm to factory, Crabtree declared that the Brand-Nye bill "ought to appeal to men and women whether they believe in government subsidies or not." The NEA *Journal* was less circumspect than had been the association's convention. The latter had called only for a study of the best method of equalization of educational opportunity for rural children. The *Journal* urged passage of the Brand-Nye bill.[27] Thus, the NEA resumed its pursuit of federal funding.

As the NEA continued its drive for a department of education, so too did the Klan wing of the Scottish Rite Masonry. In August 1929 James Vance of the *Fellowship Forum* began a widespread fund-raising campaign for the Robsion bill. Under the *Forum* letterhead, his pitch was distributed as a handbill. Playing on emotions and bigotry, Vance told readers that the Robsion bill for a department of public education would "afford every child in the United States at least a grammar school education, thereby perpetuating THE LITTLE RED SCHOOLHOUSE—the bulwark of religious and civil liberty in America." To allay fears of federal control, he contended that the proposed department would only advise and cooperate with the states "to the end that every boy and girl in America will be accorded full and equal educational opportunities. . . . Public schools must be made good enough for the best yet inexpensive enough for and in the reach of the poorest." The flyer had harsh words for the Catholic church. Vance warned that opposition to the bill was limited almost exclusively to the Catholic hierarchy. "This political institution, which operates under the name of a church would have the youth of America grow up in ignorance rather than see the American system of education completely developed and perpetuated." The only response to this threat against the American way of life was to send money. Vance urged everyone to send five to ten dollars—at the very least one dollar—and send it quickly to the *Forum*. Only as an afterthought, a postscript, did he explain why funds were needed: for postage to disperse vast amounts of American defense literature.[28]

The Hearst papers, too, joined the campaign with a well-timed barrage, delivered shortly after Senator Arthur Capper of Kansas, leader of the farm bloc, introduced the companion measure to Robsion's bill, and just as the nation's schools opened.[29] For nearly a week in early September 1929, the Hearst chain propounded the question. It challenged Hoover to lead the way for a department of education. People wanted a president who championed popular causes, not one who fought legislation initiated by Congress. For six months the nation had waited for him to take the lead in abolishing illiteracy in America. By way of absolution, the papers implied that although Hoover had originally intended to establish a department of education, his cabinet (meaning Wilbur) had convinced him otherwise.[30]

The next issue of the *Los Angeles Examiner* carried a press release from Capper, who assured readers that his bill did not violate state rights. The proposed department was intended simply as a research agency that would be as beneficial to private as to public schools. In the senator's view, the question was not whether the federal government should have a role in education; that principle had already been established. The question was whether the government's role would be made more efficient by coordinating its various educational agencies.[31]

The paper's cartoonist was then summoned to action. He drew a corps of marching school buildings being reviewed by Uncle Sam, who commented, "Very good, if there's enough of them! The more the better!" The picture bore the caption, "BUILD UP THIS ARMY!" Beneath was a brief article stating that Uncle Sam was both worried and proud because he knew that there was an insufficient number of schools to accommodate the nation's children; it added that he was "certainly thinking of the urgent need of a National Department of Education."[32] Left unsaid was how a department was to build up this army without federal aid at its disposal. The cartoon and article seemed to corroborate what opponents had been saying since 1925, namely, that advocates were seeking the department first and would seek the subsidy later. In fact, the subsidy bill was already in Congress in the form of the Brand-Nye measure for rural education.

Finally, the Hearst chain printed an open letter to Hoover. Quoting out of context a statement that Coolidge had made in his 1927 state of the union message, it alleged that the former president had endorsed the consolidation of all federal educational agencies into one department under one cabinet member, making it sound as if Coolidge favored a separate department of education. In fact, he had called for a department of education and relief. Reminding Hoover that he had been elected on a platform that pledged him to continue the policies of his predecessor, the letter urged him to create a department of education.[33]

America responded by challenging Capper's press release. His denial that a secretariat would lead to federal control flew in the face of "the well attested fact that officials and departments always seek to increase the orbit of their powers, and resist any attempt to check it." *America* argued that it was incumbent on Capper to show where the Constitution empowered the federal government to enter the field of education. In so doing, he was cautioned to avoid the tactic employed by the Hearst papers and the *Fellowship Forum.* "Pitiful appeals based on the pitiful condition of illiterate children, and grandiose panegyrics of education, are . . . irrelevant."[34] The San Francisco *Monitor* reprinted this editorial in full.[35]

Other Catholic papers warned of the political situation around the Capper-Robsion bill. The *Indiana Catholic and Record* cautioned that because the measure proposed a department of public education, its advocates could accuse antagonists of opposition to public schools. The paper further advised that although many senators were against the legislation, they were under heavy pressure to enact it.[36] The NCWC news service was more objective in its assessment but no less cautious. Noting the impossibility of forecasting Congress's attitude toward the bill, it commented that the question was "one of those controversial ones which politicians, as a rule, try to avoid and political effort will probably be directed to preventing its being brought to an issue rather than to direct opposition." Still, if proponents could force the measure out of committee, there was no telling what Congress might do. Moreover, if the bill were to be enacted, there was no guarantee that Hoover would veto it, even though his opposition was taken for granted.[37]

When the second (regular) session of Congress opened in early December 1929, department advocates launched their campaign. Republican Daniel Reed, chairman of the House Committee on Education, introduced the old Curtis-Reed bill with a speech that linked public education with the general welfare. Calling the failure to have already established a department of education "tragic and suicidal," he warned that not to create one now "would be even more short-sighted and deplorable." At stake was national development. There were parts of the country rich in natural resources, "where the lack of educational advantages has pauperized what should have been a land of plenty." Generations of children, innocent victims of these conditions, "have had their mental and spiritual lives starved and stunted." "The real loss to the Nation, therefore, is not alone one of material wealth," said Reed. "The real tragedy is in the useless sacrifice of social and spiritual force, . . . which it [the nation] can ill afford to lose." A department of education had yet to be established because "foreign opposition" had "thus far been highly effective in preventing consideration of the bill." Although Reed implied that various European governments were behind this antagonism, such a scenario was unbelievable. The objections he chose to refute were those advanced by Catholics.[38] Thus, even Reed, who hitherto had not been overtly anti-Catholic, hinted that the failure to have established a department was attributable to the Roman Church.

Outside Congress, the Southern Jurisdiction of the Scottish Rite reasserted its commitment to a department, while both Capper and Robsion circularized the Klan wing of Masonry through the *Fellowship Forum*. In an open letter printed by the paper, the two declared that America was "face to face with a crucial test of its determination to function for the benefit of mankind in general and our own children in particular." The mettle of American conviction would be proved by the establishment of a department of public education to remove the nation's deplorable illiteracy and to enthrone intelligence. Capper and Robsion called for a "Monster Appeal in behalf of the childhood of America": a petition to Congress in favor of their bill signed by 10 million people. Masons were to bring the measure to the favorable attention of their lodges, churches, and societies with a request that signed petitions be sent to the *Fellowship Forum*. The two sponsors hoped to present the appeal to Congress in January 1930.[39]

The Catholic press was shocked by Capper and Robsion's alliance with the Klan wing of Masonry. Vincent de Paul Fitzpatrick, editor of the *Baltimore Catholic Review*, felt humiliated to announce that a man of Capper's stature—he published a chain of respectable Midwestern newspapers catering to farmers—would align himself with the policies and aims of a journal like the *Fellowship Forum*.[40] Other editors were less concerned about the senator's reputation than about the implication of the coalition. They feared that discussion of the education bill, which had previously been conducted on a high plane, would be reduced to the level of prejudice. Worse, given the role of the Klan and the Scottish Rite in promoting the Oregon school law, no doubt remained about the actual intent of the Capper-Robsion bill.[41]

Although concerned, too, lest bigotry be inserted into the issue, Johnson thought that the alliance might be turned to Catholic advantage. Intelligent people would readily understand that any bill that summoned "the riff raff [*sic*]" to its support would be unhealthy for American life. In his view, the NEA was a hidden partner in this entente.[42] Airing this opinion in an editorial for the NCWC news service, Johnson noted that while the NEA never openly enlisted the forces of bigotry, the words of the prophet Habakkuk applied to the situation: "He who runs may read. Had he been running at the February [1929] meeting of the Department of Superintendence of the NEA in Cleveland, he might have read clippings from the Fellowship Forum in a scrap-book containing articles from the press in favor of its bill. Recently he might have read in the daily press an account of the appearance of Miss Charl Williams [NEA field secretary] before a convention of the Scottish Rite Masons in Washington." These two instances caused Johnson to have misgivings about the sincerity of the NEA. How could an organization dedicated to American educational ideals allow "its paid agents to fraternize with the forces of ignorance and un-Americanism"? The education question demanded dispassionate, thoughtful study, not the hatred, bitterness, and hysteria that were being invoked.[43]

Johnson's words bespoke Catholic disdain for the NEA and the Southern Scottish Rite, particularly its Klan wing. His sentiment was understandable. He had complained to the NEA about its display of anti-Catholic editorials in its scrapbook, only to receive an unsatisfactory reply.[44] Yet it was unfair to characterize Masonry and its Klan wing as the forces of ignorance. The majority of Masons were respected members of their communities. As recent historians have shown, early—and presumably present—members of the Ku Klux Klan were middle-class businessmen. Obviously, feelings ran deep on each side of the school issue. What added to the depth was a fundamental antagonism over what it meant to be an "American." One-hundred-percenters sought social cohesiveness and the promotion of traditional, Anglo-Saxon values and culture through a drive for conformity. Schools were to play an important role in inculcating and promoting that cohesiveness and those values. For 100-percenters, to be an American meant cleaving to those mores and striving to conform. Catholics, on the other hand, believed that opportunity and pluralism had made America great. To be an American meant embracing the nation's ideals and aspirations, even while clinging to some cherished ways. Catholics viewed 100-percenters as ignorant, un-American bigots; 100-percenters viewed Catholics as alien sappers, bent on subverting the nation by undermining its schools.

Johnson's subordinate, Frank Crowley, director of the NCWC's bureau of education, noted that the Capper-Robsion bill for a department of public education was so named to make opponents appear un-American. In its campaign, the NEA enjoyed the support of the Southern Jurisdiction of the Scottish Rite and the Ku Klux Klan, two organizations that opposed Catholicism. When Wilbur made it clear that the Hoover administration was unsympathetic to a department of education, the NEA shifted tactics.

Agitation for farm relief was at its height, and the recent presidential election had revealed a cleavage between rural and urban voters, one that the NEA sought to turn to its advantage. The association turned to an endorsement of federal aid for rural education in the Brand-Nye bill. Crowley reminded readers that Williams had indicated in 1925 that the NEA had simply deferred its commitment to federal aid in proposing the Curtis-Reed bill.[45] Although Crowley's assessment was accurate in the main, it missed the mark in suggesting that the administration's unsympathetic attitude toward a department had deflected efforts toward federal aid. In fact the NEA kept up its drive for both a department and aid, a position that won considerable sympathy from the NACE.

While many Catholic editors feared an escalation of bigotry over the issue, a few broke ranks. On Christmas Day 1929 the progressive *Commonweal* came out in favor of both federal aid to and federal control of education. Some school districts were so backward, commented the editor, that only federal involvement could correct them. Arguing that the improvement of decrepit systems would benefit the nation as a whole, he warned that the ideology of the advocates of a department posed danger. Most wanted nationally standardized schooling. By distinguishing between control and standardization, the *Commonweal* concluded: "If a measure of federal authority or federal money can be expended to advantage upon school systems which are not now functioning properly, that measure should be endorsed. But it must be perfectly clear that the aid is offered to the schoolroom and not to a pedagogical idea. We want to foster the children and not some particularly talkative professor."[46] The distinction advanced in this argument was almost self-contradictory. Any federal intervention would necessarily aim at bringing backward schools up to some standard, thereby bringing about a modicum of standardization.

Having already broken ranks from the church's position in 1925, the progressive Milwaukee *Catholic Citizen*, which thereafter had remained silent on the education issue, renewed its insurgency. At first discouraging his coreligionists from making the school matter a Catholic question, the editor later reprinted his original piece in support of a department of education. Within two months he also gave a qualified endorsement to federal aid for schools.[47] The positions taken by the *Catholic Citizen* and *Commonweal* indicate that the Catholic church was not monolithic on the education question. They also testify to the continued spirit of progressivism in Catholic quarters.

As the drive for a department escalated, Catholics kept tabs on Congress. Because the special session had been devoted to farm relief, most committees were not yet organized when the regular session opened in December 1929. In the first months of 1930, members had been appointed to the education committees of the House and Senate. The balance between advocates and opponents of a department of education was apparently much the same as in the previous Congress. Of the twenty-one members of the House committee, six were known to oppose the bill: Republicans Paul Kvale of Minnesota, son of the late Ole Kvale, and E. Hart Fenn of Connecticut, and four Catholic Democrats, Loring Black of New York, John Douglass of Massachusetts, Vincent Palmisano of Maryland, and René De Rouen of Louisiana. Six others were known to favor the measure: Republicans Reed, Robsion, Willis Sears of Nebraska, Elmer Leatherwood of Utah, and two Democrats, Malcolm Tarver of Georgia and La Fayette Patterson of Alabama. The positions of the remaining committeemen are unknown.[48] As in the previous Congress, opponents represented the urban wing of each party together with rural religious objectors. Proponents uniformly represented the rural wing of each party. The situation of the Senate committee was little different. Of the ten members, five had shown themselves unfavorable to previous education bills: Republicans Jesse Metcalf (chairman) of Rhode Island, William Borah of Idaho, Lawrence Phipps of Colorado, and Frederick Gillett of Massachusetts. They were joined by Democrat David Walsh of Massachusetts. Three additional members were added in the second session. Of the thirteen, only Park Trammell of Florida publicly favored the legislation.[49]

In early February 1930 William Montavon checked with Metcalf's office about the Capper-Robsion bill. Metcalf's secretary informed him that the Committee on Education and Labor had agreed not to consider the measure during the current session. Montavon inferred that as with previous committees, this one considered the issue too controversial to broach before the fall elections.[50] Within a month, an NCWC agent received a false alarm from Republican Senator Hiram Bingham's secretary, who warned that Robsion had plans for a hearing and was lining up his friends in the House to push the legislation through. Asked to confirm the story, Reed denied any knowledge of a proposed hearing. The House was too occupied with other business to pressure his committee on the education bills. In confidence, he added that the measure generated little interest outside of certain quarters.[51] No doubt the elements he had in mind were the educational trust, Masonic circles, the Klan, and the farm bloc.

This lack of interest was not for want of trying. Both Capper and Republican congressman Wilburn Cartwright of Oklahoma took the case to the airwaves. Broadcasting on Radio Station WJSV in Washington, D.C., owned by Vance of the *Fellowship Forum,* the senator explained his bill and denied that it would standardize academics, interfere with private or parochial schools, or plunge education into politics. "Twoscore [*sic*] great American organizations are supporting this bill," concluded Capper. "But it should be supported by every patriotic man and woman in the country. . . . [W]hen all the people know the facts about this bill Congress will meet the public demand. . . ."[52]

Cartwright explained why public demand had yet to be met. In the two previous Congresses, eastern representatives controlled both the steering committee and the Rules Committee in the House and blocked a vote on the Curtis-Reed measure by stacking the Education Committee against it. Despite alleged evidence that a bipartisan vote of both houses of those Congresses would have passed the bill comfortably, the legislation was never brought to the floor. The same situation prevailed in the present Congress with respect to the Capper-Robsion bill. Cartwright told his listeners that although Reed refused to call a hearing on the legislation because key witnesses from the educational community were serving on the NACE, "it is generally conceded that the real reason is the pressure from certain Republicans on the powerful steering committee and Rules Committee . . . and doubtlessly like pressure from Doctor Wilbur." To unblock the legislation, Cartwright encouraged citizens who favored the bill to find out if their representatives were urging "the political leaders in Congress opposed to it to let the bill come to a vote in the two Houses."[53]

Whether as a misunderstanding or as a ploy to force action, the *Fellowship Forum* ran an article in March 1930 stating that the House committee would hold a hearing on the Capper-Robsion bill in early April. An NCWC agent queried Reed about the information. "Confronted by the announcement in the *Forum* and alleged activities of . . . Robsion and others, Mr. Reed became cautious," reported the agent. "He realizes that the Committee may take it into their own heads to hold hearings in spite of his present opinion. It is this possibility which renders him cautious."[54] In other words, although Reed wanted no hearing, some members of his committee might take it upon themselves to force one.

Not only were Capper, Cartwright, and the *Fellowship Forum* pushing the Capper-Robsion bill, the Southern Scottish Rite was active as well. In March 1930 it sent a circular to members with word to promote the measure and furnished them with letters in its favor. Each Mason received three copies of a canned dispatch to be signed and forwarded to his state legislature, his congressman, and his senator.[55]

In early April the Hearst chain broke out with another rash of what *America* called "public-school measles."[56] An editorial entitled "The School—Earth's Noblest Monument" romanticized public education. Simplistic and misleading in its argument, it implied that common schools imparted true American virtues whereas private and parochial ones did not: "The public school typifies the spirit of the United States, the Constitution, laws and beliefs of the United States. The public school is DEMOCRATIC. The public school, like the Constitution of the United States, forbids all discrimination because of religion. It teaches THE FACTS that all intelligent human beings accept. It leaves to the homes, the churches, the private or religious schools . . . the teaching of special religious or other beliefs. The public school recognizes only social equality. . . . Fortunate the boy or girl that go to the public school. Much to be pitied are those deprived of that splendid training in American life and American thought."[57] And on it went, painting public schools in glowing terms, while damning private ones by implication.

Despite a mighty effort by Capper, Robsion, and the Southern Scottish Rite, the bill for a department of public education remained in committee. At the end of April, the Republican leadership in the House issued a special order permitting Patterson, a Democrat, to address the House in favor of the legislation, a gesture tantamount to a consolation prize. Referring to Capper and Robsion's monster appeal, he asserted that within the past year congressmen had received at least 5 million letters petitioning for passage of the bill. He further argued that the measure enjoyed the support of various national bodies, like the NEA, the AFT, the AFL, the National Congress of Parents and Teachers, the Southern Jurisdiction of the Scottish Rite, the National League of Women Voters, and other distaff societies, whose combined memberships amounted to 29 million citizens. As will be seen shortly, the AFL had cooled toward the bill and adopted a cautious attitude. Like Cartwright, Patterson claimed that although the measure had the support of a comfortable plurality in Congress, a stacked Education Committee and the Republican leadership on the steering committee prevented the bill from coming to the floor. "Let the case stand on its merits and give the people a voice," challenged Patterson, "but do not try to kill it with guile, because it is controversial, but let us vote openly on the matter." His words drew applause.[58]

A month later Capper pleaded with the Senate. Without accusing his party's leadership, he complained that the bill remained in committee even though it enjoyed widespread public support: the 5 million petitions mentioned by Patterson. Declaring that Congress could ill afford to ignore such a popular bill, the senator promised to agitate for the question during the short session in the coming December.[59] That, however, was not to be. The present drive was the last great failure to secure congressional action on the issue. In July 1930 the NEA decided to cease agitation of the matter during the NACE's study of the federal role in education.[60] In fact, this decision marked the end of the NEA's drive for congressional action on the establishment of a department during the inter-war years.

Although Capper, Cartwright, and Patterson claimed that the bill enjoyed widespread support in and out of Congress, the actual amount of support is difficult to gauge. Certainly, the applause won by Patterson's words suggested a measure of congressional agreement with his argument. To judge by the residency of known congressional advocates, the bill's support came principally from farm-bloc areas.[61] Although Capper and Robsion's monster appeal had failed to garner the hoped-for 10 million petitions, the 5 million it did fetch indicated a measure of popular support for their bill. While there is no doubt that the national conventions of the organizations named by Patterson endorsed the measure, his claim that their combined memberships of 29 million backed the bill is probably an exaggeration, not because they failed to total that amount, but because groups are rarely monolithic or unanimous. Thus, while their combined total memberships might number 29 million, it is unlikely that actual support for the bill approached that figure. If 15 to 20 million people favored the legislation, their number constituted about 20 to 25 percent of the national voting population.

Congressional proponents contended that the passage of the education bill had been thwarted for several reasons. First, the majority of opposition came from representatives of populous eastern states. Such an explanation fit well with the urban-rural tension made evident in the 1928 presidential election, a tension that had actually been present since 1920. Second, the postwar resurgence of states' rights—anti-bureaucracy—hardened as the decade wore on. This resurgence cut across both party lines and the urban-rural divide, especially in the Senate. Third, proponents argued that eastern interests had stacked the education committees of both houses against the bill. Evidence does not support the charge regarding the House committee; there was, however, truth regarding its Senate counterpart, especially in the Seventieth Congress and probably in the Seventy-first. Finally, advocates charged that the Republican leadership in Congress refused to let the issue come to the floor. This was probably true. Both Hoover and Wilbur opposed a department of education, and the matter was under study by the NACE.

With the failure to secure congressional action on the Capper-Robsion bill, the fate of the proposed department education now rested with the NACE, which resumed work in the fall of 1929. In October a steering committee of twelve provided direction in a document aimed at facilitating the NACE's task. The committee represented a cross section of the educational trust: Capen, Cody, Coffman, Crabtree, Davidson, Judd, Lamkin, Mann, Suzzallo, James E. Russell, William F. Russell, and George Zook—at least three of them (Capen, Cody, and Judd) objectors to the Smith-Towner bill. The steering committee's paper was divided into two parts, a brief preamble and a list of "questions." Upholding local control of schooling as traditional policy, the preamble recounted that throughout American history certain educational problems arose that could be solved only by the state or the federal government. Because the federal role had evolved gradually without a preconceived plan, its educational activities lacked unity and coherence. The purpose of the NACE was to make a recommendation about the proper adjustment of those responsibilities. To stimulate discussion on this issue, the committee posed ten "questions" that were actually assertions. Falling into two categories, these declarations (1) affirmed the federal duty to conduct educational research and to disseminate information, and (2) proposed federal aid to supplement the resources of colleges and universities, equalize educational opportunity at the common and high school levels, and initiate programs that might not otherwise be undertaken. For two days the NACE wrestled with these "questions." Coming to agreement on the assertions in the first category, the advisory committee remanded those in the second to the two subcommittees on federal aid. Moreover, to facilitate the entire project, the steering committee was authorized to gather information from interested groups and government agencies.[62]

During the winter of 1930 the NACE steering committee hosted conferences with thirteen organizations.[63] Several meetings simply collected data about the functions and problems of various federal offices and programs. Six organizations addressed the federal role in education directly. Three were unequivocal in their exhortations for federal participation in education. Two were cautious regarding any federal involvement, while the National Catholic Educational Association was totally unsympathetic.

Already beneficiaries of federal largesse, vocational educators naturally supported continued federal funding of their cause: "If you can't serve, you can't earn; if you can't earn, you can't buy; and if you can't buy, industry is paralyzed," they pointed out. In their view, public schools were simply interested in academics. The quickest way to overcome that sort of thinking was through federal stimulation of education on skill-building, something that was imperative because performance standards were rising rapidly. These educators viewed the friction between them and those who supported local control as arising from "the desire of each agency to get credit for any good results." Conversely, blame was too frequently laid at the doorstep of the federal agency. Like Brand and Crabtree, they claimed that the concentration of wealth in urban financial centers necessitated federal intervention in the form of permanent federal subsidies.[64]

The Southern Jurisdiction of the Scottish Rite was equally supportive of federal involvement. Grand Commander John Cowles explained that the Rite advocated a department of education with federal aid for public schools to be administered "under the absolute control of the States"; the compulsory use of English as the language of instruction; the complete separation of church and state to exclude "sectarian institutions" from federal funds; and "the American public school, non-partisan, non-sectarian, efficient, democratic, for all the children of all the people." (Clearly, the Southern Jurisdiction remained committed to compulsory public education.) The country needed a department of education to give schooling the dignity it deserved. Whether the secretary had a portfolio of authority or was simply an adviser to the president, the important thing was that there be a secretary of education. A poll of Masonic bodies in and out of the Southern Jurisdiction showed widespread support for the order's program. "Only political expedience, fear of federal control, or un-American opposition prevents some effective action on the subject," concluded the commander.[65]

Ten representatives of the NEA's Department of Superintendence also pleaded for federal aid because of the financial plight of schools. The group argued that many taxpayers protested against the increasing cost of education. Moreover, with mergers and consolidation in business, wealth was being concentrated in a few financial centers, diminishing local taxable capital, especially in rural areas. Evidence showed that federal aid stimulated local spending by as much as three to nine times the amount of the grants. Finally, as the nation's population became more mobile, "some means of insuring comparability of schooling must be found."[66] As mentioned earlier, the NEA's "vanished" commitment to federal aid had quickly reappeared.

Representatives of the National Association of State Universities and the Conference of Separate State Universities were cautious about federal intervention in education. Presidents Frank L. McVey of the University of Kentucky, David Kinley of the University of Illinois, W. A. Jessup of the University of Iowa, and others expressed concern about increased federal participation in schooling and the way it altered, if not threatened, traditional state management of higher education. They reminded the NACE that the absence of any mention of schooling in the constitution had, "in the American mind, long carried the implication that this function was left to the states." Moreover, federal agencies charged with oversight of existing aid to education "constantly tended to increase their determining influence on local management of education." On the other hand, federal income tax drained away precious state resources needed for the improvement of public schools at home. Thus, if all American children were to have equal educational opportunity, it was becoming necessary for the federal government to offer funds to schools. These presidents were skeptical, however, about federal aid on a matching basis, which seemed to them as nothing more than "a forced matching of money which originally came from the states." If the government were to grant aid, it ought to do so with no strings attached and no supervision of any sort.[67]

More cautious was the AFL. Representatives expressed concerns similar to those of Catholics. They believed that control of education should be as decentralized as possible. While favoring federal aid for education in backward communities, they warned that such aid might be used by the government to assert federal control over schooling. Their admonition was based in fact, namely, the operation of the Smith-Hughes Act for vocational education, which permitted the government to veto state plans for implementation. "Negative power of the veto may become positive power in reality," they cautioned, threatening to stifle valuable experimentation. It also threatened to standardize "the treatment of human beings," against which they expressed "grave objection." They suggested that federal money would be better spent on educational research rather than in direct aid to schools.[68]

Last to confer with the NACE was the NCEA. Representing it were Johnson, Monsignor Joseph McClancy of Brooklyn, Monsignor William McNally of Catholic High School in Philadelphia, Father F. M. Connell, S.J., of New York City, and Charles Lischka of Georgetown University. Both Johnson and Lischka also held posts in the NCWC Department of Education. According to the NACE's digest of the meeting, the delegates "heartily" encouraged the federal government to engage in educational research and the dissemination of information. Familiar with the benefits of a decentralized parochial school system, they opposed the centralization of education because it would interfere "with necessary local adjustment and experimentation." Continued the digest, "Because of the general tendency of control to follow financial aid, their attitude is one of caution, but they are not opposed to Federal financil [*sic*] aid to equalize educational opportunity among the States, if it is really needed, and Federal control is divorced from the idea." If funds were to be granted, the government should let states administer them with no strings attached; if that was impossible, subsidies should be used to foster only such matters as did not touch the educational process, like teachers' salaries, construction of buildings, and health inspection. "The attitude of the representatives present concerning making the Office of Education [the new name of the Bureau] a Department seemed one of general caution in the fear that control might follow enlargement of official position as it tends to follow financial grants," concluded the digest. "Caution, rather than opposition, was the characteristic attitude."[69]

When the five NCEA delegates read the digest, they believed that their words had been twisted. In Connell's view, the final sentence of the report had to be changed. "Our attitude cannot be defined a 'caution rather than opposition,'" he told Johnson, "but as 'opposition based on caution.'"[70] McNally registered the same objection, contending that the digest as a whole misrepresented the Catholic position. Admitting that many statements contained therein had been made during the conference, he complained: "As they appear on the Report they convey an entirely different idea than was present in our minds. We made it very clear that our opposition was based on the belief that Federal aid would of necessity imply Federal control."[71] McClancy, too, argued that the delegates had clearly rejected the idea of federal grants of any sort, regardless of their application.[72] Although evidence indicates that Johnson reconvened the NCEA representatives to formulate a supplementary report, no such document can be found.

Using information from the thirteen conferences as well as data gathered by a team of researchers, Suzzallo, who served as director of the NACE study, drafted a statement on American education. American democracy rested on the pillars of personal responsibility and local self-government. Because government resided in the people, an educated citizenry was imperative; therefore, the nation had always viewed schooling as the bulwark of freedom. The school itself was the most cherished expression of local self-determination, for "in none of our civil affairs are we as much locally self-governed as in the public management of our schools." Every departure from this long-established policy was to be regarded with the greatest skepticism and accepted "only when overwhelming evidence is presented that changes in our life require detailed modifications here and there." Suzzallo noted that since the Civil War, the federal government had increasingly inserted itself into education, first with land-grant colleges, then with agricultural experiment stations, and finally with the Smith-Lever and Smith-Hughes Acts that carried detailed regulations. At present, people were clamoring for federal aid to education in general. The time had come to determine what the national government could and should do for education.[73]

Based on Suzzallo's analysis, the steering committee established several principles to serve as guidelines. In the committee's view, (1) the federal government had an obligation to aid public education. Although (2) this help ought to consist primarily of research and information, (3) it was also appropriate that the government offer financial assistance, but in a way that would leave control of education at the local level. In so doing, (4) Congress should avoid granting money on a matching basis, as in bills of the Smith-Towner variety. Neither (5) should it determine "the social purposes to be served by schools" nor should it establish "the techniques of educational procedure." Finally, (6) any new federal participation in education must be regarded as experimental, subject to change as circumstances warranted.[74]

Guided by these principles, the steering committee recommended that Congress "create an adequate Federal Headquarters for educational research and information, so organized as to serve both as a cooperating center for all federal agencies" engaged in education and as a clearinghouse for information. The regulations attached to the Smith-Lever and Smith-Hughes Acts ought to be revoked. To promote research, existing federal educational offices ought to have their appropriations substantially increased. Finally, the federal government should grant annually to each state $2.50 for every child under the age of twenty-one, with the sole proviso that the money be used for education. Based on the 1930 census, such federal aid would have amounted to $150 million, or half again as much as the grant proposed in the Smith-Towner bill and its successors. This subsidy was to be readjusted every ten years according to census returns. If Congress accepted this plan, all existing federal aid for education ought to be repealed.[75]

In mid-June 1930 the steering committee proposed these recommendations to the NACE. They won widespread support with but few objections. Paul Chapman, R. L. Cooley, and Wesley O'Leary—representatives of the vocational education community—wanted 15 percent of the appropriation earmarked for vocational education; they argued that the Board of Vocational Education should continue to maintain minimum standards until such time as experience warranted removing them. Elliott thought the proposals lacked adequate provision for safeguarding scientific research, especially the work of the agricultural experiment stations. These objections were minor compared to the one raised by Johnson. Rejecting the philosophy beneath the proposals, he argued that far from safeguarding local control, the plan opened the way for the most far-reaching kind of federal domination. Of concern was the undefined federal headquarters, which he believed would become a sort of "super-department" that would coordinate all other federal educational agencies and act as a "mentor" to them. Through research and information, such a headquarters could profoundly influence federal, state, and local legislation. Moreover, it would be administered by the party in power and would be "used . . . albeit with the best of motives, for the perpetuation and propagation of its [the party's] own particular political, social, and economic theories." Johnson was convinced that the headquarters would become an administrative arm that exercised federal control because Congress would never grant the kind of aid envisioned without government oversight. Federal money always came with strings attached, and the headquarters would be the agency to set and enforce the regulations. "It is extremely doubtful, human nature being what it is," concluded Johnson, "if federal control can ever be divorced from federal subsidy."[76]

Disregarding Johnson's warning, the NACE voted tentative approval of the principles. To appease the opposition, it added to the steering committee three objectors, Johnson, Elliott, and O'Leary. The committee then appointed a subcommittee of five, including Johnson, to devise political mechanisms to implement the principles formulated by the NACE.[77] At the end of the summer in 1930 the subcommittee filed a report recommending three alternative structures: a department of education, a national board of education, and an expanded Office of Education under an assistant secretary in the Department of the Interior. Three new subcommittees were appointed to formulate detailed plans for each of these structures.[78]

In mid-October 1930 Johnson met with a special committee of Catholic bishops to discuss the NACE's various plans for federal participation in education. Archbishop John T. McNicholas of Cincinnati, who succeeded Archbishop Austin Dowling as chairman of the NCWC Department of Education, had formed the committee to consider problems facing his department. Although summoned as an ad hoc group, it became a standing body known as the Committee on Education during McNicholas's tenure as head of the department. In addition to McNicholas, it consisted of Archbishop Francis Beckman of Dubuque, Archbishop Samuel Stritch of Milwaukee, Bishop John Murray of Portland (Maine), Bishop Thomas O'Reilly of Scranton, Bishop Francis Howard of Covington, and Bishop Joseph Albers, auxiliary of Cincinnati. Convinced that any federal interference in or regulation of education violated the American educational tradition, the committee opposed both the creation of a department of education or the appointment of an assistant secretary in the Department of the Interior. Cautiously approving the idea that the government should engage in educational research, the bishops objected to any research agency that might achieve such political stature, thereby being able to enforce its findings.[79] In mid-November the American hierarchy concurred with the recommendations of McNicholas's committee, rejecting the notion of either a department of education or an assistant secretary. If, however, the federal government wished to gather its educational activities under a bureau that lacked both authority and aid, the hierarchy agreed to "remain neutral."[80]

When the NACE steering committee met in December 1930, Mann presented a draft of the first four chapters of what was intended to be the final report. He argued that existing federal subsidies, like those in the Smith-Hughes and Smith-Lever Acts, were "curative" in nature, that is, they had "been given to correct shortcomings in the established educational policy of the states." Such aid had a "separatist effect"; in other words, it had been granted at the behest of groups that favored a particular type of education and enjoyed the support of strong political organizations. Because these organizations were not interested in "general education," the nation's educational community lacked solidarity; it was fragmented. In Mann's view, federal subsidies, if given at all, should be "preventive," that is, they should be made "in the interest of general education, [so] that the states may be enabled to work out a fundamental educational policy calculated to produce in the pupils ability to adapt themselves to changing conditions without any continuing care on the part of the school system."[81]

The steering committee rejected the chapters outright because it thought Mann was attempting to define the social purposes of education, whereas the NACE's task was to determine the correct federal role in education and the best method of coordinating federal activities in that regard. The chapters also sparked a sharp debate over whether or not "separatism" was a good thing. The committee soon found itself in "a hopeless muddle." In order to find common ground, Mann called for a show of hands on various issues. In effect, the committee ended up reaffirming the principles agreed on in June. All believed that educational research was a proper function of the federal government, and all desired the repeal of the regulatory features and matching of funds contained in the Smith-Hughes and Smith-Lever Acts. All except O'Leary rejected the notion that the federal government should offer monetary aid for special educational purposes. The committee voted fifteen to three in favor of a general grant to education, like the one proposed in June. All agreed that the funds should not be offered on a matching basis and that the audit be restricted to determining only if the money was used for education. Because the vote was fifteen to three on the issue of a general grant, the committee decided to defer a final ballot on that issue pending further study. With respect to the federal headquarters, plans were proposed for one of the following: the establishment of a department of education, the appointment of an assistant secretary for education in the Department of the Interior, or the creation of a National Academy of Education. Johnson proposed a fourth: the enlargement of the Office of Education, with the possibility of making it a freestanding bureau independent of any department. Though most present favored a separate department of education, the committee agreed to report all four alternatives with the number of members supporting each.[82]

In the opening months of 1931 the steering committee itself examined the question of federal aid. An exhaustive study of the interrelated problems of the concentration of capital and the multiple systems of support for education reached the conclusion "that any attempt to equalize wealth or to make taxes more equitable by means of any kind of adjusted federal appropriation for education is inadequate. . . . The problem of federal versus state revenues and the just distribution thereof, are basic, economic questions—a problem for experts on taxation and for statesmanship. It is not the primary task of educators to attack or to solve directly the confused and baffling questions of taxation." In the view of this report, the only thing certain was that federal grants for such special matters as agricultural education, vocational education, and the like contributed to, rather than eased, the problems of imbalance and inequity.[83]

Baffled by the issues of taxation and the concentration of wealth, the steering committee sought the advice of experts. It sent a questionnaire to fifty-nine university economists, but only one-third responded. Denying that the concentration of capital in financial and industrial centers had resulted in an "*absolute* decrease in real wealth" in the "so-called frontier communities," the majority of respondents contended that rural areas had only "suffered a loss of *relative* ability to support education." Although most saw federal aid as an equitable means of redistributing wealth, opinion was almost evenly divided on the question of whether government grants would weaken local interest in education. In the view of the researcher who conducted the inquiry, the study was inconclusive: "On the whole, the answers. . . were products of acknowledged conjecture or theoretical deduction upon the part of economists. It is evident that, so far as the present meager materials indicate, there exist no generally accepted criteria or definitive body of factual material upon which to base conclusions concerning these problems from the standpoint of contemporary economics."[84] In short, the field of economics was not an exact science, and the economists consulted were essentially making educated guesses.

By June 1931 the steering committee had before it a draft of the final report. Written by Suzzallo and critiqued by Judd, Capen, and Zook, it was divided into two parts, the first dealing with the federal government's relation to education and the second with the federal headquarters. Part I reaffirmed that education was primarily a state and local responsibility. Though the federal government had an interest in schooling, it was "one of cooperation in fostering the education of the people under state jurisdiction." Recounting the history of federal aid, the report warned that the control exercised by government agencies under the Smith-Lever and Smith-Hughes Acts could no longer be ignored. It recommended that special programs, like grants for vocational education, be continued only for five more years and that no new ones be instituted. Moreover, the restrictions or regulations attached to existing programs ought to be revoked. In the future, Congress should grant aid on a per capita basis for schooling in general, allowing states the freedom to determine the educational purposes for which the money would be spent. The sole requirement attached to such grants should be an audit that showed that the money had been used for education. While upholding local control of schooling, the report sought to ensure that no single federal agency achieved domination of the national government's interest in education. Consequently, Part I proposed that most agencies remain scattered in their present locations in the executive branch because "a condition of dispersal among Departments most competent to exercise these functions is the part of wisdom."[85]

Instead of proposing all four alternatives for a federal headquarters as the steering committee had agreed to do, Part II of the report came out squarely in favor of a department of education because only a cabinet-level secretary would have the prestige and ability to coordinate the agencies dispersed throughout the government. Although both the Office of Education and the Board of Vocational Education were to be relocated in the proposed secretariat, "no regulatory or executive responsibilities should be vested in the Department of Education through these transfers." All other educational offices were to remain where they were, their efforts being coordinated by an interdepartmental council on education, chaired by the new secretary. The purpose of the proposed department would be to conduct research and disseminate information.[86]

In a separate vote on each part, the steering committee approved the first by a margin of fourteen to one, and the second by eleven to four. O'Leary was the only member to oppose federal aid, while Cody, Johnson, and Zook joined him in objecting to the recommendation of a department of education. The committee then unanimously agreed to submit the report to the NACE when it met in October 1931.[87]

In essence, the steering committee was recommending a sanitized version of the Smith-Towner bill. It proposed the establishment of a department of education devoted to research and devoid of any regulatory functions; it proposed regulation-free aid for education in general and an end to aid for all special purposes. Thus, the federal education question had come full circle—with a twist. The committee recommended the type of department contained in the Curtis-Reed bill, the sort envisioned by Capen and Judd. It recommended the type of aid favored by Capen, namely, federal grants with no strings attached and given without the necessity of a state's matching them. It is interesting to note that these recommendations embodied the views of the two men who had been frozen out of the framing of the original NEA measure in 1918.

Within a week of the steering committee's vote, Johnson conferred with McNicholas's Committee on Education. Much to his surprise, the bishops on the committee ordered him and Pace to cast negative ballots on every issue raised by the NACE. The deeply conservative Howard went so far as to protest against Johnson's membership on the NACE as connivance with a heretical government that opposed the principles of Catholic education. Johnson confided to Burke that the prelates seemed to be acting out of some fear. This information worried Burke because if the two Catholic representatives on the NACE took an intransigently negative stand, there would be repercussions on all church relations with the federal government. So upsetting was the situation that Burke went to speak about it with the apostolic delegate, Archbishop Pietro Fumasoni-Biondi. The latter concluded that Burke was duty-bound to express his views to McNicholas, which he did.[88]

While awaiting reply, Burke received a letter from Johnson's ordinary, Karl Alter, the new bishop of Toledo, Ohio. Johnson had sent him a copy of the NACE report and apprised him of Burke's view of it. As a new bishop, Alter had no way of knowing that the hierarchy, at its last annual convention, had gone on record as opposed to a department of education. Thus, the report impressed him favorably. He believed that as American citizens, Catholics had an obligation "to lend our efforts to securing such a form of Federal organization in the field of education as will truly advance the interests of our country." He asserted that if Congress were to entertain a bill that proposed a department of education like the one in the report without legal, financial, regulatory, and executive authority, the church could back it.[89] Agreeing that Catholics had a responsibility as citizens to promote some form of federal agency for education, Burke urged taking a "positive stand" against a department. By this he meant that the church should state that if it were possible to have a department like the one outlined in the report, it would be a good thing. But it would "be practically impossible to keep a Federal Department within those limitations, and . . . inevitably a Federal Department, once established, would fall into the very faults and evils which the report of the National Advisory Committee itself condemns." Burke favored expansion of the Office of Education with its own separate budget. He encouraged Alter to help McNicholas and his committee to a more positive expression of its opposition, such as the one just outlined.[90] Continuing to see no difficulty "in the principle of accepting a Federal Department of Education which would have the safeguards written into the law creating it," Alter confessed that "it would be far better strategy and perhaps safer" to favor an expansion of the Office of Education.[91] His underlying conviction placed him in the camp of a progressive bishop like the late John Carroll of Helena, Montana, who in 1925 had argued that Catholic bishops had a moral obligation to promote the welfare of people within and outside the Catholic fold.

Whether at Alter's urging or Burke's protestation—or both—McNicholas saw the wisdom of allowing Johnson and Pace to follow Burke's advice. When the NACE met in October 1931 to ballot on the report, it approved Part I, which recommended a per-capita, regulation-free grant for education by a vote of forty-five to six, with both Johnson and Pace casting affirmative ballots. Because neither man believed Congress would ever grant a subsidy without strings attached, both felt safe in approving the proposal. Thus, they took a positive stance toward the federal education question, regardless of their personal reservations. Opposition to such aid was limited almost exclusively to representatives of the vocational education and the black education communities, two special-interest groups that stood to lose from the proposal because it revoked aid for special purposes. For Johnson and Pace, the proposal for a department of education was a different matter. The hierarchy had expressed its opposition to the creation of such a department at its last meeting. Yet, if established, a department would undoubtedly administer the proposed federal aid and become the instrument of government control. Consequently, the two priests cast dissenting votes. The vote on Part II was thirty-eight in favor, eleven opposed (including Cody and Zook), with two abstentions. In a final vote on whether to adopt or reject the report as a whole, the tally was forty-three to eight, with both Johnson and Pace joining the representatives of vocational education and the black educational community in casting negative ballots.[92] Unlike the situation with the Smith-Towner bill in which the educational trust divided over the issue, this time, on a sanitized version of that measure, it was united, save for the representatives of special interests, including Catholic education. Unfortunately, this unity occurred at a moment when the administration, Congress, and the public at large had lost interest. As will be seen, the Atlanta *Constitution* had been correct. The discussion of federal participation in education had been academic.

Remarkably absent from the NACE's discussion had been any direct reference to the Great Depression, already well on the way to its darkest days when the committee had concluded its deliberations. To a degree, this absence is understandable because the NACE's purpose was to study the proper role of the federal government in education and recommend how best to coordinate its multiple educational agencies. Although one might have expected the NACE to have addressed the Depression in its consideration of federal aid, its deliberations were framed in terms of the Brand-Nye bill, that is, the NACE saw federal aid as a means of redressing the concentration of wealth in major urban centers. Given the depressed nature of agriculture throughout the decade—hence the depressed nature of rural America—the committee focused on redressing the woes that had beset the agricultural portion of the nation for the past ten years, while minimizing the crisis that was beginning to cripple urban America as well.

With regard to federal participation in education, Johnson thought that most members of the NACE operated on the assumption that the establishment of a department of education was the best way to avoid federal control. They believed that a cabinet secretary would be able to monitor federal educational legislation and intervene to scrap it if it tended toward federal autocracy. In short, the best way to avoid federal control was to appoint a secretary to guard against it. Johnson found this line of thinking "quite gratuitous" because the office was a political one, and the incumbent "would in all likelihood respond to the pressure and become amenable to the wishes of those groups who heretofore have been active in the cause of a greater measure of federal domination in education and who will not experience any particular change of heart because of the issuance of this report."[93] The groups he had in mind, no doubt, were the Southern Jurisdiction of the Scottish Rite, the Ku Klux Klan, and the NEA.

In mid-October Johnson and Pace filed a joint minority report arguing the underlying inconsistency in the NACE's recommendations. While upholding education as a local concern and rejecting federal control as undesirable, the NACE proposed the establishment of a department of education. By nature, a secretariat was an administrative agency; as such, the proposed department would eventually bring about the very centralization and control feared by the NACE, despite the limitations that Congress might write into the act of establishment. An inevitable lever of control would be any federal grant to education, which would not be regulation-free, but would certainly come with strings attached to be administered by the secretary of education. Besides, Johnson and Pace argued that as a political appointee, the secretary would be subject to party and lobby pressure. By the same token, his selection of research projects would be guided by the philosophy of the party in power and might result in propaganda. In the view of the two priests, Congress would do better to expand the Office of Education and let it carry out the functions of the proposed department.[94]

Publication of the NACE report evoked a unanimous response from Catholic editors. All followed Johnson and Pace in pointing out the inconsistency in the report. A few even suggested that the committee may have been aiming at federal control all along. For example, Vincent de Paul Fitzpatrick of the *Baltimore Catholic Review* declared that the committee members either failed to comprehend the implication of their approval of a department or lacked the courage to express their true design regarding education.[95] Without giving the NACE the benefit of the doubt, the editor of the *Indiana Catholic and Record* stated bluntly, "The point we would like to make is that what is on foot is merely a movement to federalize education, thus making private education impossible."[96] Probably the wisest comment appeared in the Omaha *True Voice.* Convinced that politicians and the public had wearied of bureaucracy, the editor quoted the *Washington Post*'s opinion: "Instead of starting a new fight for representation of education in the President's Cabinet, the report probably will be placed in a safe pigeon-hole where it can gather dust until even the committee members have forgotten about it. . . . It appears that the public does not even take the report seriously."[97] As it turned out, this was an accurate assessment, but the fears of Catholic officials were not to be so easily allayed.

Catholic papers were not alone in opposing the recommendations of the NACE. In addition to the *Washington Post,* at least sixty other secular journals in all parts of the nation came out against the report, including the *New York Times,* the *New York Herald-Tribune,* the *Philadelphia Inquirer,* the Baltimore *Sun,* the *Detroit Free Press,* the *Chicago Daily News,* the Saint Louis *Post-Dispatch,* the Kansas City *Times,* the Atlanta *Journal,* the Milwaukee *Journal,* and the Saint Paul *Pioneer Press.* Like Catholics, more than half of the editors pointed out that the recommendations in the report were inconsistent. The editor of the *Miami News* voiced this criticism most succinctly and colorfully:

"The committee might as well have favored swallowing boiling water while opposing getting burned." Without referring to the inconsistency, the other half rejected the report because its recommendations would lead to federal control of education. Resurrecting an earlier tactic, the NCWC gathered the negative opinions of secular papers and published them in pamphlet form.[98]

On 10 November 1931, the day before the annual convention of the American hierarchy in Washington, D.C., McNicholas informed Burke that several bishops thought it would be a good idea to find out where Hoover stood regarding the report. Without wishing to make a public protest against the establishment of a department of education, the hierarchy might be compelled to do so unless he was opposed to the plan. Rather than approach the president directly, Burke spoke with Mr. Burlew, Wilbur's administrative secretary. Burke reminded Burlew that the Catholic church had consistently opposed the establishment of a department because it would "trammel free thought among our people in its educational channels." The priest advised Burlew that if Hoover were to favor the NACE report, the bishops might have to issue a public protest, something they hoped to avoid unless absolutely necessary because such an action might be given political significance by the public. If assurance could be given that there was no immediate plan to implement the report, the air might be cleared. With the understanding that any word would be confidential, Burlew promised to telephone Wilbur, who was in Texas. That evening, Burlew informed Burke that although Wilbur could not speak for Hoover, the matter would definitely be settled without the creation of a department of education. Wilbur hoped that Burke would be able to forestall precipitate action by the hierarchy; a public protest would only make Wilbur's own opposition to the report more difficult.[99]

Burke took the appropriate steps, immediately going to the Shoreham Hotel to inform McNicholas of the news. The archbishop seemed satisfied and thought that the minority report filed by Johnson and Pace sufficiently expressed the Catholic position. If difficulties arose during the hierarchy's convention, he would telephone Burke for further assistance. The priest then visited Cardinal Patrick Hayes of New York to confer about the situation. In the cardinal's view, Wilbur's assurance was sufficient to prevent a protest.[100] And so it was. When McNicholas made his report to the hierarchy the next morning, the bishops accepted Wilbur's word. No protest was forthcoming and no further action was contemplated.[101]

For Catholics, the best news came several days later. Archbishop Edward Hanna, chairman of the NCWC Administrative Committee, called on Hoover about Welfare Conference business. During the conversation, Hoover indicated that he would not follow the NACE's recommendation. In his view, education and public welfare both properly belonged in the Department of the Interior, and he intended to reorganize government agencies accordingly.[102] In effect, the *Washington Post* had been correct: the NACE report was a dead letter to be pigeonholed and forgotten by all but history. This was the final chapter in the department of education battle, which ended with a whimper rather than a bang. The government ignored the NACE's report and the educational trust let the matter drop.

Even though the department of education battle was virtually over, shell-shocked Catholics still reacted nervously to the fading sound of arms. The Seventy-second Congress, which opened in December 1931, was a house divided: Republicans controlled the upper chamber, while Democrats had captured the lower. Most important, both Education Committees were chaired by long-standing opponents of a department of education: Republican Metcalf at the helm of the Senate Committee on Education and Labor, and Catholic Democrat Douglass with the gavel of the House counterpart. In what turned out to be a parting shot, Reed introduced a bill to create a department of education, the last such measure for many years. A mixture of the Capper-Robsion and Curtis-Reed legislation, it followed more or less the recommendations outlined in the NACE report.[103] No one sponsored a companion measure in the Senate. Still, the NCWC kept a wary watch. In early February 1932 William Montavon advised Burke that the NEA Department of Superintendence was to meet in the national capital at the end of the month. "In past years the holding of this convention in the City of Washington has been invariably accompanied by an intensified campaign for the enactment of legislation establishing a Federal Department of Education," warned Montavon. "I feel that steps should be taken at once to be prepared for any emergency that may arise."[104] The superintendents met and departed, and still no companion measure was introduced and no hearings were held, despite the activity of proponents.[105] Congress had lost interest in a department of education.

Reorganization was the watchword, and, as Burke cautioned the NCWC Administrative Committee, "in the various measures proposed . . . a sub-Secretary of Education might be appointed and this would be a definite step towards the creating of a Department of Education." His superiors ordered him to be vigilant and to oppose any move for an under secretary of education.[106] None was forthcoming.

Even after the drive for a department of education had collapsed, Catholic tensions were high. Prevented by Depression-depleted finances from sending an official representative to the 1932 Democratic convention in Chicago, Burke had to depend on an NCWC news service reporter, Burke Walsh, to look after the education question. Shortly after Walsh's arrival in the Windy City, fellow newsmen assured him that no one had appeared before the first session of the platform committee to plead for an endorsement of a department of education. The next day Walsh himself checked with committee members, including former Senator Gilbert Hitchcock, the chairman, Catholic Senator David Walsh of Massachusetts, A. Mitchell Palmer, and Michael Igoe, Catholic national committeeman from Illinois. None had heard anything about a possible plank for a department, and all "plainly indicated that any such proposals would hardly receive favorable consideration." According to Senator Walsh, moreover, the draft of the platform was virtually finished. Burke Walsh telephoned this information to his superior, Frank Hall, director of the NCWC Press Department. Hall told him "to keep this matter upper most[*sic*], and at the first word or even rumor of a hearing on the subject or plans to ask a hearing, to wire us."[107]

A conflicting report began a scramble that bordered on hysteria. Margaret Lynch of the NCCW had contacted Mrs. George McIntyre of Chicago, first vice president of that organization, to encourage vigilance for any attempt to insert a plank in favor of a department of education. Having supported Igoe's rival in the recent Democratic primary for governor, McIntyre did not wish to approach Igoe directly, so she asked one of his supporters to inquire if there was any truth to a report that an endorsement was being sought for such a department. On receiving affirmative word—something that was incorrect—McIntyre relayed it to John Burke. The latter telephoned Father James Callaghan of Saint Malachy's Church, Chicago, and urged him to see Igoe. Burke then contacted David Walsh, who assured him that the platform was already complete and contained no such plank. Fearing that the senator might be mistaken, Burke drafted a protest in Johnson's name to be presented to the platform committee by Burke Walsh if necessity demanded. Burke then spent hours trying to telephone Johnson for his approval of the statement. With Johnson's approbation, Hall then telephoned the protest to Burke Walsh. Throughout the night the latter was in contact with Callaghan, Igoe, and Senator Walsh. None knew anything about an attempt to insert an endorsement of a department of education in the platform.[108] In fact there was none. This incident serves to show the skittishness of NCWC officials after years of battle.

Shell shock was not the only legacy of the long combat. Intransigence was another. As mentioned above, the Administrative Committee had ordered Burke to fight every attempt to appoint an under secretary for education.

When the hierarchy met in November 1932, it reinforced this resolve. Disregarding the fear of some bishops that antagonism against the establishment of such a position might be misunderstood, the hierarchy commanded the NCWC secretariat "to oppose every advance toward a Department of Education," including the appointment of an under secretary.[109] The more remote a department became, the more unyielding and uncompromising became Catholics. Unknown to them, little was to be feared. The administrative progressives of the educational trust had shelved the idea of a department, not to be seriously entertained again until the 1960s. During the intervening period, their attention focused on securing federal aid for education.[110] The long struggle had come to an end.

Conclusions

The battle over the establishment of a department of education and a national system of public schools was emblematic of a cultural and religious conflict, closely related to Prohibition and immigration restriction. On one side were old-stock, Protestant Americans mainly in the South, Midwest, and West, who sought to reinforce white, Anglo-Saxon, Victorian culture in the face of a massive influx of foreigners, most of them Catholics from eastern and southern Europe, the majority of whom settled in eastern and midwestern cities. Many came for economic advantage, either to earn money to send home or to amass it for their own return, and they retained close ties to the old country. Some incorporated into the political structure of America through citizenship and participation in the machine politics of urban bosses. Others incorporated through ethnic societies and the Catholic church. Their retention of loyalty to their native land made them seem resistant to the melting pot. Most Americans considered them racially inferior, posing a threat to dilute native stock, the "Nordic" type. Old-stock Americans, moreover, considered the Catholic church itself a foreign, political institution intent on subverting the nation in the name of papal Rome. They viewed Catholics as intruders on an experiment in Protestant civilization.

On the other side were American Catholics themselves. In the mid-1890s the pope had urged them to hold themselves apart from Protestant society. During the early twentieth century they sought to stem the leakage of professionals from the church by reinforcing the bonds of faith. In order to preserve and promote their religious identity within the American Protestant milieu, they began to create a Catholic subculture with organizations paralleling those in the host society. The CEA (NCEA), for instance, was the Catholic counterpart of the NEA. Similarly, the NCWC was to serve as the counterweight to the Federal Council of Churches. Catholics not only wanted to maintain their identity, they also believed that they had something special to offer the host culture. They viewed religion—their religion—as the sure foundation on which morality, citizenship, and democracy rested. They conducted parochial schools to inculcate civic values in their children while preserving their religious and cultural heritage. They ran settlement houses to incorporate Catholic newcomers into national life. Through organizations like the CEA and the NCWC, Catholics sought to protect and promote their view. The NCWC and its ancillaries, the NCCM and NCCW, engaged the faithful in the social and political issues of the day, especially the education question. As a result, the more militant that Catholics became, the more Anglo-Saxon Americans sought to tame them.

The focal point of the cultural conflict was the school. Since the 1840s, old-stock Americans had viewed public schools as melting pots for the assimilation of immigrants, places where newcomers learned Americanism: the traditions of the nation and the values of democracy. With the influx of eastern and southern Europeans, the understanding of Americanization became transformed. Increasingly, to Americanize immigrants meant to Anglicize them. The tool of this enculturation process was the public school, not just for immigrants, but for the children of Catholic hyphenates because parochial schools instilled an intolerable dual allegiance between Rome and Washington. Public education would ensure that all children imbibed Anglo-Saxon, Protestant values and culture. Catholics, on the other hand, had early established their own system of education to preserve the faith of their children by providing an alternative to the public schools, which were pan-Protestant. For them, Americanization meant adherence to the ideals and political institutions of the nation, while leaving people free to retain religious beliefs and cherished ethnic ways. Thus, each side vested its school system with the preservation of its religious and cultural identity, hence the bitterness of the clash. For one, the issue was schooling for conformity; for the other, it was schooling for pluralism. Each side viewed its position as the truly American one.

America's entry into the First World War revealed significant weaknesses in the public school system, which the educational trust sought to remedy through the establishment of a department of education. Administrative progressives at Columbia University, working closely with the NEA, wanted to go further and establish a standardized, national system of education. Seeking to capitalize on the wartime emergency and patriotic fervor, the NEA introduced a bill for the creation of a department of education with an aid package of $100 million for the promotion of such a national system, complete with an Americanization component. Although George Strayer, architect of the measure, had declared that the country ought to "conscript" all children into public schools, the NEA never specifically advocated such a course.

There were groups that did. The Southern Jurisdiction of Scottish Rite Masonry and the resurgent Ku Klux Klan espoused both the NEA bill and compulsory public education. With the majority of their members hailing from the South, Midwest, and West, both organizations viewed with alarm the separateness of the Catholic church, especially its school system. The two were soon joined by the Sunday School Council of Evangelical Denominations, the International Sunday School Association, and the Council of Church Boards of Education, which considered public schools to be Protestant ones, at least in clientele. Like the Southern Scottish Rite and the Klan, the National Patriotic Council, an offshoot of the Evangelical Protestant Society, advocated compulsory public education.

Considering political Romanism to be tyranny and therefore antithetical to democracy, supporters of compulsory public education believed that parochial schools were the seedbeds of foreignism, nurturing children on principles antagonistic to the country. To safeguard national institutions, all children must be Americanized in public schools. Catholics, on the other hand, viewed their schools as the bulwarks of democracy and traditional Americanism. For them, the secularization rampant in public education represented the abandonment of the founding fathers' vision of the country as a Christian nation and the forsaking of their understanding that religion formed an integral part of schooling. America's first schools were religious schools. As Catholics saw it, the nationalist philosophy of education propounded by the NEA, the Southern Jurisdiction, the Klan, and other likeminded groups made the child a creature of the state, a theory that attacked both parental rights and the freedom of education. Considering this philosophy to be un-American and out of harmony with America's educational past, Catholics viewed themselves as defenders of the nation's educational tradition.

Though a handful of Catholic progressives gave on-and-off support to some sort of federal participation in education, the majority of Catholics viewed the establishment of a department of education, especially one with federal aid at its disposal, as a peril to their schools. Federal money was the inducement to accept federal standards. Because organizations that supported compulsory education also endorsed the NEA bill, Catholics surmised that the real intent of that measure was the abolition of parochial education, despite the NEA's disclaimers.

The church's protection of its educational interests took institutional form in the NCWC, a Catholic lobby complete with national organizations of men and women. Originally intended as a Catholic counterpart of the Federal Council of Churches, the NCWC had been established in large measure in response to the NEA's legislation; consequently, for more than a decade the NCWC acted as the counterbalance, not of the Federal Council, but of the NEA. The activism of the NCWC in thwarting the creation of a department of education was taken as proof by some educators, the Ku Klux Klan, the Southern Jurisdiction of the Scottish Rite, and many rural Americans that the church was in politics and intent on subverting the nation.

During the first seven years of the conflict, Catholics benefited by a division within the educational trust itself, a significant portion of which opposed federal aid for one reason or another. To unite the trust and win the support of opponents, especially Catholics, the NEA dropped the subsidy from its bill, a ploy that failed to garner Catholic backing. Catholics rightly believed that the association was simply seeking to implement its program piecemeal. However, the church's continued antagonism to the establishment of a now apparently innocuous department of education—one shorn of aid—made it seem all the more clear to proponents that Catholics were not merely being protective of their schools, but were actually opposed to public education.

It would be too much to claim that Catholics were responsible for the victory of home rule in education in the period under study. A postwar wave of anti-statism acted as a brake on further federalization, especially with the return to power of the Republican Old Guard. Given the latter's pro-business policies, the NEA's legislation faced a serious obstacle because taxes from eastern and midwestern urban industrial centers were to be used to pay for the education of rural children. After the Sixty-sixth Congress, bipartisan opposition from the urban wing of each party kept the bill in committee, despite bipartisan support from farm-bloc progressives. As the decade wore on, it became more and more apparent that Catholics were, at least on the federal education question, in the political mainstream. Perhaps it was for this reason that the administration of Herbert Hoover requested Catholic participation on the NACE. Certainly, the church's representatives on that committee were the most sympathetic to the administration's objectives. Though their views were disregarded by the educational trust, which was to be expected, their position was actually the one followed, not because they had convinced the administration of its worth, but because the administration was antecedently committed to it.

As the goal of establishing a department of education, to say nothing of achieving a national school system, receded further from the grasp of the educational trust, the federal education question became increasingly religiously charged. That Catholics found themselves in sympathy with current antistatism was both a blessing and a curse: a blessing because it sometimes permitted them to hide beneath the coattails of administrations opposed to increased federal participation in education; a curse because they, more than the administrations, bore the brunt of the proponents' disappointment over their inability to advance their cause. The cultural conflict surrounding the department of education battle left a legacy of anti-Catholicism in the countryside that lingered for decades.

Appendix

The Percentage of Difference between the Percentage of the Smith-Towner Subsidy Received by a State and the Percentage of the Subsidy Contributed by a State

	Sum Allotted According to Terms of Smith-Towner	*Percent of Total Smith-Towner Subsidy*	*Contribution to Annual Revenue of Federal Government*	*Difference between Contribution and Portion Received*
New York	$ 9,172,838	9.173%	24.49%	−62.545%
Michigan	$ 3,166,211	3.166%	5.93%	−46.607%
Rhode Island	$ 514,081	0.514%	0.92%	−44.122%
Massachusetts	$ 3,163,927	3.164%	5.66%	−44.100%
Illinois	$ 5,469,021	5.469%	8.46%	−35.354%
Pennsylvania	$ 6,898,424	6.898%	10.64%	−35.165%
California	$ 2,665,256	2.665%	3.97%	−32.865%
Maryland	$ 1,103,178	1.103%	1.57%	−29.734%
Connecticut	$ 1,102,956	1.103%	1.56%	−29.298%
Delaware	$ 183,947	0.184%	0.26%	−29.251%
Ohio	$ 4,799,012	4.799%	6.22%	−22.845%

(Continued)

	Sum Allotted According to Terms of Smith-Towner	*Percent of Total Smith-Towner Subsidy*	*Contribution to Annual Revenue of Federal Government*	*Difference between Contribution and Portion Received*
New Jersey	$ 2,668,489	2.668%	3.12%	−14.472%
North Carolina	$ 2,574,185	2.574%	2.72%	−5.361%
Kentucky	$ 1,119,887	1.120%	1.10%	1.808%
Missouri	$ 3,124,690	3.125%	2.75%	13.625%
Colorado	$ 948,719	0.949%	0.74%	28.205%
Oregon	$ 856,407	0.856%	0.61%	40.395%
Minnesota	$ 2,619,890	2.620%	1.69%	55.023%
Wisconsin	$ 2,512,381	2.512%	1.62%	55.085%
West Virginia	$ 1,496,583	1.497%	0.91%	64.460%
Indiana	$ 2,818,931	2.819%	1.70%	65.819%
Virginia	$ 2,289,451	2.289%	1.35%	69.589%
Washington	$ 1,363,504	1.364%	0.79%	72.595%
Utah	$ 453,565	0.454%	0.23%	97.202%
Louisiana	$ 1,859,473	1.859%	0.87%	113.733%
New Hampshire	$ 474,293	0.474%	0.22%	115.588%
Kansas	$ 2,034,957	2.035%	0.84%	142.257%
Maine	$ 967,683	0.968%	0.39%	148.124%
Texas	$ 4,261,343	4.261%	1.70%	150.667%
Florida	$ 932,070	0.932%	0.36%	158.908%
South Carolina	$ 1,751,922	1.752%	0.62%	182.568%
Oklahoma	$ 1,811,123	1.811%	0.60%	201.854%
Tennessee	$ 2,268,829	2.269%	0.75%	202.511%
Vermont	$ 426,088	0.426%	0.14%	204.349%
Nebraska	$ 1,681,329	1.681%	0.52%	223.333%
Nevada	$ 97,630	0.098%	0.03%	225.433%
Wyoming	$ 229,596	0.230%	0.07%	227.994%
Arizona	$ 307,767	0.308%	0.09%	241.963%
Georgia	$ 2,918,115	2.918%	0.81%	260.261%
Iowa	$ 3,479,468	3.479%	0.82%	324.325%
Idaho	$ 461,522	0.462%	0.10%	361.522%
Alabama	$ 2,325,294	2.325%	0.40%	481.324%
New Mexico	$ 347,356	0.347%	0.04%	768.390%

(*Continued*)

	Sum Allotted According to Terms of Smith-Towner	*Percent of Total Smith-Towner Subsidy*	*Contribution to Annual Revenue of Federal Government*	*Difference between Contribution and Portion Received*
Arkansas	$ 2,132,005	2.132%	0.23%	826.959%
Mississippi	$ 2,111,559	2.112%	0.20%	955.780%
South Dakota	$ 1,424,892	1.425%	0.11%	1195.356%
North Dakota	$ 1,055,218	1.055%	0.07%	1407.454%
Montana	$ 3,629,461	3.629%	0.12%	2924.551%

Notes

Preface

1. David B. Tyack and Elisabeth Hansot, *Managers of Virtue: Public School Leadership in America, 1820–1980* (New York: Basic Books, 1982), 28–39; David B. Tyack, *The One Best System: A History of American Urban Education* (Cambridge, Mass.: Harvard University Press, 1974), 15–21; Robert L. Church and Michael Sedlak, *Education in the United States: An Interpretive History* (New York: Free Press, 1976), 3–21; Lloyd P. Jorgenson, *The State and the Non-Public School, 1825–1925* (Columbia: University of Missouri Press, 1987), 1–19.

2. Tyack and Hansot, *Managers of Virtue*, 28–72; Jorgenson, *State and Non-Public School*, 20–54; Church and Sedlak, *Education in the United States*, 55–81; Ellwood Cubberley, *Public Education in the United States: A Study and Interpretation of American Educational History* (Boston: Houghton Mifflin, 1934), 165–66, 231–34. As Timothy Smith notes, the transition to nonsectarian schools had been accomplished in New York City long before any of the above-named reformers appeared on the scene ("Protestant Schooling and American Nationality," *Journal of American History* 53 [March 1967]: 681–88).

3. Ray Allen Billington, *The Protestant Crusade, 1800–1860: A Study of the Origins of American Nativism* (New York: Rinehart and Company, 1952), 53–76; Gustavus Myers, *History of Bigotry in the United States*, ed. and rev. by Henry M. Christman (New York: Capricorn Books, 1960), 84–91; Jorgenson, *State and Non-Public School*, 69–110; David J. O'Brien, *Public Catholicism* (New York: Macmillan, 1988), 42–44; Michael Feldberg, *The Turbulent Era: Riot and Disorder in Jacksonian America* (New York and Oxford: Oxford University Press, 1980), 34–37.

4. For the development of Catholic education during this period, see Harold A. Buetow, *Of Singular Benefit: The Story of Catholic Education in the United States*

(London: Macmillan, 1970); Jorgenson, *State and Non-Public School,* 75–76, 98–99, 101.

5. Vincent P. Lannie, "Alienation in America: The Immigrant Catholic and Public Education in Pre–Civil War America," *Review of Politics* 32 (October 1970): 507–11; Vincent P. Lannie, *Public Money and Parochial Education: Bishop Hughes, Governor Seward, and the New York School Controversy* (Cleveland: Case Western University, 1968), 21–244; Joseph J. McCadden, "Bishop Hughes Versus the Public School Society of New York," *Catholic Historical Review* 50 (July 1964): 188–207; Jay P. Dolan, *The Immigrant Church: New York's Irish and German Catholics, 1815–1865* (Notre Dame: University of Notre Dame Press, 1983), 100–104; O'Brien, *Public Catholicism,* 44–46; Vincent P. Lannie and Bernard Diethorn, "For the Honor and Glory of God: The Philadelphia Bible Riots of 1840 [*sic* for 1844]," *History of Education Quarterly* 8 (Spring 1968): 44–95; Billington, *Protestant Crusade,* 220–34. Michael Feldberg masterfully explores the socioeconomic background of the Philadelphia trouble and gives a blow-by-blow account of the violence in *The Philadelphia Riots of 1844: A Study of Ethnic Conflict* (Westport, Conn.: Greenwood Press, 1975).

6. Jorgenson, *State and Non-Public School,* 69–110; Lannie and Diethorn, "For the Honor and Glory of God," 62–63; Billington, *Protestant Crusade,* 380–430; Tyler Anbinder, *Nativism and Slavery: The Northern Know-Nothings & the Politics of the 1850s* (New York and Oxford: Oxford University Press, 1992), 10–51.

7. *Concilium Plenarium Totius Americae Septentrionalis Foederatae, Baltimore Habitum Anno 1852* (Baltimore: John Murphy and Sons, 1853), 47; Peter Guilday, *The National Pastorals of the American Hierarchy* (Washington, D.C.: NCWC, 1923), 191; *Concilii Plenarii Baltimorensis II, In Ecclesia Metropolitana Baltimorensi, a die VII ad diem XXI Octobris, A.D., MDCCCLXVI, Habita, et a Sede Apostolica Recognita, Acta et Decreta* (Baltimore: John Murphy and Sons, 1868), 222; Bernard Meiring, *Educational Aspects of the Legislation of the Councils of Baltimore, 1829–1884* (New York: Arno Press, 1978), 183–85. Dolan, *Immigrant Church,* 103–7; Buetow, *Of Singular Benefit,* 117–46.

8. *Sadlier's Catholic Directory, Almanac, and Ordo for the Years of Our Lord 1868 and 1876* (New York: D. and J. Sadlier, 1868 and 1876). The figures for parishes were adjusted by the author to reflect only those with a resident pastor.

9. *Acta et Decreta Concilii Plenarii Baltimorensis Tertii, A.D. MDCCCLXXXIV* (Baltimore: John Murphy et Sociorum, 1894), 100–101; Francis P. Cassidy, "Catholic Education in the Third Plenary Council of Baltimore," *Catholic Historical Review* 34 (October 1948): 292–97; Meiring, *Educational Legislation,* 231.

10. *The Official Catholic Directory* (New York: Kennedy and Sons, 1890–1920).

11. Thomas Wangler, "Emergence of John J. Keane as a Liberal Catholic and Americanist (1878–1887)," *American Ecclesiastical Review* 166 (September 1972): 457–78; "John Ireland and the Origins of Liberal Catholicism in the United States," *Catholic Historical Review* 56 (January 1971): 617–29; "John Ireland's Emergence as a Liberal Catholic and Americanist: 1875–1887," *Records of the American Catholic Historical Society of Philadelphia* 81 (June 1970): 67–82; "The Birth of Americansim: 'Westward the Apocalyptic Candlestick,' " *Harvard Theological Review* 65 (July 1972): 415–36; James H. Moynihan, *The Life of Archbishop John Ireland* (New York: Harper and Brothers, 1953), 49; John T. Farrell, "Archbishop Ireland and Manifest Destiny," *Catholic Historical Review* 33 (October 1947): 295–301; Patrick Ahern, *The Life of John Keane,*

1839–1918 (Milwaukee: Bruce Publishing Company, 1955), 94; Gerald P. Fogarty, S.J., *The Vatican and the Americanist Crisis: Denis J. O'Connell, American Agent in Rome, 1885–1903* (Rome: Universita Gregoriana, 1974), 266–67; Margaret Mary Reher, "The Church and the Kingdom of God in America: The Ecclesiology of the Americanists" (Ph.D. diss., Fordham University, 1972), 48–91, 123–24, 126–29.

12. Daniel F. Reilly, *The School Controversy, 1891–1893* (Washington, D.C.: The Catholic University of America Press, 1943), 75–86, 111–13; Gerald P. Fogarty, S.J., *The Vatican and the American Hierarchy from 1870 to 1965* (Wilmington, Del.: Michael Glazer, 1985), 65–80; Buetow, *Of Singular Benefit*, 170–74; Margaret Mary Reher, *Catholic Intellectual Life in America: A Historical Study of Persons and Movements* (New York: Macmillan, 1989), 69–71; R. Emmett Curran, *Michael Augustine Corrigan and the Shaping of Conservative Catholicism in America, 1878–1902* (New York: Arno Press, 1978), 339–83; La Vern J. Rippley, "Archbishop Ireland and the School Language Controversy," *U.S. Catholic Historian* 1 (Fall 1988): 1–16; Coleman J. Barry, O.S.B., *The Catholic Church and German Americans* (Milwaukee: Bruce Publishing Company, 1953), 184–85; Marvin R. O'Connell, *John Ireland and the American Catholic Church* (St. Paul: Minnesota Historical Society Press, 1988), 289–90; Jorgenson, *State and Non-Public School*, 187–201; C. Joseph Nuesse, "Thomas Joseph Bouquillon (1840–1902), Moral Theologian and Precursor of the Social Sciences in The Catholic University of America," *Catholic Historical Review* 72 (October 1986): 603–6; John Tracy Ellis, *The Life and Times of James Cardinal Gibbons, Archbishop of Baltimore, 1834–1921*, 2 vols. (Milwaukee: Bruce Publishing Company, 1952), 1:667–707; Joseph H. Lackner, S.M., "Bishop Ignatius Horstmann and the School Controversy of the 1890s," *Catholic Historical Review* 75 (January 1989): 73–90; Frederick Zwierlein, *The Life and Letters of Bishop McQuaid*, 3 vols. (Rochester, N.Y.: Art Print Shop, 1925–1927), 3:160–98.

13. Leo XIII, "*Longinqua Oceani*," in John Tracy Ellis, ed., *Documents of American Catholic History* (Milwaukee: Bruce Publishing Company, 1962), 504; Fogarty, *Vatican and the American Hierarchy*, 263–71, and *Vatican and the Americanist Crisis*, 153–80; Reher, *Catholic Intellectual Life in America*, 74–86, and " Kingdom of God in America," 81–4; Thomas McAvoy, *The Great Crisis in American Catholic History* (Chicago: H. Regnery Co., 1957), 189–280; Neil T. Storch, "John Ireland's Americanism *After* 1899: The Argument from History," *Church History* 51 (December 1982): 434–44; Alfred J. Ede, *The Lay Crusade for a Christian America: A Study of the American Federation of Catholic Societies, 1900–1919* (New York: Garland, 1988), 160–71.

14. William J. Kerby, "Reinforcement of the Bonds of Faith," *Catholic World* 84 (January and February 1907): 508–22, 591–606; Elizabeth McKeown, "Catholic Identity in America," in Thomas McFadden, ed., *America in Theological Perspective* (New York: Seabury Press, 1976), 56–57; O'Brien, *Public Catholicism*, 124–28; John J. Burke to Kerby, 7 November 1906, ACUA, Kerby Papers; Douglas J. Slawson, "John J. Burke, C.S.P.: The Vision and Character of a Public Churchman," *Journal of Paulist Studies* 4 (1995–1996): 50–51.

15. William M. Halsey, *The Survival of American Innocence: Catholicism in an Era of Disillusionment, 1920–1940* (Notre Dame, Ind.: University of Notre Dame Press, 1980), 26–28, 39 (the quotation is on 39); Louis J. Rogier et al., eds., *The Christian Centuries*, 5 vols. (London: Darton, Longman, and Todd, 1964–1978), vol. 5, Roger Aubert et al., *The Church in a Secularized Society*, 186–97; Hubert Jedin, ed., *History of the Church*,

10 vols. (New York: Seabury Press, 1980–1981), vol. 9, Roger Aubert, *The Church in the Industrial Age*, 307–18, 384–93, 431–59; Bernard M. G. Reardon, *Roman Catholic Modernism* (Stanford: Stanford University Press, 1970), 9–63; Gabriel Daly, O.S.A., *Immanence and Transcendence: A Study of Catholic Modernism and Integralism* (Oxford: Oxford University Press, 1980); McKeown, "Catholic Identity in America," 64–65; R. Scott Appleby, *"Church and Age Unite!": The Modernist Impulse in American Catholicism* (Notre Dame, Ind.: University of Notre Dame Press, 1992); Reher, "Kingdom of God in America," 239–97; Michael V. Gannon, "Before and After Modernism: The Intellectual Isolation of the American Priest," in John Tracy Ellis, ed., *The Catholic Priest in the United States: Historical Investigations* (Collegeville, Minn.: Saint John's University Press, 1971), 293–383; Henry F. May, *The End of American Innocence: A Study of the First Years of Our Own Time, 1912–1917* (New York: Alfred A. Knopf, 1959), 3–51; Daniel Howe Walker, "American Victorianism as a Culture," *American Quarterly* 27 (December 1975): 507–32; William Leuchtenburg, *The Perils of Prosperity, 1914–1932* (Chicago: University of Chicago Press, 1965), chapters 8 and 9; Richard M. Abrams, *The Burdens of Progress, 1900–1929* (Glenview, Ill.: Scott, Foresman, 1978), chapters 1 and 2; Frederick Lewis Allen, *Only Yesterday: An Informal History of the Nineteen-Twenties* (New York: Harper and Row, 1964), chapters 5 and 9; Gilman M. Ostrander, "The Revolution in Morals," in John Braeman, Robert H. Bremner, and David Brody, eds., *Change and Continuity in Twentieth-Century America: The 1920s* (Columbus: Ohio State University Press, 1968), 323–49; Norman Furniss, *The Fundamentalist Controversy, 1918–1931* (Hamden, Conn.: Archon Books, 1963); Lawrence W. Levine, *Defender of the Faith: William Jennings Bryan: The Last Decade, 1915–1925* (London: Oxford University Press, 1965); Lynn Dumenil, *The Modern Temper: American Culture and Society in the 1920s* (New York: Hill and Wang, 1995), 145–47.

16. Halsey, *Survival of American Innocence*, 37–83; Douglas J. Slawson, *The Foundation and First Decade of the National Catholic Welfare Council* (Washington, D.C.: The Catholic University of America Press, 1992), 10–23, hereafter cited as *NCWC*. See also Philip Gleason, "In Search of Catholic Unity: American Catholic Thought 1920–1960," *Catholic Historical Review* 65 (April 1979): 185–205.

17. Arthur S. Link, "What Happened to the Progressives in the 1920's?" *American Historical Review* 64 (July 1959): 833–51.

18. Christopher J. Kauffman, *Faith and Fraternalism: The History of the Knights of Columbus, 1882–1982* (New York: Harper & Row, Publishers, 1982), 170–71 (the so-called bogus oath of the Knights is quoted here); C. Vann Woodward, *Tom Watson: Agrarian Rebel* (New York: Oxford University Press, 1963), 416–30; John Higham, *Strangers in the Land: Patterns of American Nativism, 1860–1925* (New York: Atheneum, 1975), 178–86; Myers, *Bigotry in the United States*, 192–99, 211–13.

19. Maisie Ward, *Unfinished Business* (New York: Sheed and Ward, 1964), 81–95, 254; Frank Ward, *The Church and I* (New York: Sheed and Ward, 1974), 41–53; Stephen A. Leven, *Go Tell It in the Streets: An Autobiography of Bishop Stephen A. Leven* (Edmond, Okla.: M. Leven, 1984), 23–27; Patrick McKenna, C.M., "The Catholic Motor Missions in Missouri," *Vincentian Heritage* 7, no. 2 (1986): 97–104; Douglas J. Slawson, "Thirty Years of Street Preaching: Vincentian Motor Missions, 1934–1965," *Church History* 62 (March 1993): 63–64; Debra Campbell, " 'I Can't Imagine Our Lady on an Outdoor

Platform': Women in the Catholic Street Propaganda Movement," *U.S. Catholic Historian* 3 (Spring–Summer 1983): 103–14; Campbell, "A Catholic Salvation Army: David Goldstein, Pioneer Lay Evangelist," *Church History* 52 (September 1983): 322–32; Campbell, "David Goldstein and the Rise of the Catholic Campaigners for Christ," *Catholic Historical Review* 72 (January 1986): 33–50; Campbell, "Part-Time Female Evangelists of the Thirties and Forties: The Rosary College Catholic Evidence Guild," *U.S. Catholic Historian* 5 (Summer–Fall 1986): 371–83; Christopher J. Kauffman, *Mission to Rural America: The Story of W. Howard Bishop, Founder of Glenmary* (New York: Paulist Press, 1991); 150–53, 167–68, 170–72; Edgar Schmiedler, O.S.B., "Motor Missions," *Homiletic and Pastoral Review* 38 (March 1938): 585–88; Schmiedler, "Churches-on-Wheels," ibid. 40 (January 1940): 62, 366–72; Schmiedler, "Motor Missions, 1940," ibid. 41 (January 1941): 388–400; Schmiedler, "Trailing Trailer-Chapels," ibid. 42 (December 1941): 237–47; Schmiedler, "Little Motor Mission Rationing," ibid. 43 (March 1943): 502–16.

1. A National Program for Education

1. Alfred D. Chandler, Jr., "The Origins of Progressive Leadership," in Elting Morison et al., eds., *Letters of Theodore Roosevelt* (Cambridge, Mass.: Harvard University Press, 1954) 8:1462–64; George E. Mowry, "The California Progressive and His Rationale: A Study in Middle Class Politics," *Mississippi Valley Historical Review* 34 (September 1949): 239–50; J. Joseph Huthmacher, "Urban Liberalism in the Age of Reform," ibid. 44 (September 1962): 321–41; Richard Hofstadter, *The Age of Reform: From Bryan to F.D.R.* (New York: Vintage Books, 1955), 131–73; Arthur S. Link, *Woodrow Wilson and the Progressive Era, 1910–1917* (New York: Harper and Row, 1954), 1–24; George Mowry, *The Era of Theodore Roosevelt and the Birth of Modern America, 1900–1912* (New York: Harper and Row, 1958), 85–105; Otis Graham, Jr., *The Great Campaigns: Reform and War in America, 1900–1928* (Englewood Cliffs, N.J.: Prentice-Hall, 1971), 1–52; Arthur S. Link and Richard L. McCormick, *Progressivism* (Arlington Heights, Ill.: Harlan Davidson, 1983), 1–25; Samuel P. Hays, "The Politics of Reform in Municipal Government in the Progressive Era," *Pacific Northwest Quarterly* 55 (October 1964): 157–69; Herbert Croly, *The Promise of American Life* (New York: Macmillan, 1909), 22–23, 138–40, 148–54, 272–76; Theodore Roosevelt, *Progressive Principles: Selections from Addresses Made During the Presidential Campaign of 1912*, ed. by Elmer H. Youngman (New York: Progressive National Service, 1913), 141–52, 216–17; Woodrow Wilson, *The New Freedom: A Call for the Emancipation of Generous Energies*, ed. by William Bayard Hale (New York: Doubleday, Page and Company, 1913), 192–222.

2. Dumenil, *Modern Temper*, 16–21 (the quote is on 20), 203–7, 226–35; Martin E. Marty, *Pilgrims in Their Own Land: 500 Years of Religion in America* (New York: Penguin Books, 1988), 374–77.

3. Link, *Wilson and the Progressive Era*, 60–61; Higham, *Strangers in the Land*, 162–64, 186–93, 202–4, 234–50; Herbert Asbury, *The Great Illusion: An Informal History of Prohibition* (Garden City, N.Y.: Greenwood Press, 1950), 121–29; Edward Behr, *Prohibition: Thirteen Years that Changed America* (New York: Arcade Publications, 1996), 7–61; Sean Dennis Cashman, *Prohibition: The Lie of the Land* (New York: Free

Press, 1981), 6–15; John Kobler, *Ardent Spirits: The Rise and Fall of Prohibition* (London: Putnam, 1973), 198–206; Charles Merz, *The Dry Decade* (Garden City, N.Y.: Doubleday, 1931), 1–24; Andrew Sinclair, *Era of Excess: A Social History of the Prohibition Movement* (New York: Harper and Row, 1964), 63–82.

4. Lawrence A. Cremin, *The Transformation and the School: Progressivism in American Education, 1876–1957* (New York: Alfred A. Knopf, 1961), 1–273.

5. Tyack and Hansot, *Managers of Virtue*, 105–121; Tyack, *One Best System*, 126–76; Raymond E. Callahan, *Education and the Cult of Efficiency: A Study of the Social Forces That Have Shaped the Administration of the Public Schools* (Chicago: University of Chicago Press, 1964), 1–220; Morton Keller, *Regulating a New Society: Public Policy and Social Change in America, 1900–1933* (Cambridge, Mass.: Harvard University Press, 1994), 46–48.

6. Quoted in Tyack and Hansot, *Managers of Virtue*, 129–44 (the quotation appears on 132), Callahan, *Education and the Cult of Efficiency*, 117.

7. William A. Link, *The Paradox of Southern Progressivism, 1880–1930* (Chapel Hill: University of North Carolina Press, 1992), 124–42 (the quotation is on 131); Dewey W. Grantham, Jr., *Southern Progressivism: The Reconciliation of Progress and Tradition* (Knoxville: University of Tennessee Press, 1983), 246–74; C. Vann Woodward *Origins of the New South, 1877–1913* (Baton Rouge: University of Louisiana Press, 1951), 396–406; Charles Lee Lewis, *Philander Priestly Claxton: Crusader for Public Education* (Knoxville: University of Tennessee Press, 1948), 112–25, 150–68.

8. Dewey W. Grantham, Jr., *Hoke Smith and the Politics of the New South* (Baton Rouge: Louisiana State University Press, 1958), 254–64; H. G. Good, *A History of American Education*, 2d ed. (New York: Macmillan, 1964), 303–4; Freeman R. Butts and Lawrence Cremin, *A History of Education in American Culture* (New York: Holt, Rinehart, and Winston, 1953), 427–29.

9. Grantham, *Hoke Smith*, 264–67; Lewis, *Philander Priestly Claxton*, 217; Church and Sedlak, *Education in the United States*, 304–8; Good, *History of American Education*, 304–5; Cubberley, *Public Education*, 644–45.

10. Woodrow Wilson, "Address delivered at Joint Session of the Two Houses of Congress, 2 April 1917," in Henry Steele Commager, ed., *Documents of American History*, 2 vols. (New York: Appleton-Century-Crofts, 1963), 2:128–32 (quotations are on 131–32); Daniel M. Smith, *The Great Departure: The United States and World War I, 1914–1920* (New York: John Wiley and Sons, 1965), 1–82; August Heckscher, *Woodrow Wilson* (New York: Charles Scribner and Sons, 1992), 359–439; Thomas J. Knock, *To End All Wars: Woodrow Wilson and the Quest for a New World Order* (New York and Oxford: Oxford University Press, 1992), 48–122. For the economic dimensions behind America's entry into the war see, for instance, Emily Rosenberg, *Spreading the American Dream: American Economic and Cultural Expansionism, 1890–1945* (New York: Hill and Wang, 1982), 63–86; William Appleman Williams, *The Tragedy of American Diplomacy* (Cleveland and New York: The World Publishing Company, 1959), 23–76. For a blend of both interpretations, see Walter Lafeber, *The American Age: United States Foreign Policy at Home and Abroad since 1750* (New York and London: W. W. Norton & Company, 1989), 253–58, 268–80.

11. David M. Kennedy, *Over Here: The First World War and American Society* (New York and Oxford: Oxford University Press, 1980), 53.

12. Dean R. Esslinger, "American German and Irish Catholic Attitudes toward Neutrality, 1914–1917: A Study of Catholic Minorities," *Catholic Historical Review* 53 (July 1967): 194–216; Edward Cuddy, "Pro-Germanism and American Catholicism, 1914–1917," ibid., 54 (October 1968): 427–54; Philip Gleason, *Conservative Reformers: German American Catholics and the Social Order* (Notre Dame, Ind.: University of Notre Dame Press, 1968), 160–71.

13. "Chronicle," *America* 17 (21 April 1917): 25–26; Minutes of the Board of Archbishops, 18 April 1917, AASL, RG 9, U.S.A. Hierarchy; Halsey, *Survival of American Innocence*, 45; Bureau of the Census, *Historical Statistics of the United States: Colonial Times to 1957* (Washington, D.C.: U.S. Government Printing Office, 1961), 7, 228; James Hennesey, S.J., *American Catholics: A History of the Roman Catholic Community in the United States* (Oxford and New York: Oxford University Press, 1983), 225–26; Slawson, *NCWC*, 16.

14. Kauffman, *Faith and Fraternalism*, 190–200; Michael Williams, *American Catholics in the War: The National Catholic War Council, 1917–1921* (New York: Macmillan, 1921), 110–12; Elizabeth McKeown, *War and Welfare: American Catholics and World War I* (New York: Garland, 1988), 72–73.

15. Williams, *Catholics in War*, 105–53, and chapters 9–18; McKeown, *War and Welfare*, 72–88, and chapters 3 and 4; John B. Sheerin, *Never Look Back: The Career and Concerns of John J. Burke* (New York: Paulist Press, 1975), 38–46; Evelyn Savidge Sterne, "Beyond the Boss: Immigration and American Political Culture from 1880 to 1940," in Gary Gerstle and John Mollenkopf, eds., *E Pluribus Unum? Contemporary and Historical Perspectives on Immigrant Political Incorporation* (New York: Russell Sage Foundation, 2001), 51–57; Slawson, *NCWC*, 26–33; Slawson, "John J. Burke," 51–58; O'Brien, *Public Catholicism*, 128–29; Lynn Dumenil, "The Tribal Twenties: 'Assimilated' Catholics' Response to Anti-Catholicism in the 1920s," *Journal of American Ethnic History* 11 (Fall 1991): 21–25.

16. Kennedy, *Over Here*, 45–53, 117–24; Graham, *Great Campaigns*, 97–111; Dumenil, *Modern Temper*, 20–21.

17. Church and Sedlak, *Education in the United States*, 353–58; Joel H. Spring, "Psychologists and the War: The Meaning of Intelligence in the Alpha and Beta Tests," *History of Education Quarterly* 12 (Spring 1972): 3–15; Cremin, *The Transformation of the School*, 187–92; Cubberley, *Public Education*, 92.

18. Lewis P. Todd, *Wartime Relations of the Federal Government and the Public Schools, 1917-1918* (New York: Teachers College, Columbia University, 1945) 197–201.

19. Bureau of Census, *Historical Statistics*, 91; *The Towner Education Bill, H.R. 15400, 65th Congress, Third Session* (Washington, D.C.: NEA, undated [early 1919]); Albertina Adelheit Abrams, "The Policy of the National Education Association toward Federal Aid to Education, (1857–1953)" (Ph.D. diss.: University of Michigan, 1954), 58–62; Todd, *Wartime Relations*, 197–201.

20. *A National Program for Education—A Statement Issued by the National Education Association Commission on the Emergency in Education and the Program for Readjustment During and After the War*, printed in U.S. Congress, Senate, Committee on Education and Labor, *Department of Education: Hearing on S. 4987*, 65th Cong., 3rd sess., 1918, 44–57 (hereafter cited as *Senate Hearing, 1918*); Professor R. C. Brooks,

"Teachers' Salaries and the Cost of Living, 1918," printed in *Senate Hearing, 1918*, 76-95; Todd, *Wartime Relations*, 197–99; A. Abrams, "Policy of NEA," 71.

21. *National Program for Education*, in *Senate Hearing, 1918*, 44–57 (the quotation is on 48); Brooks, "Teachers' Salaries," 76–95; Todd, *Wartime Relations*, 197–99; A. Abrams, "Policy of NEA," 71.

22. *National Program for Education*, in *Senate Hearing, 1918*, 50-51; A. Abrams, "Policy of NEA," 71–72.

23. Cubberley, *Public Education*, 615–16; Todd, *Wartime Relations*, 118–19.

24. *National Program for Education*, in *Senate Hearing, 1918*, 53–54; Frank J. Munger and Richard F. Fenno, Jr., *National Politics and Federal Aid to Education* (Syracuse: University of Syracuse Press, 1962), 4; A. Abrams, "Policy of NEA," 57–58. The figures were compiled from Bureau of Census, *Historical Statistics*, 7, 10, and 214, and *Senate Hearing, 1918*, 57. The Census Bureau classified as illiterates those who were unable to read or write in any language. Thus immigrants who could read or write in their native tongue were considered literate even though they were not so in English.

25. Bureau of Census, *Historical Statistics*, 615–16; R. M. Abrams, *Burdens of Progress*, 53; Gary Gerstle, "Liberty, Coercion, and the Making of Americans," *Journal of American History* 84 (September 1997): 534–39; Ewa Morawska, "Immigrants, Transnationalism, and Ethnicization: A Comparison of This Great Wave and the Last," in *E Pluribus Unum?*, 179–87; Reed Ueda, "Historical Patterns of Immigrant Status and Incorporation in the United States," ibid., 295.

26. *National Program for Education*, in *Senate Hearing, 1918*, 53; Cubberley, *Public Education*, 591; Barry, *German Americans*, 1–12; Gleason, *Conservative Reformers*, 29–31; Dolores Liptak, R.S.M., *Immigrants and Their Church* (New York: Macmillan, 1989), 92-101; E. Clifford Nelson, *The Lutherans in North America* (Philadelphia: Fortress Press, 1975), 295–98, 367, 398–99.

27. "Statement of Dr. Robert Kelly of the American Council on Education," *Senate Hearing, 1918*, 23–24.

28. Todd, *Wartime Relations*, 22–27, 34–37; Kennedy, *Over Here*, 53–75.

29. Paul L. Campbell (secretary of the Executive Committee *ad interim* of the Emergency Council on Education) to Francis Howard, 9 February 1918, with undated memorandum to Franklin K. Lane, copies, ACUA, Rector files; Donald J. Cowling, "The Emergency Council on Education," *Journal of the Proceedings of the National Education Association* 56 (1918): 200–201.

30. Campbell to Howard, 9 February 1918, with undated memorandum to Franklin K. Lane, copies, ACUA, Rector files; Cowling, "Emergency Council on Education," 200–201.

31. Campbell to Howard, 9 February 1918, with undated memorandum to Lane, copies, ACUA, Rector files; Report of the Organization of the Emergency Council on Education, Organized at Washington, 24–25 January 1918, ACUA, Edward A. Pace Papers; Cowling, "Emergency Council on Education," 201.

32. Minutes of the Meetings of Executive Committees of National Education Associations, Washington, 29–30 January 1918, ACUA, Pace Papers; Campbell to Howard, 9 February 1918, copy, ACUA, Rector files; Cowling, "Emergency Council on Education," 201–202; U.S. Congress, Senate, S. 18, 65th Cong., 2d sess., 1917.

33. Harry Pratt Judson, John H. McCracken, and Paul L. Campbell, "Creation of a National Department of Education: Statement Presented to the Chairman of the Senate Committee on Education," undated [1918], ACUA, Rector files.

34. NEA, *An Education Platform,* in *Senate Hearing, 1918,* 40–43; "The National Emergency in Education," *NEA Bulletin* 6 (April 1918): 16–17; A. Abrams, "Policy of NEA," 68–70; Edgar B. Wesley, *NEA: The First Hundred Years, The Building of the Teaching Profession* (New York: Harper & Brothers, 1957), 300.

35. Minutes of the Emergency Council on Education, Washington, 26–27 March 1918, ACUA, Pace Papers; A. Abrams, "Policy of NEA," 68–74.

36. Minutes of the Emergency Council on Education, Washington, 26–27 March 1918, ACUA, Pace Papers.

37. Herman Ames to Thomas Shahan, 12 February 1918, ACUA, Rector files; Shahan to Ames, 11 March 1918, copy, ACUA, Department of Philosophy files. The quotation is from the latter.

38. Minutes of the Meeting of the Emergency Council on Education, University of Pennsylvania, 17 May 1918, ACUA, Pace Papers; Lewis, *Philander Priestly Claxton,* 204–9; Keller, *Regulating a New Society,* 52. On German Catholics and anti-German sentiment, see Higham, *Strangers in the Land,* 194–212; Gleason, *Conservative Reformers,* 159–171; Esslinger, "American German and Irish Attitudes Toward Neutrality," 194–216; Cuddy, "Pro-Germanism and American Catholicism" 427–54.

39. Charles Phillips, "Not German, but Germanism," *America* 18 (16 March 1918): 585–86; Joseph Schrembs, "Catholic Education and After-the-War Problems," *CEA Bulletin* 15 (November 1918): 475; Gleason, *Conservative Reformers,* 172; Frederick C. Luebke, *Bonds of Loyalty: German Americans and World War I* (De Kalb: Northern Illinois University Press, 1974), 292.

40. "No God for the Child," *America* 18 (6 April 1918): 656; Schrembs, "After-the-War Problems," 474; Higham, *Strangers in the Land,* 204–7.

41. Louis Walsh to Howard, 19 March 1918, ANCEA, Francis Howard Papers. A fellow member of the Advisory Committee who shared Walsh's suspicion that the proponents of federalization were using the war as a pretext for securing their aim was Brother John Waldron of Chaminade College (John Waldron to Howard, 30 March 1918, ANCEA, Howard Papers).

42. Lewis, *Philander Priestly Claxton,* 229; "Federal Government Should Cooperate with the States, Says Commissioner of Education," *School Life* 3 (16 August 1919): 1–2; Arthur C. Monahan to William Devlin, S.J., 22 June 1921, copy, ACUA, NCWC Department of Education files; William P. Burris, "A Federal Department of Education," in National Education Association, *Addresses and Proceedings of the Fifty-eighth Annual Meeting Held at Salt Lake City, Utah, 4–10 July 1920,* 58 (1920): 446 (hereafter cited as *Addresses and Proceedings of NEA*).

43. Walsh to Howard, 19 March 1918, 12 June 1918, 4 January 1919, ANCEA, Howard Papers; Walsh to Howard, 14 January 1921, copy, ADP Walsh Papers.

44. Howard to CEA Advisory Committee, 26 March 1918, ACUA, Pace Papers; Howard to Shahan, 27 March 1918, copy, ANCEA, Howard Papers; Ambrose A. Clegg, Jr., "Federal Aid to Education: A Study of the Interest of Church and Labor Groups in Proposals for Federal Aid to Elementary and Secondary Schools, 1890–1945" (Ph.D. diss., University of North Carolina, 1963), 74–76; A. Abrams, "Policy of NEA," 83–84.

45. Shahan to Howard, 4 April 1918, ANCEA, Howard Papers.

46. "Meetings of the Executive Board: Report of the Advisory Committee," *CEA Bulletin* 15 (November 1918): 16–17. There are no minutes of the meeting.

47. No draft of the bill is available. The points outlined here are taken from the version presented to Congress (U.S. Congress, Senate, S. 4987, 65th Cong., 2d sess.). Details have been omitted because several committees reviewed the measure and may have changed it. Available evidence indicates that the draft certainly contained the above elements (cf. *Senate Hearing, 1918*); *Addresses and Proceedings of NEA,* 56 (1918): 23–26.

48. Minutes of the Meeting of the Emergency Council on Education, University of Pennsylvania, 17 May 1918, ACUA, Pace Papers; Ames to Shahan, 6 June 1918, ACUA, Rector files (in which Ames quotes Judson's entire letter).

49. *Addresses and Proceedings of NEA,* 56 (1918): 24–26.

50. See *Senate Hearing, 1918,* 21–29.

51. George Strayer, "The National Emergency in Education," *Addresses and Proceedings of NEA,* 56:129–31. Emphasis in original.

52. William C. Bagley, "Education and Our Democracy," ibid., 55–58.

53. "Report of the Committee on Resolutions," ibid., 24.

54. A. Abrams, "Policy of NEA," 74–76.

55. Dumenil, *Modern Temper,* 40–42, 46–47; Lynn Dumenil, " 'The Insatiable Maw of Bureaucracy': Antistatism and Education Reform in the 1920s," *Journal of American History* 77 (September 1990): 508–9.

56. Callahan, *Education and the Cult of Efficiency,* 79–94.

57. Strayer, "Plan to Meet Emergency in Schools Due to the War," *New York Times,* 21 July 1918, section 3.

58. "The Catholic Educational Association," *Fortnightly Review* 25 (1 July 1918): 201–2; Howard, "Report of the Secretary General," *CEA Bulletin* 15 (November 1918): 11; "Meetings of the Executive Board," ibid., 15.

59. Richard Tierney to Howard, 16 April 1918, ANCEA, Howard Papers.

60. L. F. Happel, "Education: A New Danger for Our Schools," *America* 19 (27 July 1918): 389.

61. Link, *Paradox of Southern Progressivism,* 268–83; George B. Tindall, *The Emergence of the New South, 1913–1945* (Baton Rouge: Louisiana State University Press, 1967), 258–76.

62. U.S. Congress, Senate, S. 4987, 65th Cong., 2d sess., 1918. According to the consumer price index, $100 million in 1918 had a purchasing power equivalent to $1.248 billion in 2004.

63. U.S. Congress, Senate, S. 4987, 65th Cong., 2d sess., 1918.

64. James Burns to Howard, 30 September 1918, ANCEA, Howard Papers.

65. Pace to James Gibbons, 27 January 1919, copy, ACUA, Vice-Rector files; Fogarty, *Vatican and American Hierarchy,* 153 and 158; Gannon, "Before and After Modernism," 317 and 327; Reher, *Catholic Intellectual Life in America,* 91.

66. Walsh to Shahan, 4 February 1919, ACUA, Vice-Rector files; Pace to Walsh, 6 February 1919, copy, ACUA, Vice-Rector files; Robert Trisco, "The Church's History in the University's History," *Catholic Historical Review* 75 (October 1989): 659–61; Patrick Henry Ahearn, *The Catholic University of America, 1887–1896: The Rectorship of*

John J. Keane (Washington, D.C.: The Catholic University of America Press, 1948), 121–30; Peter E. Hogan, S.S.J., *The Catholic University of America, 1896–1903: The Rectorship of Thomas J. Conaty* (Washington, D.C.: The Catholic University of America Press, 1949), 11, 155–57.

67. Howard to the CEA Executive Board, 25 October 1918, ACUA, Pace Papers; "Meetings of the Executive Board," *CEA Bulletin* 16 (1919): 9–10. Present were Shahan, Pace, Burns, Howard, Peterson, Francis Moran, John Fenlon, John Chidwick, B. P. O'Reilly, M. A. Schumacher, and Albert Fox.

68. Waldron to Howard, 24 November 1918 and 3 January 1919, ANCEA, Howard Papers. The quotation is from the first, in which Waldron also states that he was taking steps to contact Reed and Clark. The second indicates that he had made contact with both legislators and received acknowledgments from them.

69. "A Plot Against Our Schools," *Fortnightly Review* 25 (15 October 1918): 306; San Antonio-Dallas *Southern Messenger*, 7 November 1918; Detroit *Michigan Catholic*, 6 February 1919.

70. Press Bulletin of the Central Bureau of the Central Verein, 19 November 1918, copy in ACUA, Vice-Rector files; Gleason, *Conservative Reformers*, 5–13, 23–29.

71. Cincinnati *Catholic Telegraph*, 23 January 1919.

72. Saint Paul *Catholic Bulletin*, 25 January 1919.

73. "The Smith Bill," *America* 20 (30 November 1918): 184–85; "Our Proposed Prussian Schools," ibid. (28 December 1918): 296; Cincinnati *Catholic Telegraph*, 27 January 1919; Detroit *Michigan Catholic*, 31 October 1918.

74. Boston *Pilot*, 16 November 1918 and 11 January 1919.

75. "Our Proposed Prussian Schools," *America* 20 (28 December 1918): 296. See also Paul L. Blakely, S.J., "Education: What is the Smith Bill?" ibid. (18 January 1919): 374–75. Several Catholic papers made the same point. See the Boston *Pilot*, 16 November 1918 and 11 January 1919; San Antonio-Dallas *Southern Messenger*, 23 January 1919; Detroit *Michigan Catholic*, 13 October 1918; *Our Sunday Visitor*, 23 February 1919.

76. "The Smith Bill and an American Institution," *America* 20 (15 February 1919): 475. Writing in the Los Angeles *Tidings*, Reverend Z. J. Maher, S.J., echoed the theme. The Smith bill, he said, was "one of the most un-American bills which has ever been presented, and, incidentally one of the most anti-Catholic which has ever been successfully disguised" (31 January 1919).

77. Lynn Dumenil, *Freemasonry and American Culture, 1880–1930* (Princeton, N.J.: Princeton University Press, 1984), 115–47; Dumenil, *Modern Temper*, 16–23, 203–7, 226–34.

78. "The Peril of the Public Schools," *New Age* 26 (October 1918): 450. Emphasis in original.

79. See Detroit *Michigan Catholic*, 9 January 1919.

80. Nebraska State Legislature, 37th Sess., House Roll No. 4 (a copy of the bill is in ACUA, NCWC Committee on Special War Activities, Executive Secretary files); Orville B. Zabel, *God and Caesar in Nebraska: A Study of the Legal Relationship of Church and State, 1854–1954* (Lincoln: University of Nebraska Press, 1955), 135.

81. Link, *Woodrow Wilson*, 54–80, 223–51; David Burner, *The Politics of Provincialism: The Democratic Party in Transition, 1918–1932* (New York: Alfred A.

Knopf, 1968), 28–41; Leuchtenburg, *Perils of Prosperity,* 50–51; David A. Shannon, *Between the Wars: America, 1919–1941* (Boston: Houghton Mifflin Company, 1965), 8.

82. *Senate Hearing, 1918,* 29–37.

83. Ibid., 135.

84. U.S. Congress, House, H.R. 15400, 65th Cong., 3rd sess., 1919; San Antonio-Dallas *Southern Messenger,* 27 March 1919; NCWC news service, 22 October 1923. The information about the equalization fund was provided to the news service by William P. Burris, dean of Teachers College in the University of Cincinnati, who handled the negotiations with the NEA for the American Federation of Teachers. He later foreswore the idea of a department because he claimed that it had the support of "all the educational Bolsheviks" and aimed at a federal control that would "imperil our orderly educational process" (NCWC news service, 22 October 1923). Regarding the NEA's intention that money from the equalization fund be used to increase teachers' salaries, see *Senate Hearing, 1918,* 48–50, 73–95.

85. U.S. Congress, Senate, S. 5635, 65th Cong., 3d sess., 1919.

86. A. Abrams, "Policy of the NEA," 76.

87. Hugh S. Magill, *A National Program for Education: Address at a Conference on Reconstruction under the auspices of the National League of Popular Government in the Auditorium of the Department of the Interior, Washington, D.C.* (Washington, D.C.: NEA, undated [late 1918 or early 1919]).

88. *The Towner Education Bill: H.R. 15400, 65th Congress, Third Session* (Washington, D.C.: NEA, undated [early 1919]).

89. Pace to Gibbons, 27 January 1919, copy, ACUA, Vice-Rector files; Ellis, *Gibbons,* 2:543.

90. Pace, memorandum for Cardinal Gibbons, 27 January 1919, copy, ACUA, Vice-Rector files.

91. Ibid.

92. Walsh to Shahan, 4 February 1919, ACUA, Vice-Rector files.

93. Merz, *Dry Decade,* 1–42; Asbury, *Great Illusion,* 121–34; Behr, *Prohibition,* 7–75; Cashman, *Prohibition,* 6–23; Kobler, *Ardent Spirits,* 198–212; Sinclair, *Era of Excess,* 63–82, 116–28, 152–66; Douglas J. Slawson, "Wine for the Gods: Negotiations for the Sacramental Use of Alcohol during Prohibition," in Joseph C. Link, C.O., and Raymond J. Kupke, eds., *Building the Church in America: Studies in Honor of Monsignor Robert F. Trisco on the Occasion of His Seventieth Birthday* (Washington, D.C.: The Catholic University of America Press, 1999), 161–95.

94. Pace to Walsh, 6 February 1919, copy, ACUA, Vice-Rector files. See also [Pace], Memorandum on the School Question, undated [February 1919], ACUA, Vice-Rector files. This memorandum elaborates more fully on the arguments made to Walsh.

95. Walsh to Pace, 14 February 1919, ACUA, Vice-Rector files.

96. "Meeting of the Executive Board: Report of the Advisory Committee," CEA *Bulletin* 16 (November 1919): 16–17.

97. Slawson, *NCWC,* 45–47.

98. Quoted in Diary of Muldoon, 20 February 1919, ACUA, microfilm; *Baltimore Catholic Review,* 1 March 1919; Ellis, *Gibbons,* 2:298–99; Slawson, *NCWC,* 47.

99. Minutes of the Meeting of the Archbishops, 1919, AABa, 128 H7; William Russell, *Confidential Communication to the Archbishops* [By Request of His Eminence

James Cardinal Gibbons], undated [February 1919], ACUA, NCWC General Secretary flles; Slawson, NCWC, 47–48.

100. Minutes of the Meeting of the Archbishops, 1919, AABa, 128 H7; Slawson, *NCWC,* chapter 3.

101. Boston *Pilot,* 24 May 1919. See also Cincinnati *Catholic Telegraph,* 8 May 1919, for concern about federalization.

102. Los Angeles *Tidings,* 16 May 1919; *Denver Catholic Register,* 22 May 1919. Much of the latter editorial was reprinted in the *Indiana Catholic and Record,* 30 May 1919.

103. Blakely, "Hit the Smith Bill Hard," *America* 21 (19 April 1919): 52, and "The Smith Bill and Sherman's Pigs," ibid. (31 May 1919): 211; Boston *Pilot,* 17 May 1919; Davenport *Catholic Messenger,* 24 April 1919. The quotation is from the last named. See also Robert K. Murray, *The Harding Era: Warren G. Harding and His Administration* (Minneapolis: University of Minnesota Press, 1969), 172.

104. Figures are based on *Congressional Record,* 66th Congress, 1st Sess., vol. 58, pt. 4, 3241; ibid., 3rd Sess., vol. 60, pt. 2, 2061–62; U.S. Congress, Senate, Committee on Education and Labor, *To Create a Department of Education and to Encourage the States in the Promotion and Support of Education: Hearings on S. 1337,* 68th Cong., 1st sess., 1924, 330 and 353.

105. Pace to Russell, 29 April 1919, SAB, RG 10, box 23; copy of same in ACUA, Vice-Rector files.

106. Russell to Edward Dyer, 30 April 1919, SAB, RG 10, box 23; Slawson, *NCWC,* 49–53.

107. Gibbons to the General Committee on Catholic Interests and Affairs, 5 May 1919, ACUA, NCWC Chairman files; Slawson, *NCWC,* 54.

108. Gibbons to the Hierarchy, 17 May 1919, AABa, 126 B8; Slawson, *NCWC,* 54–55.

2. The Making of a Religious Issue

1. Dumenil, *Modern Temper,* 26–31, and "'Insatiable Maw of Bureaucracy,'" 510–24; Shannon, *Between Wars,* 15–25; Robert K. Murray, *The Politics of Normalcy: Governmental Theory and Practice in the Harding-Coolidge Era* (New York: W. W. Norton & Company, 1973), 36–40.

2. Leuchtenburg, *Perils of Prosperity,* 66–83.

3. Leuchtenburg, *Perils of Prosperity,* 142–69; Dumenil, *Modern Temper,* 145–59; R. M. Abrams, *Burdens of Progress,* 182–86; Allen, *Only Yesterday,* 188–203; Hofstadter, *Age of Reform,* 288–89.

4. Bureau of Census, *Historical Statistics,* 12–14; R. M. Abrams, *Burdens of Progress,* 20–22, 159; Leuchtenburg, *Perils of Prosperity,* 7, 43; Dumenil, *Modern Temper,* 45, 11; Harvey Green, *The Uncertainty of Everyday Life, 1915–1945* (New York: Harper Perennial, 1993), 6, 62–69, 187–206, 243 n. 64.

5. Bureau of Census, *Historical Statistics,* 56; Gerstle, "Liberty, Coercion, and the Making of Americans," 534–39; Morawska, "Immigrants, Transnationalism, and Ethnicization," 179–87; Ueda, "Historical Patterns of Immigrant Status and Incorporation in the United States," 295; Sterne, "Beyond the Boss," 37–57; Desmond King, "Making Americans: Immigration Meets Race," in ibid., 143–51, 158–66; R. M.

Abrams, *Burdens of Progress,* 45, 51–53; Kirschner, *City and Country,* 14–16, 27–38, 117; Mowry, *Era of Theodore Roosevelt,* 13–14, 92–94; Dumenil, *Modern Temper,* 204–7, 260–63; Higham, *Strangers in the Land,* 150–63; Leuchtenburg, *Perils of Prosperity,* 66–83; Allen, *Only Yesterday,* 38–62; Green, *Uncertainty of Everyday Life,* 33–36.

6. Bureau of Census, *Historical Statistics,* 46–47, 285, 510; Dumenil, *Modern Temper,* 59–97; Green, *Uncertainty of Everyday Life,* 17–40, 215–19; Kirschner, *City and Country,* 12–21; Burner, *Politics of Provincialism,* 4–5, 101.

7. Church and Sedlak, *Education in the United States,* 358–70; Howard K. Beale, *Are American Teachers Free? An Analysis of Restraints upon the Freedom of Teaching in American Schools* (New York: Octagon Books, 1972, originally published in 1936), 1–319, 523–658; Allen, *Only Yesterday,* 73–101; Leuchtenburg, *Perils of Prosperity,* 169–74; Dumenil, *Modern Temper,* 134–38; Green, *Uncertainty of Everyday Life,* 128–34; Beth L. Bailey, *From Front Porch to Back Seat: Courtship in Twentieth-Century America* (Baltimore: Johns Hopkins University Press, 1988), 13–96 passim; Tyack, "School for Citizens," 344–50.

8. U.S. Congress, House, H.R. 7, 66th Cong., 1st sess., 1919; U.S. Congress, Senate, S. 1017, 66th Cong., 1st sess., 1919; U.S. Congress, Senate, Committee on Education and Labor, *Education Bill: Joint Hearings before the Committees on Education and Labor, Sixty-sixth Congress, First Session, on S. 1017 and H.R. 7* (Washington: U.S. Government Printing Office, 1919), 32, 114–16 (hereafter cited as *Joint Hearings, 1919*). With regard to state constitutional prohibitions against aid to parochial schools, see Cubberley, *Public Education,* 239.

9. As will be seen, Burris, who helped the American Federation of Teachers to negotiate for the outright admission that moneys in the Smith-Towner equalization fund should be used to increase teachers' salaries, would soon break with the NEA over the bill. He came to consider the appropriations in the bill a pure "bribe" to win the support of poorly paid teachers for the administrative portion of the measure. He quoted, moreover, one of the NEA drafters of the original bill, turned opponent, as stating that the measure was ardently promoted by "a little group who have for the past several years dominated the N.E.A." and had the support of "all the educational bolsheviks [*sic*]" (NCWC news service, 22 October 1923). More recently, historian Lynn Dumenil has contended that the NEA's withdrawal was simply a tactical maneuver to curry support for the measure ("'Insatiable Maw of Bureaucracy,'" 564–65). On the other hand, Charles Judd, dean of the Cleveland Conference, inferred from this waffling on the part of the NEA that the framers of the bill "were without any real policy in the matter" and were prepared to "reverse themselves on cardinal issues to secure support" ("Discussion," in *The Relation of the Federal Government to Education: Installation of David Kinley as President of the University of Illinois, December 1 and 2, 1921* [Urbana: University of Illinois Press, 1922], 98).

10. *Congressional Record,* 67th Congress, 4th sess., 1923, 64, pt. 5, 4432. Layton conducted his survey in February 1921 at the height of the Smith-Towner controversy. He did not choose to report his findings, however, until just before his retirement from Congress in 1923.

11. Blakely, "'The Same Old Bill,'" *America* 21 (7 June 1919): 222. See also Detroit *Michigan Catholic,* 5 June 1919; Thomas E. Shields, "The Towner Bill and the Centralization of Education Control," *Catholic Educational Review* 17 (June 1919): 326.

12. Boston *Pilot,* 28 June 1919.

13. Central Bureau of the Central Verein, *For the Freedom of Education: An Argument against the Towner and Smith Bills* (Saint Louis: Central Bureau of the Central Verein, 1919); Los Angeles *Tidings,* 1 August 1919.

14. *Congressional Record,* 66th Cong., 1st sess., 1919, 58, pt. 1, 512–15; 622–23.

15. Abstract and Questions on Proposed Legislation, [June 1919], copy, ACUA, Rector files.

16. John Fenlon to Muldoon, 10 June 1919, ACUA, NCWC Chairman files; Pace to Muldoon, 10 June 1919, copy, ACUA, Vice-Rector files.

17. Muldoon, Russell, Schrembs, and Joseph Glass, C.M., to the hierarchy, 13 June 1919, ACUA, Vice-Rector files.

18. "Meetings of the Executive Board," *CEA Bulletin* 16 (November 1919): 14–19; undated [June 1919], untitled tally sheet in Fenlon's hand recording the vote of the CEA Executive Board on the Pace questionnaire, SAB, RG 13, box 8. Although there were eleven people present on the first night of the convention when the Executive Board met, the tally sheet makes it clear that at least on one question only seven voted the next evening. Because evidence on either side of the meeting clearly indicates that both Pace and Shahan favored in some measure the federalization of education, one is forced to assume that they abstained from voting. Fenlon, too, who was also a faculty member at Catholic University and in sympathy with Pace and Shahan on many other issues of national Catholic importance, probably also abstained.

19. "Sermon of the Most Reverend John J. Glennon, D.D., Archbishop of St. Louis," *CEA Bulletin* 16 (November 1919): 44–46.

20. William O'Connell, "The Reasonable Limits of State Activity," ibid., 62–76. The address was delivered by John B. Peterson.

21. On this point, see Halsey, *Survival of American Innocence,* 37–83; McKeown, *War and Welfare,* 45–59; Arnold J. Sparr, "From Self-Congratulation to Self-Criticism: Main Currents in American Catholic Fiction, 1900–1960," *U.S. Catholic Historian* 6 (1987): 213–22; Joseph M. McShane, S.J., "Mirrors and Teachers: A Study of Catholic Periodical Fiction Between 1930 and 1950," ibid., 181–98; Fayette Breaux Veverka, *"For God and Country": Catholic Schooling in the 1920s* (New York: Garland, 1988), 69–91; Slawson, *NCWC,* chapter 1; Dumenil, "Tribal Twenties," 27–30.

22. Muldoon to Fenlon, 30 July 1919, SAB, RG 13, box 8.

23. Gibbons to James Flaherty, 1 August 1919, copy, SAB, RG 13, box 8. Although the letter was probably written by Fenlon, the views expressed are essentially those of Gibbons, who opposed the Smith-Towner bill (see Muldoon's list of responses to Pace's questionnaire, ACUA, NCWC Chairman files).

24. *Joint Hearing, 1919,* 10–90 passim.

25. Ibid., 10–17, 53–54, 143–44.

26. Ibid., 30–33, 54.

27. Correspondence from the bishops is not extant. Both Muldoon and Fenlon, however, compiled lists of the responses, located in ACUA, NCWC Chairman files, and in SAB RG 13, box 8. While Muldoon's list names the bishops, it is crude in tallying their responses, stating simply whether they opposed the legislation or favored it, with conditions. Fenlon's list gives no names, but accurately registers the number of responses to each of the eight questions on Pace's questionnaire. Unfortunately, his list contains

only the first thirty-four responses, not the final eleven, which must have come to Muldoon late.

28. Gibbons to the hierarchy, 1 August 1919, ACUA, NCWC Chairman files.

29. *Joint Hearings, 1919*, 152–54: Tyack and Hansot, *Managers of Virtue*, 136–40. The Old Guard had initially opposed the formation of the Emergency Council on Education on the ground that it would only add to the proliferating number of organizations and would hinder the war service of the universities. Once it was established, however, they supported it (Carol Gruber, *Mars and Minerva: World War I and the Uses of Higher Education* [Baton Rouge: Louisiana State University Press, 1975], 98–100).

30. Lewis, *Philander Priestly Claxton*, 229; Burris, "Federal Department of Education," *Addresses and Proceedings of the NEA, 1920*, 58: 445–6; Monahan to William Devlin, 22 June 1921, copy, ACUA, NCWC Department of Education files.

31. "Federal Government Should Cooperate with States, Says Commissioner of Education," *School Life* 3 (16 August 1919): 1; Lewis, *Philander Priestly Claxton*, 229; Monahan to Devlin, 22 June 1921, copy, ACUA, NCWC Department of Education files. Monahan, as will be seen below, was a Catholic career educator in the public schools who knew Claxton. He rightly believed that Catholics had grossly misunderstood the commissioner's position on the Smith-Towner bill. In Monahan's view, Claxton's proposal would have given the federal government "no control of the State systems."

32. "Federal Government Should Cooperate with States," 1–2.

33. Lewis, *Philander Priestly Claxton*, 214–17, 222–23.

34. "Education: The American Spirit and the Smith Bill," *America* 21 (23 August 1919): 505.

35. *Congressional Record*, 66th Cong., 1st sess., 1919, 58, pt. 4, 3238.

36. Ibid., 3239–41.

37. *New York Times*, 29 July 1919.

38. Gibbons to the hierarchy, 1 August 1919, ACUA, NCWC Chairman files. A typed draft of the letter is in the Fenlon Papers, SAB, RG 13, box 8.

39. [Pace], Memorandum in re Smith-Towner Bill, undated [late August or early September 1919], ACUA, National Catholic War Council, Chairman files.

40. Slawson, *NCWC*, 57–69.

41. Diary of Walsh, 25 September 1919, ADP; *Minutes of the First Annual Meeting of the American Hierarchy, September 1919*, 14 and 16, ACUA, NCWC General Secretary files; Slawson, *NCWC*, 67. For the Smith-Bankhead bill, see U.S. Congress, Senate, S. 17, 66th Cong., 1st sess., 1919. The companion measure was H.R. 1204.

42. Diary of Walsh, 26 September 1919, ADP.

43. *Congressional Record*, 66th Cong., 1st sess., 1919, 58, pt. 7, 6748, 7018–20; Blakely, "Education: School Legislation, Proposed and Debated," *America* 22 (10 January 1920): 261; Grantham, *Hoke Smith*, 336.

44. Blakely, "Education: School Legislation," 261–62; Grantham, *Hoke Smith*, 334–35.

45. Joseph E. Ransdell to Russell, 17 February 1920, ADCh, 717.50.

46. Slawson, *NCWC*, 73–74, 80.

47. Callahan, *Education and the Cult of Efficiency*, 188–96.

48. Burris, "A Federal Department of Education," 444–49; Wesley, *NEA: The First Hundred Years*, 247–48.

49. Charles Judd, "The Federal Department of Education," *School and Society* 11 (5 June 1920): 661–74.

50. Strayer, "National Leadership and National Support for Education," ibid., 674–81.

51. Charles Eliot, "Discussion," ibid., 681–83.

52. Strayer, "A National Program for Education: A Final Report of the Commission on the Emergency in Education, Presented by Its Chairman," *Addresses and Proceedings of NEA, 1920*, 58: 41–48; "Legislative Commission Begins Work," *NEA Bulletin* 9 (October 1920): 41.

53. "Reorganization of the National Education Association," *NEA Bulletin* 9 (September 1920): 7–10; "By-Laws of the National Education Association: Complete as Amended at the Annual Business Session, Salt Lake City, July 9, 1920," ibid., 10–19; Wesley, *NEA: The First Hundred Years*, 328–32; Upton Sinclair, *The Goslings: A Study of the American Schools* (Pasadena: Upton Sinclair, 1924), 234–69.

54. Slawson, *NCWC*, 70–72; Slawson, "John J. Burke," 62–63, 66–68; Sheerin, *Never Look Back*, 60–61.

55. Slawson, "John J. Burke," 47–62; Slawson, *NCWC*, 70–71.

56. Slawson, *NCWC*, 72–81.

57. Austin Dowling to Fenlon, 6 February 1920, SAB, RG 13, box 7; Slawson, *NCWC*, 81.

58. Dowling to Fenlon, 6 February 1920, SAB, RG 13, box 7; Slawson, *NCWC*, 81–82.

59. Blakely, "School Legislation," 262–63; Zabel, *God and Caesar*, 135–37.

60. James Hamilton, "Michigan School Amendment," *New Age* 28 (October 1920): 459–60; Frank T. Martin, "The Michigan School Controversy" (M.A. thesis, The Catholic University of America, 1949), 11–31; Timothy Mark Pies, "The Parochial School Campaigns in Michigan, 1920–1924: The Lutheran and Catholic Involvement," *Catholic Historical Review* 72 (April 1986): 222–31; Slawson, *NCWC*, chapter 4.

61. David M. Chalmers, *Hooded Americanism: The History of the Ku Klux Klan* (New York: Franklin Watts, 1976); Charles C. Alexander, *The Ku Klux Klan in the Southwest* (Lexington: University of Kentucky Press, 1966); Kenneth T. Jackson, *The Ku Klux Klan in the City, 1915–1930* (New York and London: Oxford University Press, 1967), 3–255; Higham, *Strangers in the Land*, 86–99; Robert Moats Miller, "The Ku Klux Klan," in *Change and Continuity in Twentieth-Century America: The 1920s* (Columbus: Ohio State University Press, 1968): 215–56; Grantham, *South in Modern America*, 101–4; Graham, *Great Campaigns*, 125. Revisionist interpretations apparent in the above text are from Robert Alan Goldberg, "Hooded Empire: The Ku Klux Klan in Colorado, 1921–1932" (Ph.D. diss., University of Wisconsin, 1977); Shawn Lay, ed., *The Invisible Empire in the West: Toward a New Historical Appraisal of the Ku Klux Klan in the 1920s* (Urbana and Chicago: University of Illinois Press, 1992); Leonard Moore, *Citizen Klansmen: The Ku Klux Klan in Indiana, 1921–1928* (Chapel Hill: University of North Carolina Press, 1991); Dumenil, *Modern Temper*, 235–44.

62. Perhaps the best catalog of the deeds expressive of this side of the Klan is given in Chalmers, *Hooded Americanism*, and Alexander, *Klan in the Southwest*.

63. "The Smith-Towner Bill and the Catholics," *New Age* 28 (February 1920): 69–70. Like the Ku Klux Klan, Masonry shared in the racism, anti-Semitism, and

anti-Catholicism of the time, though its racism and anti-Semitism were manifested covertly (Dumenil, *Freemasonry*, 120–26).

64. "The Supreme Council," *New Age* 28 (July 1920): 322.

65. "Is the Parochial School the Place for Patriotism?" *New Age* 28 (May 1920): 212–14.

66. Cleveland *Catholic Universe*, 4 June 1920; Detroit *Michigan Catholic*, 1 July 1920; Denver *Catholic Register*, 17 June 1920. Sharp's comments were made at the above-mentioned meeting of the Harvard Teachers' Association, featuring the debate between Judd and Strayer.

67. Quoted in The Catholic Information League of Philadelphia, *Bulletin* 1 (July 1920): [2], copy in ACUA, Pace Papers.

68. Wesley M. Bagby, *The Road to Normalcy: The Presidential Campaign and Election of 1920* (Baltimore: Johns Hopkins University Press, 1962), 22–52; Francis Russell, *The Shadow of Blooming Grove: Warren G. Harding in His Times* (New York: McGraw-Hill, 1968), 313–54; Burner, *Politics of Provincialism*, 41–58; Leuchtenburg, *Perils of Prosperity*, 88–89; Murray, *Politics of Normalcy*, 2–6.

69. Blakely, "The Republican Party and the Smith Bill," *America* 23 (12 June 1920): 190; Russell, *Shadow of Blooming Grove*, 367.

70. Dowling to Hanna, telegram, undated, AASF, NCWC files; Dowling to Burke, telegram, 4 June 1920, ACUA, NCWC General Secretary files. Punctuation has been added and spelling corrected.

71. Pace to the Committee on Resolutions, undated draft in pencil, ACUA, Vice-Rector files.

72. See printed copy in *NCWC News Sheet*, 14 June 1920.

73. The code names were as follows: Ogden Mills was Henry, Will Hays was Brown, William Sproul was Jones, Hiram Johnson was West, Warren Harding was Ford, William Lowden was Hudson, Medill McCormick was Mack, Archbishop Dowling was Darlington, Cardinal Gibbons was Simmons, and Archbishop George Mundelein was Casey (Code Words, undated [1920], ACUA, NCWC General Secretary files).

74. Burke to Mundelein, telegram, 5 June 1920, copy, ACUA, NCWC General Secretary files; Michael Slattery to Burke, 7 June 1920, ACUA, NCWC General Secretary files; Slattery to Burke, 9 June 1920, ACUA, NCWC General Secretary files; James Watson to James Hugh Ryan, 5 November 1920, ACUA, Pace Papers; Russell, *Shadow of Blooming Grove*, 367; Bagby, *Road to Normalcy*, 31–33.

75. Slattery to Burke, 7 June 1920, ACUA, NCWC General Secretary files; Bagby, *Road to Normalcy*, 87; Russell, *Shadow of Blooming Grove*, 368.

76. Burke to Mills, 6 June 1920, copy, with enclosure, Gibbons [to Republican Platform Committee, undated, June 1920], ACUA, NCWC General Secretary files; Burke to Slattery, telegram, 7 June 1920, ACUA, NCWC General Secretary files. The telegram actually reads: "Have had my letter to Henry handed to him personally this morning."

77. Slattery to Burke, 7 June 1920, ACUA, NCWC General Secretary files; Slattery to Burke, telegram, 8 June 1920, ACUA, NCWC General Secretary files.

78. Slattery to Burke, telegram, 8 June 1920, ACUA, NCWC General Secretary files; Slattery to Burke, 9 June 1920 (4 A.M.), ACUA, NCWC General Secretary files.

79. Slattery to Burke, 9 June 1920 (4 A.M.), ACUA, NCWC General Secretary files; *New York Times*, 9 June 1920; Russell, *Shadow of Blooming Grove*, 368.

80. Slattery to Burke, 9 June 1920 (second of the day), ACUA, NCWC General Secretary files; Dowling to Ryan, 11 June 1920, copy, ACUA, Vice-Rector files.

81. Slattery to Burke, 9 June 1920 (second of the day), ACUA, NCWC General Secretary files.

82. *NCWC News Sheet*, 14 June 1920; *New York Times*, 11 June 1920; Karl Schriftgiesser, *This Was Normalcy: An Account of Party Politics During the Twelve Republican Years, 1920–1932* (Boston: Little, Brown and Company, 1948), 9.

83. Dowling to Pace, 19 June 1920, ACUA, Vice-Rector files.

84. Bagby, *Road to Normalcy*, 25–42, 83–96; Russell, *Shadow of Blooming Grove*, 325–40, 355–96; Schriftgiesser, *This Was Normalcy*, 3–19; Leuchtenburg, *Perils of Prosperity*, 85–86.

85. Pace to Bourke Cockran, 24 May 1920, copy, ACUA, Vice-Rector files; Pace to Dowling, 16 June 1920, copy, ACUA, Vice-Rector files; Pace to Ryan, 19 June 1920, copy, ACUA, Pace Papers.

86. Pace to Dowling, 21 June 1920, copy, ACUA, Vice-Rector files; Slattery to Burke, 3 July 1920, ACUA, NCWC General Secretary files.

87. Catholic Information League of Philadelphia, *Bulletin* 1 (July 1920): [1-2], copy in ACUA, Pace Papers; NCWC news service, 12 July 1920.

88. Slattery to Burke, 24 July 1920, ACUA, NCWC General Secretary files. The proposed plank is quoted within.

89. Ibid.; Bagby, *Road to Normalcy*, 102 and 120.

90. Slattery to Kerby (acting general secretary for Burke), 29 June 1920, ACUA, NCWC, General Secretary files; Slattery to Burke, 24 July 1920, ACUA, NCWC General Secretary files. For a detailed account of Walsh's position on the bill, see Thomas Walsh to Mrs. E. J. English, 28 February 1921, MCLC, Walsh-Erickson Papers.

91. Slattery to Kerby, 29 June 1920, ACUA, NCWC General Secretary files; Slattery to Burke, 24 July 1920, ACUA, NCWC General Secretary files; William E. Ellis, "Catholicism and the Southern Ethos: The Role of Patrick Henry Callahan," *Catholic Historical Review* 69 (January 1983): 41–50.

92. Slattery to Kerby, 29 June 1920, ACUA, NCWC General Secretary files; Slattery to Burke, 3 July 1920, ACUA, NCWC General Secretary files.

93. Slattery to Kerby, 29 June 1920, with enclosure, Arguments against the Federalization of Education, ACUA, NCWC General Secretary files; Slattery to Burke, 3 July, ACUA, NCWC General Secretary files.

94. Slattery to Burke, 24 July 1920, ACUA, NCWC General Secretary files.

95. Slattery to Kerby, 29 June 1920, ACUA, NCWC General Secretary files; Slattery to Burke, 3 July 1920, ACUA, NCWC General Secretary files; Slattery to Burke, 24 July 1920, ACUA, General Secretary files; *New York Times*, 28 and 29 June 1920; Bagby, *Road to Normalcy*, 104. Again, Slattery offered two different numbers of telegrams. In his letter to Kerby, he stated that within twenty-four hours he had received 122. In his later report to Burke, he stated that he had received 400. Because the fight over the platform ultimately stretched over three days, I believe the number Slattery gave to Kerby on the first day of the platform fight is correct. It is possible that over the next two days he received enough to reach the 400.

96. Slattery to Kerby, 29 June 1920, ACUA, NCWC General Secretary files; Slattery to Burke, 24 July 1920, ACUA, NCWC General Secretary files; NCWC news service, 12 July 1920.

97. Slattery to Burke, 3 July 1920, ACUA, NCWC General Secretary files; Slattery to Burke, 24 July 1920, ACUA, NCWC General Secretary files. Schrembs, Report of the Chairman of the Department of Lay Organizations, September 1920, ADCl, Schrembs Papers; Dowling to Edward Hanna, undated, AASF, NCWC files; *New York Times*, 29 June 1920; Bagby, *Road to Normalcy*, 104.

98. Bagby, *Road to Normalcy*, 54–78, 110–20; Burner, *Politics of Provincialism*, 41–73.

99. Dowling, [Confidential Report to the Hierarchy], 21 July 1920, ACUA, Vice-Rector files.

100. Minutes of the Meeting of the NCWC Department of Education, 27 July 1920, copy, ACUA, Pace Papers.

101. "The Public School Question Once More," *New Age* 28 (September 1920): 406–7.

102. "The Smith-Towner Bill: A Paramount Necessity," *Addresses and Proceedings of NEA, 1920* 58 (1920): 24.

103. "Legislative Commission Begins Work," *NEA Bulletin* 9 (October 1920): 41; "Statements of Presidential Candidates," ibid., 42–43. See also *New York Times*, 3 September 1920.

104. Will Hays to Burke, 7 September 1920, ACUA, NCWC General Secretary files; Russell, *Shadow of Blooming Grove*, 407–8; Burner, *Politics of Provincialism*, 57; Schriftgiesser, *This Was Normalcy*, 73–78.

105. "Presidential Candidates Endorse Federal Aid," *NEA Bulletin* 9 (November 1920): 47.

106. "Legislative Commission Begins Work," ibid., 41; Fred M. Hunter, "The Reorganized Association and Its Work," *NEA Bulletin* 9 (October 1920): 23; "Women's Organizations Urge Smith-Towner Bill," *NEA Bulletin* 9 (November 1920): 47; *Washington Post*, 15 November 1920; Dumenil, *Modern Temper*, 47.

107. Pace to Ryan, 15 November 1920, copy, ACUA, Pace Papers. The ACE would not complete the referendum until February (see below).

108. Pace to Dowling, 9 November 1920, copy, ACUA, Vice-Rector files. On Watson and Ryan, see Ryan to Haberlin, 4 October 1922, copy, ACUA, NCWC J. H. Ryan Papers; Ryan to Watson, 25 October 1920, in *Indiana Catholic and Record*, 29 October 1920; Watson to Ryan, 5 November 1920, ACUA, Pace Papers.

109. Ryan to Pace, 15 November 1920; Ryan to Pace, Wednesday [17 or 24 November 1920]; Ryan to Pace, Monday night [29 November 1920]; Pace to Ryan, 7 December 1920, copy—all in ACUA, Pace Papers.

110. Pace to Ryan, 15 November 1920, copy, ACUA, Pace Papers; NCCW, *News Sheet to Organizations*, 1 January 1921, ACUA, NCWC NCCW files.

111. Minutes of the General National Committee, 13 January 1921, ACUA, NCWC Department of Education files.

112. *A Federal Department of Education* and *Federal Aid for Education in the United States: Action of the Senate of the University of Chicago, November 20, 1920* (Chicago: University of Chicago, 1920).

113. U.S. Congress, Senate, S. 4542, 66th Cong., 3d sess., 1920; Pace to Dennis Dougherty, 22 April 1921, copy, ACUA, Vice-Rector files; Bagby, *Road to*

Normalcy, 159–61; Russell, *Shadow of Blooming Grove,* 418; Murray, *Politics of Normalcy,* 1–2.

114. *Congressional Record,* 66th Cong., 3d sess., 1920, 60, pt. 1, 450; Murray, *Harding Era,* 414–15.

115. Ryan to Pace, 26 November 1920, ACUA, Pace Papers.

116. George Johnson to Shahan, undated [probably 7 or 8 December, 1920], ACUA, Pace Papers; Pace to Johnson, 9 December 1920, copy, ACUA, Pace Papers; Pace to Dowling, 3 January 1920 [*sic* for 1921], copy, ACUA, Vice-Rector files; Dowling to Pace, 20 January 1921, ACUA, Vice-Rector files; Ryan to Pace, 6 January 1921, ACUA, Pace Papers; Slawson, *NCWC,* 100–101.

117. "Welfare Council Issues Statement on the Smith-Towner Bill," *NCWC Bulletin* 2 (February 1921): 24; Slawson, *NCWC,* 102–3.

118. Minutes of the General National Committee, 13 January 1921, ACUA, NCWC Department of Education files; Minutes of the Meeting of the NCWC Department of Education, 2 February 1921, ACUA, Pace Papers; Slattery to the clergy, 19 January 1921, ACUA, Pace Papers.

119. Pace to Ryan, 12 January 1921, copy, ACUA, Pace Papers; Pace to Dowling, 14 January 1921, copy, ACUA, Vice-Rector files; Minutes of the General National Committee, 13 January 1921, ACUA, NCWC Department of Education files.

120. U.S. Congress, House, Committee on Education, *Create a Department of Education: Report to Accompany H.R. 7,* 66th Cong., 3d sess., 1921, H. Rept. 1201.

121. *NCWC News Sheet,* 17 January 1921; Blakely, "The Unamendable Smith Bill," *America* 24 (22 January 1921): 341–42; Boston *Pilot,* 22 January 1921.

122. Russell to Dyer, 27 January 1921, SAB, RG 10, box 23. During the 1930s, bills for federal aid to education ran into difficulties when the NAACP demanded that segregated black schools in the South receive equal funding (Gilbert E. Smith, *The Limits of Reform: Politics and Federal Aid to Education 1937–1950* [New York: Garland, 1982]).

123. David I. Walsh to Pace, 27 January 1921, ACUA, Vice-Rector files; Grantham, *Hoke Smith,* 335. Smith had been defeated in the Georgia Democratic primary by Tom Watson, the Great Populist turned 100-percenter.

124. Minutes of the General National Committee, 3 and 24 February 1921, ACUA, NCWC Department of Education files; Monahan to Dowling, 15 March 1921, with enclosure from the Washington *Star,* 13 March 1921, ACUA, Pace Papers. On the result of the ACE referendum, see "The Referendum on a Federal Department of Education," *Educational Review* 2 (April 1921): 43–51.

125. *Congressional Record,* 66th Cong., 3d sess., 1921, 60, pt. 2, 2102–3.

126. Ibid., 2183.

127. Pace to D. I. Walsh, 28 January 1921, copy, ACUA, Vice-Rector files.

128. Burke to O'Connell, 8 February 1921, copy, ACUA, NCWC General Secretary files.

129. *Congressional Record,* 66th Cong., 3d sess., 1921, 60, pt. 3, 2890; Minutes of the General National Committee, 11 February 1921, ACUA, NCWC Department of Education files.

130. *Congressional Record,* 66th Cong., 3d sess., 1921, 60, pt. 3, 3038–49.

131. Minutes of the Administrative Committee, 26 January 1922, ACUA, NCWC General Secretary files; "Editorial Comment: Menace of Federalization," *NCWC Bulletin* 3 (February 1922): 15–16.

132. Burke to O'Connell, 16 February 1921, copy, ACUA, NCWC General Secretary files; Circular from Slattery, 19 February 1921, AGU, Sullivan Papers.

133. Circular of William McGinley to Grand Knights, 15 February 1921, AKC, Education file; Slawson, *NCWC*, 106–8.

134. *New York Times*, 13 February 1921. On the Knights' troubled relationship with the NCWC, see Kauffman, *Faith and Fraternalism*, chapter 8; Slawson, *NCWC*, 26–37, 106–12.

135. Burke to Hanna, 23 February 1921, copy, ACUA, NCWC General Secretary files.

136. *New York Times*, 20 February 1921.

137. Burke to Hanna, 23 February 1921, copy, ACUA, NCWC General Secretary files.

138. U.S. Congress, Senate, Committee on Education and Labor, *Creating a Department of Education: Report to Accompany S. 1017*, 66th Cong., 3d sess., 1921, S. Rept. 824; D. I. Walsh to Fenlon, 8 March 1921, SAB, RG 13, box 7. Walsh noted that only five of the six signers actually favored the legislation.

139. Higham, *Strangers in the Land*, 300–11; Link, *Wilson and Progressive Era*, 60–61.

3. "Those Who Do Not Answer to the Lash"

1. U.S. Congress, House, H.R. 7, 67th Cong., 1st sess., 1921; U.S. Congress, Senate, S. 1252, 67th Cong., 1st sess., 1921; Monahan to Dowling, 15 March 1921, ACUA, Pace Papers; J. H. Ryan to Pace, 21 March 1921, ACUA, Pace Papers.

2. U.S. Congress, House, H.R. 7, 67th Cong., 1st sess., 1921; U.S. Congress, Senate, S. 1252, 67th Cong., 1st sess., 1921; George Strayer, "Report on Legislative Committee," *Addresses and Proceedings of NEA, 1921* 59 (1921): 161–62.

3. *Congressional Record*, 67th Cong., 1st sess., pt. 1, 169–73; Murray, *Harding Era*, 125–26, 170–72, 408, and *Politics of Normalcy*, 46–47; Eugene P. Trani and David L. Wilson, *The Presidency of Warren G. Harding* (Lawrence: Regents Press of Kansas, 1977), 47–48; John D. Hicks, *The Republican Ascendancy, 1921–1933* (New York: Harper and Row, 1960), 50–72; George H. Mayer, *The Republican Party, 1854–1966* (New York: Oxford University Press, 1967), 383.

4. Pace to Dowling, 4 April 1921, copy, ACUA, Vice-Rector files.

5. Blakely, "Judge Towner Rewrites His Bill," *America* 25 (23 April 1921): 21–22.

6. Blakely to Towner, enclosure with Blakely to William McGinley, 18 April 1921, copy, AKC, SC3–3–31. A local council of the Knights in Towner's district had protested the bill, prompting Towner to reply. The council forwarded his letter to Blakely for ideas about a response. His answer offers enlightenment about his thinking, captured in brief above.

7. Detroit *Michigan Catholic*, 9 June 1921.

8. *Denver Catholic Register*, 9 June 1921.

9. *Baltimore Catholic Review*, 23 April 1921.

10. New York *Catholic News*, 14 April 1921.

11. Cited in "The Fight Against Federal Aid for Schools," *Literary Digest* 69 (16 April 1921): 26.

12. Quoted in ibid.

13. Hicks, *Republican Ascendancy*, 53–55, 62–63; Mayer, *Republican Party*, 381–83; Leuchtenburg, *Perils of Prosperity*, 98, 101–2; Murray, *Politics of Normalcy*, 47–58. Murray portrays the bloc as simply a greedy, self-serving group. Hofstadter, *Age of Reform*, 285, considered farm-bloc progressivism "fake," apparently because they declined to bolt the major parties. Theodore Saloutos and John D. Hicks, *Agricultural Discontent in the Middle West, 1900–1939* (Madison: University of Wisconsin Press, 1951), 321–43, sympathize with the view that the bloc had of itself, i.e., that it was seeking to promote national interest and prosperity through legislation that benefited a weakened sector of the economy, one that had long suffered from the favoritism and special privilege shown to commerce and industry (see Arthur Capper, *The Agricultural Bloc* [New York: Harcourt, Brace and Company, 1922]). Graham, *Great Reforms*, 115–16, views the bloc as a progressive element in the politics of the 1920s. See also, Elmer D. Graper, "The American Farmer Enters Politics," *Current History* 19 (February 1924): 817–26; Kirschner, *City and Country*, 1–12.

14. Higham, *Strangers in the Land*, 308–11.

15. U.S. Congress, Senate, S. 1607, 67th Cong., 1st sess., 1921; Pace to Dougherty, 22 April 1921, copy, ACUA, Vice-Rector files; Lewis, *Philander Priestly Claxton*, 229–31; Murray, *Harding Era*, 125–26, 408; *Congressional Record*, 67th Cong., 2d sess., vol. 62, pt. 8, 8072.

16. Pace to Dougherty, 22 April 1921, copy, ACUA, Vice-Rector files.

17. Burke and Pace interview with Dougherty, 23 April 1921, ACUA, NCWC General Secretary files; Pace to Dowling, 25 April 1921, copy, ACUA, Vice-Rector files.

18. Pace to Dowling, 25 April 1921, copy, ACUA, Vice-Rector files; Dowling to NCCM, 4 May 1921, copy, ACUA, Vice-Rector files.

19. Slawson, *NCWC*, 111–12. The file of the Knights of Columbus on the education bill contains no entries after 1921, and as an organization the Knights ceased to agitate the issue.

20. Minutes of the General National Committee, 5 May 1921, ACUA, NCWC Department of Education files.

21. Minutes of the General National Committee, 12 May 1921, ACUA, NCWC Department of Education files; Pace to Howard, 9 May 1921, copy, ACUA, Pace Papers; Pace to Dowling, 9 May 1921, ACUA, Vice-Rector files; A. Abrams, "Policy of NEA," 93–94.

22. U.S. Congress, Senate, Committee on Education and Labor, *Department of Public Welfare: Joint Hearings on S. 1607 and H.R. 5837*, 67th Cong., 1st sess., 1921, 49–51; Strayer, "Report on Legislative Committee," 162.

23. For McCracken's altered view, see "Discussion," in *University of Illinois Bulletin* 19 (6 February 1922): 75.

24. U.S. Congress, Senate, Committee on Education and Labor, *Department of Public Welfare: Joint Hearings on S. 1607 and H.R. 5837*, 67th Cong., 1st sess., 1921, 90–101.

25. Ibid., 125–27.

26. Ibid., 131.

27. Frederick Kenkel to Shahan, 11 June 1921, with enclosure, Paul H. Vieth, general secretary, to the Executive Committee of the Missouri Sunday School Association, 2 June 1921, copy, recounting Walter Athearn's warning, ACUA, Nolan-Dixon Collection; Circular from Dowling, 4 May 1921, ACUA, Vice-Rector files; Press Bulletin of the

Central Bureau of the Central Verein, 4 June 1921, copy in ACUA, Nolan-Dixon Collection. The Central Bureau of the Central Verein discovered Athearn's drive and alerted church officials and the faithful.

28. Pace to Joseph Denning, 25 May 1921, copy, ACUA, Vice-Rector files. On the rumors about Harding's commitment to the bill, see Pace to Howard, 9 May 1921, copy, ACUA, Pace Papers; Pace to Dowling, 9 May 1921, ACUA, Vice-Rector files.

29. Murray, *Harding Era,* 408 and 415; and see below.

30. Monahan to Pace, 15 March 1921, with enclosure, "Position of the Roman Catholic Church in Educational Matters," ACUA, Pace Papers; NCWC Department of Education, Report on the Bureau of Education, 2 February 1921, ACUA, Pace Papers; NCWC Bureau of Education, Monthly Report for March 1921, ACUA, Vice-Rector files.

31. *NCWC News Sheet,* 13 June 1921; Monahan, "Work of the N.C.W.C. Bureau of Education," *NCWC Bulletin* 3 (June 1921): 18. For Monahan's true attitude on foreign languages, see Saint Paul *Catholic Bulletin,* 25 June 1921.

32. *Buffalo Catholic Union and Times,* 26 May 1921; Burke to Hanna, 25 July 1921, AASF, NCWC files; Minutes of the NCWC Department of Education, 25 June 1921, ACUA, NCWC Department of Education files; Martin, "Michigan School Controversy," 34–40.

33. Bishop Vincent Wehrle of Bismarck to the NCWC, 22 June 1921, copy, AUND, Kenkel Papers, part 2, box 8; Adolph Suess to Charles Korz, 30 June 1921, AUND, Kenkel Papers, part 2, box 8; Burke to Hanna, 25 July 1921, AASF, NCWC files.

34. Schrembs to Burke, 1 July 1921, ACUA, NCWC General Secretary files.

35. Burke to Hanna, 25 July 1921, AASF, NCWC files. Although the NCWC made no public disclaimer, its agents were apparently advised to deny that Monahan's views reflected organizational policy (see Robert Drady to Hanna, 12 July 1921, AASF, NCWC files).

36. Monahan, "Catholic Attitude Toward Public Schools," *Journal of Education* 94 (7 July 1921): 8–9; A. E. Winship to Monahan, 6 August 1921, copy, ACUA, Pace Papers; Winship to Monahan, 21 August 1921, copy, ACUA, Pace Papers.

37. For J. A. Ryan's pronouncements, see Slawson, *NCWC,* 100–105, 112–18.

38. Burke to Hanna, 25 July 1921, AASF, NCWC files; "Sterling-Towner Bill a Move Towards Paternalism," *NCWC Bulletin* 3 (September 1921): 25–26.

39. Minutes of the Administrative Committee, 19 September 1921, ACUA, NCWC General Secretary files; Dowling to Monahan, 30 September 1921, copy, ACUA, Pace Papers.

40. Detroit *Michigan Catholic,* 9 June 1921.

41. "The Sterling-Towner Bill," *America* 25 (18 June 1921): 206–7.

42. Dowling to Monahan, 30 September 1921, copy, ACUA, Pace Papers.

43. "Conference on the Towner-Sterling Bill," *Journal of the NEA* 2 (February 1922): 79; Murray, *Harding Era,* 415.

44. William Cochran (director of NCWC Legal Department) to Bishop Edmund Gibbons, 1 October 1921, ADA, NCWC files.

45. *New York Times,* 2 August 1921.

46. Warren G. Harding, "A Proclamation," *Supplement to the Messages and Papers of the Presidents Covering the Term of Warren G. Harding, March 4, 1921, to August 2, 1923,*

and the First Term of Calvin Coolidge, August 3, 1923, to March 4, 1925 (Washington D.C.: Bureau of National Literature, 1925), 9015–16.

47. Charl Ormond Williams et al., "A Petition for a Department of Education," October 1921, *Journal of the NEA* 10 (December 1921): facsimile reproduction on reverse of frontispiece; "A Notable Petition," ibid., 194; NEA *Press Bulletin*, 15 March 1922, ACUA, Vice-Rector files; *Washington Post*, 1 November 1921; *New York Times*, 1 November 1921.

48. Judd to Monahan, 25 November 1921, copy, with enclosures, "The Federal Government and Education" (editorial) and undated circular letter, both copies, ACUA, Pace Papers.

49. Capen, "Review of Recent Federal Legislation on Education," *University of Illinois Bulletin* 19 (6 February 1922): 20–27.

50. David Kinley, "Relation of State and Nation in Educational Policy," ibid., 31–46.

51. "Discussion," ibid., 68–69, 71–73.

52. Ibid., 98–100.

53. Towner, "Federal Aid to Education: Its Justification, Degree, and Method," ibid., 77–88. Towner's admission about the intention of congressional leaders was apparently given off the cuff, for it was not contained in his formal remarks. Judd, however, alluded to it during the discussion and later publicized it ("Discussion," 98; "Judge Towner's Announcement," *Elementary School Journal* 22 [January 1922]: 321).

54. Thomas Sterling, "Constitutional and Political Significance of Federal Legislation on Education," *University of Illinois Bulletin* 19 (6 February 1922): 89–97.

55. "Discussion," 69–71, 74–76.

56. Ibid., 69, 102–3.

57. "Judge Towner's Announcement," 321–24; Charles Judd, "Federal Standards of Educational Administration," *Elementary School Journal* 22 (February 1922): 415–26; same, "Federal Participation in Education," ibid. (March 1922): 494–504. For Monahan's view on a department of education see, Monahan to Dowling, 15 March 1921, copy, ACUA, Pace Papers; Monahan to Simeon Fess, chairman of the House Committee on Education, undated draft [May 1921], ACUA, NCWC Department of Education files.

58. Judd, "Federal Participation in Education," 503–4.

59. "Conference on the Sterling-Towner Bill," 79. According to the consumer price index, $125,000 in 1922 had a purchasing power equivalent to about $880,000 in 1988.

60. Burke to Muldoon, 4 March 1922, copy, ACUA, NCWC General Secretary files; Monahan to Burke, 9 March 1922, copy, ACUA, NCWC Department of Education files; Monahan to Dowling, 9 March 1922, copy, ACUA, Pace Papers.

61. "The Masons and Federal Education," *America* 26 (11 March 1922): 497.

62. Burris to [J. H. Ryan], "Confidential," undated, copy, ACUA, Vice-Rector files. A copy of the correspondence between the secretary general and Judd is supposed to be on file in the NCWC papers, but the writer has been unable to locate it. In March 1922 Magill spoke at a public school rally sponsored by the Masons (*New York Times*, 13 March 1922).

63. Murray, *Harding Era*, 415–16.

64. Burke to Hanna, telegram, 12 February 1922, AASF, NCWC files; Burke to Muldoon, 4 March 1922, copy, ACUA, NCWC General Secretary files; NEA Bulletin on

the Towner-Sterling Bill, 15 March 1922, copy, ACUA, Vice-Rector files; *Congressional Record,* 67th Cong., 2d sess., vol. 62, pt. 8, 8072; Circular from Edward Day, sovereign grand inspector general of the Southern Jurisdiction of the Scottish Rite in Montana, 18 February 1922, ACUA, Vice-Rector files. Encouraging members to telegraph congressmen, the circular included NEA materials and reminded brothers that Grand Commander John Cowles, head of the Southern Jurisdiction, wanted all members to cooperate with the supreme council's educational program, especially the Sterling-Towner bill. "The Rite everywhere is alert and active," said Day. "Shall we not make an especial effort to show our response to this call." On Republican factiousness and insurgency, see Murray, *Harding Era,* 128, 190, 314–16, and *Politics of Normalcy,* 43–47. Although Burke believed that Fess did not wish the Sterling-Towner bill reported because he opposed the measure, the evidence argues that Fess supported the measure but wanted to await the report of the Reorganization Committee before pressing for it (*Congressional Record,* 67th Cong., 2d sess., vol. 62, pt. 8, 8072–73).

65. Kirschner, *City and Country,* 1–7.

66. On the voting of committee members, see *Congressional Record,* 67th Cong., 1st sess., vol. 61, pt. 1, 327–55, 1033–46; pt. 2, 2035; pt. 3, 3043–44; pt. 5, 4556–57, 5341–60; pt. 6, 5449; vol. 62, pt. 3, 253–55; pt. 4, 3458–63.

67. Burke to Archbishop Michael Curley of Baltimore, 13 March 1922, AAB, B2048; NEA Bulletin on the Sterling-Towner Bill, 15 March 1922, copy, ACUA, Vice-Rector files. The NEA Bulletin indicated that at the February meeting with the House committee, Fess and his colleagues argued persuasively that the bill could not be reported until the Reorganization Committee issued its findings, but Burke's contention that Fess intended to report the measure is more compelling, borne out too by evidence external to his correspondence. In short, the NEA seemed to be glossing over the fact that it had almost carried the day, but ultimately failed to do so by a hair's breadth.

68. Burke to Hanna, telegram, 1 March 1922, AASF, NCWC files; Burke to Michael Curley, 13 March 1922, AAB, B2048; NEA Bulletin on the Sterling-Towner Bill, 15 March 1922, copy, ACUA, Vice-Rector files.

69. "Department of Superintendence," *Addresses and Proceedings of NEA, 1922* 60 (1922): 1322–46; Burke to Curley, 13 March 1922, copy, ACUA, NCWC General Secretary files. Winship gave the measure qualified support. "We confess to having had some sympathy with the non-bureaucratic arguments, with the feeling that possibly a compromise was wise . . ." ("Towner-Sterling bill," *Journal of Education* 95 [23 March 1922]).

70. "Federal Participation in Education," *School Review* 30 (April 1922): 244–45.

71. Quoted in Burke to Curley, 13 March 1922, AAB, B2048; Burke to Muldoon, 4 March 1922, copy, ACUA, NCWC General Secretary files; J. H. Ryan to Muldoon, 13 March 1922, copy, ACUA, NCWC Department of Education files; NEA Bulletin on the Towner-Sterling Bill, 15 March 1922, copy, ACUA, Vice-Rector files; *Congressional Record,* 67th Cong., 2d sess., 1922, 62, pt. 4, 3627.

72. Murray, *Harding Era,* 318, and *Politics of Normalcy,* 86.

73. William Bankhead to Fess, 9 March 1922, printed in *Congressional Record,* 67th Cong., 2d sess., 1922, 62, pt. 5, 4821; Fess to Bankhead, 11 March 1922, printed in ibid.

74. Hugh Magill, "Report of the Field Secretary," *Addresses and Proceedings of NEA, 1922* 60 (1922): 117–18; *Congressional Record,* 67th Cong., 2d sess., 1922, 62, pt. 5, 4821–22. According to a sworn deposition given in 1928 by David C. Stephenson, former Grand

Dragon of Indiana, who was at that time in prison for the rape and murder of Madge Oberholtzer, the Sterling-Towner bill "had to do with the abolition of parochial schools, and had for its objective some other purposes." Said Stephenson, the Klan "schemed against the men it could not use to its purposes, and invariably the weapon used in the effort to destroy these men was a woman of loose morals." As payback for his obstruction of the Sterling-Towner bill, the hooded empire attempted "to involve the name of . . . Fess with a woman" (*New York Times,* 2 April 1928; U.S. Congress, House, Committee on Education, *Proposed Department of Education: Hearing on H.R. 7,* 70th Cong., 1st sess., 1928, 291. Hereafter cited as *House Hearing, 1928.* The quotations are taken from both sources).

75. Burke to Muldoon, 4 March 1922, copy, ACUA, NCWC General Secretary files; *Congressional Record,* 67th Cong., 2d sess., 62, 1922, pt. 4, 4192; J. H. Ryan to Muldoon, 22 March 1922, copy, ACUA, NCWC J. H. Ryan Papers; *NCWC News Sheet,* 27 March 1922.

76. Burke to E. Gibbons, 21 March 1922, ADA, NCWC files.

77. *New York Times,* 6 April 1922.

78. San Francisco *Monitor,* 5 August 1922.

79. Slawson, *NCWC,* 123–46, 156–61; Douglas J. Slawson, " 'The Boston Tragedy and Comedy': The Near-Repudiation of Cardinal William O'Connell," *Catholic Historical Review* 77 (October 1991): 616–43.

80. The Administrative Committee of the NCWC, *Report to His Holiness, Pope Pius XI, On the Work of the Administrative Committee of the National Catholic Welfare Council,* 14, copy in ACUA, NCWC General Secretary files. For a more extended and complete treatment of the report, see Elizabeth McKeown, "Apologia for an American Catholicism: The Petition and Report of the National Catholic Welfare Council to Pius XI, April 25, 1922," *Church History* 43 (December 1974): 514–28.

81. Slawson, *NCWC,* 143–49, 160, 329 n. 70. For the debilitating effect that public knowledge of the decree had on the NCWC's effectiveness, see Burke to Muldoon, 8 and 9 June 1922, copies, ACUA, NCWC General Secretary files; Slawson, "John J. Burke," 72–73.

82. Strayer and Magill to Deans and Professors of Education, 15 March 1922, ACUA, Vice-Rector files.

83. Burris to Strayer, 20 March 1922, copy, ACUA, Vice-Rector files.

84. "Discussion of Federal Participation in Education," *Elementary School Journal* 22 (April 1922): 561–63.

85. "A Large Appropriation to Education," *School Review* 30 (May 1922): 321–25.

86. John Cowles to Burris, 26 April 1922, copy, ACUA, Vice-Rector files.

87. J. H. Ryan to Giovanni Bonzano, 1 May 1922, copy, ACUA, NCWC General Secretary files; J. H. Ryan to Burke, Thursday afternoon, [1 June 1922], second under that date, ACUA, NCWC General Secretary files; Slawson, *NCWC,* 152, 162, 163.

88. "The Issue Stated," *New Age* 30 (May 1922): 281–82. Emphasis in original.

89. *NCWC News Sheet,* 29 May 1922.

90. Magill, "Report of Field Secretary," *Addresses and Proceedings of NEA, 1922* 60 (1922): 118; J. H. Ryan to L. S. Walsh, 9 May 1922, copy, ACUA, NCWC J. H. Ryan Papers; *Conressional Record,* 67th Cong., 2d sess., vol. 62, pt. 8, 8072.

91. Magill, "Report of Field Secretary," 118.

92. *Congressional Record,* 67th Cong., 2nd sess., vol. 62, pt. 8, 8073.

93. J. H. Ryan to Walsh, 9 May 1922, ACUA, NCWC J. H. Ryan Papers.

94. Dowling to Burke, 21 August 1922, ACUA, NCWC General Secretary files; Slawson, *NCWC,* 156–78.

95. Slawson, *NCWC,* 175–91.

4. A Battle on Two Fronts

1. Dumenil, "Tribal Twenties," 30.

2. David B. Tyack, "The Perils of Pluralism: The Background of the Pierce Case," *American Historical Review* 74 (October 1968): 75–76.

3. Quoted in Paul M. Holsinger, "The Oregon School Bill Controversy, 1922–1925," *Pacific Historical Review* 37 (August 1968): 330.

4. Pace, Memorandum on the School Question, undated [February 1919], ACUA, Vice-Rector files; Pace to Walsh, 6 February 1919, copy, ACUA, Vice-Rector files; Russell to Burke, 6 May 1922, ACUA, NCWC, General Secretary files.

5. Tyack, "Perils of Pluralism," 77; Lloyd P. Jorgenson, "The Oregon School Law of 1922: Passage and Sequel," *Catholic Historical Review* 54 (October 1968): 456–58; Jorgenson, *State and Non-Public School,* 207–8; Dumenil, *Freemasonry,* 143–45; Cincinnati *Catholic Telegraph,* 10 February 1921; NCWC news service, 2 October 1922; Chalmers, *Hooded Americanism,* 50, 180; Charles C. Alexander, *The Ku Klux Klan in the Southwest* (Lexington: University of Kentucky Press, 1966), 112.

6. Chalmers, *Hooded Americanism,* 32–33, 88–89; Jackson, *Klan in the City,* 204–5; Jorgenson, "Oregon School Law," 455–59; Tyack, "Perils of Pluralism," 74–85; Veverka, *"For God and Country,"* 56–60.

7. Veverka, *"For God and Country,"* 73–82.

8. J. H. Ryan, "Education in a Democracy," in Minutes of Second Annual Convention of the NCCW, 21–22 November 1922, 2: 119–36, ACUA, NCWC NCCW files.

9. Tyack, "Perils of Pluralism," 86–88; Jorgenson, "Oregon School Law," 458–60; J. G. Shaw, *Edwin Vincent O'Hara: American Prelate* (New York: Farrar, Straus and Company, 1956), 98–105; Thomas J. Shelley, "The Oregon School Case and the National Catholic Welfare Conference," *Catholic Historical Review* 75 (July 1989): 441–43; Minutes of the meeting of the Department of Education, 11 December 1922, ACUA, NCWC George Johnson Papers.

10. Minutes of the Fourth Annual Meeting of the American Hierarchy, 27 September 1922, ACUA, NCWC General Secretary files; J. H. Ryan to Patrick Henry Callahan, 3 December 1923, copy, ACUA, NCWC J. H. Ryan Papers.

11. Alvin M. Owsley, "The Peace-Time Program of the American Legion," *Addresses and Proceedings of NEA, 1922* 60 (1922): 220–24; J. M. Gwinn, "Report of Committee to Cooperate with the American Legion," *Addresses and Proceedings of NEA, 1923* 61 (1923): 399–402; Higham, *Strangers in the Land,* 256.

12. "National Conspiracy Against the Catholic Schools," *NCWC Bulletin* 4 (October 1922): 21–22. The Northern Jurisdiction refused to involve itself in politics on grounds that it would be "exceedingly dangerous to the unity, strength, usefulness and

welfare of the order . . ." (quoted in J. H. Ryan, "Opposition to the Catholic School in the United States," *Ecclesiastical Review* 70 [February 1924]: 126).

13. Minutes of the Administrative Committee, 11 January 1923, ACUA, NCWC General Secretary files. Burris shared Ryan's view. So too did one of the drafters of the original Smith bill, who turned opponent to its successors. Without identifying the person, Burris quoted a letter from him, stating that the measure was ardently promoted by "a little group who have for the past several years dominated the N.E.A." This unidentified educator voiced the same objections against the bill as did Catholics: it would lead to federal control of education despite the alleged safeguards written into it. Another issue that "as much as anything else alienated my interest," continued the anonymous schoolman, was that the bill was "supported so ardently by all the educational bolsheviks," an unflattering reference to the Ku Klux Klan and the Scottish Rite. "The logic, if it may be called that, used [by the NEA] in urging the passage of resolutions [in the bill's favor]," he continued, "has been that of the camp meeting exhorter." He denounced the NEA's membership drive as simply a tactic to "attract teachers into the association in order to get money with which to hire walking delegates to go out and to get more money with which to support this bill" (NCWC, news service, 22 October 1923).

14. Notes on meeting held Monday, 11 December 1922, ACUA, Vice-Rector files.

15. Ibid.

16. Shelley, "Oregon School Case," 443.

17. Minutes of the Administrative Committee, 11 and 12 January 1923, ACUA, NCWC General Secretary files; Guthrie to Burke, 5 January 1923, ACUA, NCWC General Secretary files; Slawson, *NCWC*, 192–94.

18. *NCWC News Sheet*, 1 and 22 January 1923 and 26 February 1923; Burke to Hanna, 24 January 1923, copy, ACUA, NCWC General Secretary files.

19. Jackson, *Klan in the City*, 237; Mark Hurley, *Church-State Relationships in Education in California* (Washington, D.C.: The Catholic University of America Press, 1948), 103–7; Reynold Blight, editor of the *New Age*, to B. W. Reed, 26 November 1923, copy, ACUA, NCWC General Secretary files; Scott to Burke, undated, ACUA, NCWC General Secretary files.

20. Hurley, *Church-State Relationships*, 107-8; NCWC news service, 23 April 1923.

21. *NCWC News Sheet*, 7 April 1924 and 19 May 1924.

22. *Los Angeles Examiner*, 13 May 1923; *Washington Times*, 14 May 1923.

23. J. H. Ryan, "Catholic School Defense League Explained: Voluntary Organization to Defend Right of Catholic Education," *NCWC Bulletin* 4 (March 1923): 27–28; "Assembled Prelates Approve Reports of N.C.W.C.: Hierarchy Lauds Splendid Work of Past Year," ibid. (October 1923): 9.

24. Minutes of the Administrative Committee, 11 January 1923, ACUA, NCWC General Secretary files.

25. J. H. Ryan to Joseph Smith, 30 April 1923, copy, ACUA, Pace Papers; Dowling to J. H. Ryan, 12 May 1923, copy, ACUA, Pace Papers; Burke to Muldoon, 22 May 1923, copy, ACUA, NCWC General Secretary files; Burke to Hanna, 23 May 1923, AASF, NCWC files.

26. Pace to Hanna, 19 June 1923, AASF, Catholic University of America file.

27. Minutes of the Administrative Committee, 12 April 1923, ACUA, NCWC General Secretary files.

28. "Declaration of Principles," *CEA Bulletin* 20 (November 1923): 39–40.

29. Zabel, *God and Caesar,* 140–47.

30. William D. Guthrie and Bernard Hershkopf, *Brief of Amici Curiae: Supreme Court of the United States, October Term 1922, Robert T. Meyer v. State of Nebraska,* printed copy in AASF, NCWC files; Slawson, *NCWC,* 195. The brief was actually written by Garret McEnerney, perhaps with the help of Hershkopf, Guthrie's partner.

31. Burke to Muldoon, 2 March 1923, copy, ACUA, NCWC General Secretary files; Burke to John Kavanaugh, 1 March 1923, ACUA, NCWC General Secretary files; Burke to Hanna, 27 February 1923, AASF, NCWC files.

32. *Meyer* v. *Nebraska,* 262 U.S. 390 (1923); Zabel, *God and Caesar,* 147–48.

33. *Oregon School Cases: Complete Record* (Baltimore: Belvedere, 1925), 36–58.

34. Quoted in *House Hearing, 1928,* 290–91.

35. *Congressional Record,* 68th Cong., 1st sess., 1924, 65, pt. 10, 10248, 10262–63, 10267.

36. Truman R. Clark, *Puerto Rico and the United States, 1917–1933* (Pittsburgh: University of Pittsburgh Press, 1975), 48–75; Arturo Morales Carrión, *Puerto Rico: A Political and Cultural History* (New York: W. W. Norton & Company, 1983), 204–10.

37. *New York Times,* 16 February 1923; Ann Gibson Buis, "An Historical Study of the Role of the Federal Government in the Financial Support of Education, with Special Reference to Legislation Proposals and Action" (Ph.D. diss., Ohio State University, 1953), 82–87; Report of the chairman of the Department of Education, submitted 22 January 1924, ACUA, Vice-Rector files. The tally of the referendum can be found in U.S. Congress, House, Committee on Education, *To Create a Department of Education and to Authorize Appropriations of Money to Encourage the States in the Promotion and Support of Education: Hearings on H.R. 3923,* 68th Cong., 1st sess., 1924, 408, 419–20.

38. "A Set-back for Paternalism," *NCWC Bulletin* 4 (March 1923): 13.

39. *New York Times,* 15 March 1923.

40. *Massachusetts* v. *Mellon,* 262 U.S. 447 (1923).

41. Ibid.

42. Paul Murphy, *The Constitution in Crisis Times, 1918-1968* (New York: Harper and Row, 1972), 53–54.

43. Mayer, *Republican Party,* 385, 388; Burner, *Politics of Provincialism,* 103–6; Hicks, *Republican Ascendancy,* 88–89; Schriftgiesser, *This Was Normalcy,* 100–101, 127–30; Murray, *Politics of Normalcy,* 84–85.

44. Burner, *Politics of Provincialism,* 3–27, 74–102.

45. Strayer, "Report of the Legislative Commission," *Addresses and Proceedings of NEA, 1923* 61 (1923): 398–99.

46. Williams, "Report of the Field Secretary, Legislative Division," ibid., 398.

47. A. Abrams, "Policy of NEA," 96.

48. *Washington Post,* 25 October 1923.

49. NCWC news service, 25 February 1924.

50. Omaha *True Voice,* 26 October 1923.

51. Chalmers, *Hooded Americanism,* 34, 67–68, 123, 138, 191, 219, 283; Alexander, *Klan in the Southwest,* 94–95; Jackson, *Klan in the City,* 95, 161–62, 219, 259.

52. Reynold Blight to B. W. Reed, 26 November 1923, copy, ACUA, NCWC General Secretary files; Chalmers, *Hooded Americanism,* 283.

53. Minutes of the Department of Education, 22 November [1923], ACUA, NCWC Department of Education files; Burke to Muldoon, 23 December 1923, copy, ACUA, NCWC General Secretary files.

54. Paul Johnson, "Calvin Coolidge and the Last Arcadia," in *Calvin Coolidge and the Coolidge Era: Essays on the History of the 1920s*, ed. John Earl Haynes (Washington, D.C.: Library of Congress, 1998), 4–7; Claude M. Fuess, *Calvin Coolidge: The Man From Vermont* (Hamden, Conn.: Archon Books, 1965), 381–82; Robert H. Ferrell, *The Presidency of Calvin Coolidge* (Lawrence: University Press of Kansas, 1998), 167–75; Robert Sobel, *Coolidge: An American Enigma* (Washington, D.C.: Regnery Publishing, 1998), 310–13.

55. Calvin Coolidge, "First Annual Message Delivered at a Joint Session, 6 December 1923," in *The State of the Union Messages of the Presidents,* ed. Fred Israel (New York: Chelsea House–Robert Hector Publishers, 1966), 3:2650–51, 2652; George B. Nash, "The 'Great Enigma' and the 'Great Engineer,'" in *Calvin Coolidge and the Coolidge Era: Essays on the History of the 1920s*, ed. John Earl Haynes (Washington, D.C.: Library of Congress, 1998), 149; William Allen White, *A Puritan in Babylon: The Story of Calvin Coolidge* (New York: Macmillan, 1938), 267–68.

56. John McCracken, "Report of the Committee on Federal Legislation," *Educational Record* 5 (July 1924): 155–56.

57. U.S. Congress, Senate, S. 1337, 68th Cong., 1st sess., 1923; U.S. Congress, House, H.R. 3923, 68th Cong., 1st sess., 1923.

58. "Address delivered by Hon. David I. Walsh before the Southern Berkshire County Teachers' Convention, 5 October 1923," in *Congressional Record,* 68th Cong., 1st sess., 1923, 65, pt. 1, 308; Blakely, "'The Same Old Bill,'" *America* 30 (5 January 1924): 290–91; Davenport *Catholic Messenger,* 10 April 1924; Los Angeles *Tidings,* 11 April 1924; San Antonio-Dallas *Southern Messenger,* 17 January 1924; *Denver Catholic Register,* 13 March 1924.

59. *Denver Catholic Register,* 3 April 1924.

60. Chicago *New World,* 14 March 1924.

61. Jackson, *Klan in the City,* 290 n. 13.

62. J. H. Ryan, "Opposition to the Catholic Schools," *Ecclesiastical Review* 70 (February 1924): 125–31.

63. *Congressional Record,* 68th Cong., 1st sess., 1924, 65, pt. 1, 536–60.

64. U.S. Congress, Senate, Committee on Education and Labor, *To Create a Department of Education and to Encourage the States in the Promotion and Support of Education: Hearings on S. 1337*, 68th Cong., 1st sess., 1924, 106–16, 123–67 (hereafter cited as *Senate Hearing, 1924*); Minutes of meeting of the Department of Education, 22 January 1924, ACUA, NCWC Johnson Papers; J. H. Ryan to William Borah, 23 January 1924, copy, ACUA, NCWC Johnson Papers.

65. *Senate Hearing, 1924*, passim.

66. U.S. Congress, House, H.R. 6562, 68th Cong., 1st sess., 1924.

67. J. H. Ryan, "The Dallinger Education Bill," *NCWC Bulletin* 5 (March 1924): 9–10.

68. Chicago *New World,* 4 April 1924.

69. Wiley Marshall, "An Amended Sterling-Reed Bill," *America* 30 (23 February 1924): 459–60.

70. "The Dallinger Bill," *School Review* 32 (April 1924): 243–44.

71. Report of the Chairman of the Department of Education, Submitted 28 January 1925, ACUA, Vice-Rector files; Minutes of the Administrative Committee, 1 March 1924, ACUA, NCWC General Secretary files; *House Hearings, 1924*, 408, 419–20, 560–76.

72. *Congressional Record,* 68th Cong., 1st sess., vol. 65, pt. 10, 10520–21; *House Hearings, 1924*, 31 and 54.

73. McCracken, "Report of the Committee on Federal Legislation," 154.

74. Higham, *Strangers in the Land,* 312–24 (the first quote is from here); Slawson, *NCWC,* 237–38; Marty, *Pilgrims in Their Own Land,* 391 (the second quote is from here).

75. *Congressional Record,* 68th Cong., 1st sess., 1924, 65, pt. 10, 10248 and 10256; Burke to Hanna, 9 February 1924, copy, ACUA, NCWC General Secretary files.

76. Burke to Muldoon, 24 February 1924, copy, ACUA, NCWC General Secretary files; J. H. Ryan to Burke, 9 May 1924, ACUA, NCWC General Secretary files; Report of the Chairman of the Department of Education of the National Catholic Welfare Conference for the Year Ending 30 June 1924, ACUA, Vice-Rector files.

77. [Burke], Memorandum, 2 February 1924, ACUA, NCWC General Secretary files. It is important to note that Strayer viewed the NEA in the same way. The association was a public-interest lobby that would be ruined by political alliances. "We believe . . . that the professional organization of teachers should not affiliate with any other body," he had told the NEA convention in 1920. "Those who serve the whole public cannot afford to serve parties or groups within that public. We know of no surer way to destroy confidence in the American public-school system and our profession . . ." (Strayer, "A National Program for Education," 47).

78. Minutes of the Administrative Committee, 1 May 1924, ACUA, NCWC General Secretary files.

79. Ryan to Burke, telegram, 25 May 1924, ACUA, NCWC General Secretary files; Burke to Ryan, telegram, 25 May 1924, ACUA, NCWC General Secretary files.

80. Burke to Ryan, 3 June 1924, ACUA, NCWC General Secretary files; Minutes of the Administrative Committee, 1 May 1924, ACUA, NCWC General Secretary files.

81. Burke to J. H. Ryan and W. J. Cochran, 1 March 1924, copy, ACUA, NCWC General Secretary files; J. H. Ryan to Muldoon, 7 May 1924, ACUA, NCWC J. H. Ryan Papers; Minutes of the Departmental Meeting, 9 May 1924, ACUA, NCWC Department of Education files; J. H. Ryan to E. Gibbons, copy, 2 June 1924, ACUA, NCWC J. H. Ryan Papers.

82. J. H. Ryan to Muldoon, 7 May 1924, copy; J. H. Ryan to Muldoon, 20 May 1924, copy ACUA, NCWC J. H. Ryan Papers; J. H. Ryan to Gibbons, 2 June 1924, copy, ACUA, NCWC J. H. Ryan Papers; Robert K. Murray, *The 103rd Ballot: Democrats and the Disaster in Madison Square Garden* (New York: Harper & Row, 1976), 52–53; Burner, *Politics of Provincialism,* 107–35; Hicks, *Republican Ascendancy,* 94–97.

83. J. H. Ryan to Muldoon, 6 June 1924, copy, ACUA, NCWC J. H. Ryan papers; J. H. Ryan to Schrembs, 6 June 1924, copy, ACUA, NCWC J. H. Ryan papers; J. H. Ryan to Muldoon, 18 June 1924, copy, ACUA, NCWC J. H. Ryan papers; J. H. Ryan to Gibbons, 8 July 1924, copy, ACUA, NCWC J. H. Ryan papers; *New York Times,* 12 June 1924.

84. Quoted in Schriftgiesser, *This Was Normalcy,* 175; Hicks, *Republican Ascendancy,* 90–91.

85. J. H. Ryan to Burke, 24 June 1924, ACUA, NCWC General Secretary files; J. H. Ryan to Gibbons, 8 July 1924, copy, ACUA, NCWC J. H. Ryan Papers; J. H. Ryan to Muldoon, 8 July 1924, copy, ACUA, NCWC J. H. Ryan Papers; J. H. Ryan to Burke, 30 June 1924, ACUA, NCWC General Secretary files; Chalmers, *Hooded Americanism,* 204–12; Murray, *103rd Ballot,* 90, 104, 108–9.

86. *New York Times,* 29 June 1924.

87. J. H. Ryan to Burke, 30 June 1924, ACUA, NCWC General Secretary files.

88. Murray, *103rd Ballot,* 164–218; Burner, *Politics of Provincialism,* 107–35; Hicks, *Republican Ascendancy,* 94–97.

89. Strayer, "Report of the Committee on Legislation," *Addresses and Proceedings of NEA, 1924,* 62 (1924): 264–65; Williams, "Report of Legislative Service," ibid., 128–29.

90. James W. Crabtree, "Report of the Secretary of the Association," ibid., 103–5.

91. J. A. C. Chandler, "Report of Committee on Resolutions," ibid., 54.

92. Bagley, "Report of the Editorial Council," ibid., 294.

93. Washington *Evening Star,* 2 July 1924; *Washington Post,* 3 July 1924.

94. San Francisco *Monitor,* 19 July 1924.

95. J. H. Ryan to Burke, 3 July 1924, ACUA, NCWC General Secretary files.

96. *New York Times,* 5 July 1924.

97. Ibid.; [Judd], "The National Education Association," *School Review* 32 (September 1924): 485–86.

98. Hicks, *Republican Ascendancy,* 97–101; Mayer, *Republican Party,* 397–99; Burner, *Politics of Provincialism,* 128–30; Schriftgiesser, *This Was Normalcy,* 189–97.

99. Hicks, *Republican Ascendancy,* 101–2; Burner, *Politics of Provincialism,* 136–41, 150, 161, 165–78 (the quotation is on 175); Schriftgiesser, *This Was Normalcy,* 204–14; Murray, *103rd Ballot,* 221–56.

100. *NCWC News Sheet,* 7 April 1924.

101. *NCWC News Sheet,* 21 January 1924; J. Grant Kinkle, secretary of state, *A pamphlet Containing Copies of All Measures "Proposed by Initiative Petition," Measures "Passed by the Legislature and Referred to the People," and "Laws Passed by the Legislature and Referred to the People by Petition," together with "Amendments to the Constitution Proposed by the Legislature"* (Olympia: Frank M. Lamborn, 1924), 3; Chalmers, *Hooded Americanism,* 86, 90, 216.

102. Chalmers, *Hooded Americanism,* 218; Jackson, *Klan in the City,* 195; *NCWC News Sheet,* 18 February and 26 May 1924; NCWC news service, 7 July 1924.

103. Quoted in Martin, "Michigan School Controversy," 47–48; Pies, "Parochial School Campaigns," 231–36.

104. Charles N. Lischka, "Americanism Argument a Mere Pretext in School Crisis: Natural and Guaranteed Rights, Not Nationalism, the Real Issue," *NCWC Bulletin* 6 (September 1924): 18–20. Emphasis in original.

105. NCWC news service, 6 October 1924.

106. "Our Respects to Joseph Scott," *New Age* 32 (October 1924): 588–89.

107. W. P. O'Connell, executive secretary of the Educational Committee of Seattle, to J. H. Ryan, 4 October 1924, copy, ACUA, NCWC General Secretary files; Martin,

"Michigan School Controversy," 57; Pies, "Parochial School Campaigns," 231–37; Jackson, *Klan in the City*, 195.

108. Saint Paul *Catholic Bulletin*, 28 February 1925.

109. C. H. Kutz to Mary Monahan, 6 June 1924, copy, ACUA, NCWC Department of Education files; Slawson, *NCWC*, 82.

110. Quoted in Burke to Thomas O'Mara, 23 March 1925, copy, ACUA, NCWC General Secretary files.

111. *Pierce* v. *Society of Sisters*, 268 U.S. 510 (1925).

112. Ibid.

113. NCWC news service, 6 October 1924.

114. "The Education Bill," *New Age* 32 (November 1924): 645–46. This issue also contained the most sustained anti-Catholic editorial, running fifteen pages, published by the magazine in the period under study.

115. Strayer, "Report of the Legislative Committee," *Addresses and Proceedings of NEA, 1925* 63 (1925): 244; Williams, "Report of the Legislative Division NEA," ibid., 1049–51; NCCW, "News Sheet to Organizations," November-December 1924, ACUA, NCWC NCCW files; Dumenil, *Modern Temper*, 47, 108–10.

116. Minutes of the Fourth Annual Convention of the NCCW, November 1924, 491–93, ACUA, NCWC NCCW files.

117. *New York Times*, 12 December 1924.

118. NCCW, "News Sheet to Organizations," November-December 1924, ACUA, NCWC NCCW files.

119. Charles R. Mann to Members of ACE, 10 November 1924, ACUA, Pace Papers; Minutes of the Executive Committee of the ACE, 2 January 1925, ACUA, Pace Papers; John McCracken, "Report of the Committee on Federal Legislation," *Educational Record* 6 (July 1925): 194–95.

120. W. J. Cochran, "The Proposed Department of Education and Relief," *NCWC Bulletin* 6 (January 1925): 5–6; Saint Paul *Catholic Bulletin*, 24 January 1925.

121. Minutes of the Department of Education, 28 January 1925, ACUA, NCWC Johnson Papers.

122. *Congressional Record*, 68th Cong., 2d sess., 1925, 66, pt. 3, 2707–09.

5. Guardians or Betrayers of Catholic Interests?

1. Strayer, "A New Education Bill," *Journal of the NEA* 14 (May 1925): 171.

2. Strayer, "A Federal Department of Education," *Educational Record* 6 (July 1925): 227–33. The bill was clearly drafted at this meeting because it is printed in full here, bearing the date 10 March 1925.

3. Minutes of the Administrative Committee, 23 April 1925, ACUA, NCWC General Secretary files.

4. Minutes of the Departmental Meeting, 30 April 1925, ACUA, NCWC Department of Education files.

5. The draft of the bill is printed in Strayer, "Report of the Legislative Commission," *Addresses and Proceedings of NEA, 1925* 63 (1925): 246–48, and in

Educational Record 6 (July 1925): 230–33. According to the consumer price index, $1.5 million in 1925 had a purchasing power equivalent to $16.15 million in 2004.

6. [Judd], "A New Bill Providing for a Federal Department of Education," *School Review* 33 (September 1925): 482. See also Judd, "A New Bill Providing for a Federal Department of Education," *Elementary School Journal* 26 (September 1925): 13–15.

7. Strayer, "New Education Bill," 171.

8. Williams, "Report of Legislative Division, NEA," *Addresses and Proceedings of NEA, 1925*, 63 (1925): 1049–50.

9. Mabel Wilson, "Federal Aid to Education," *Addresses and Proceedings of NEA, 1925*, 63 (1925): 388–94.

10. Minutes of the Departmental Meeting, 3 July 1925, ACUA, NCWC Department of Education files.

11. Programme for the Meeting of the Bishops at The Catholic University of America, 16–17 September 1925, ACUA, Annual Meeting of Hierarchy file.

12. Minutes of the Administrative Committee, 14 September 1925, ACUA, NCWC General Secretary files.

13. *Minutes of the Hierarchy, 1925*, 5–7, ACUA, NCWC General Secretary files. See also Bishop Howard's account of the conversation: enclosure with Howard to Curley, 17 December 1925, AABa, H1418.

14. *Minutes of the Hierarchy, 1925*, 7–8, ACUA, NCWC General Secretary files; Howard, enclosure with Howard to Curley, 17 December 1925, AABa, H1418.

15. *Minutes of the Hierarchy, 1925*, 8, ACUA, NCWC General Secretary files.

16. Minutes of the Administrative Committee, 17 September 1925, ACUA, NCWC General Secretary files.

17. Hartford *Catholic Transcript*, 17 September 1925; New York *Catholic News*, 5 September 1925.

18. Blakely, "The Nose of the Smith Towner Camel," *America* 33 (3 October 1925): 583–84; John Wiltbye [Blakely], "Politics and the Old Smith-Towner," ibid. (13 June 1925): 215–16; "Federal Control of Schools," ibid. (1 August 1925): 378; Boston *Pilot*, 10 October 1925; Cleveland *Catholic Bulletin*, 28 November 1925.

19. Cleveland *Catholic Bulletin*, 28 November 1925.

20. Milwaukee *Catholic Citizen*, 14 November 1925.

21. *New York Times*, 2 July 1925.

22. "The Oregon Law," *New Age* 33 (July 1925): 389–90; A Special Contributor, "The Oregon School Law Decision," ibid., 401–2.

23. Quoted in "Masonry and Americanism," *America* 33 (11 July 1925): 305.

24. Ibid., 305–6.

25. "Majorities and Minorities," *New Age* 33 (August 1925): 453–54.

26. Mark O. Shriver, "Federal Education Again," *Commonweal* 2 (26 August 1925): 363–64.

27. Los Angeles *Tidings*, 28 August 1925.

28. *Denver Catholic Register*, 10 September 1925; Detroit *Michigan Catholic*, 6 August 1925; Saint Paul *Catholic Bulletin*, 31 October and 7 November 1925; San Antonio-Dallas *Southern Messenger*, 12 November 1925.

29. "Where the Fault Lies," *New Age* 33 (October 1925): 581–82.

30. "Priestly Condemnation," *New Age* 33 (December 1925): 709–11; NCWC news service, 16 November 1925.

31. Keller, *Regulating Society,* 26–30.

32. Hiram Wesley Evans, "The Klan's Fight for Americanism," *North American Review* 223 (March–April–May 1926): 33–63.

33. Circular from Burke, 12 October 1925, copy, ACUA, NCWC General Secretary files.

34. Felix Pitt, secretary of the diocesan school board of Louisville, to Burke, 20 October 1925; Michael Keyes of Savannah to Burke, 17 October 1925, with enclosure Joseph Mitchell to Keyes, 17 October 1925; Edward Allen of Mobile to Burke, 24 October 1925, with enclosure C. B. to Allen; John Baptist Morris of Little Rock to Burke, 19 October 1925; J. M. Kirwin, rector of Saint Mary's Seminary, LaPorte, Texas, to Burke, 20 October and 27 October 1925; Francis Kelley of Oklahoma City to Burke, 15 October 1925; James Cantwell, chancellor of San Francisco, to Burke, 28 October 1925; Edward Kelley, chancellor of Baker City, Oregon, to Burke, 17 October 1925; John O'Hara, editor of the *Catholic Sentinel* (Portland), to Burke, 19 October 1925; Edward P. Ryan, state deputy of Knights of Columbus, Spokane, to Burke, 18 October 1925; Edward O'Dea of Seattle to Burke, 20 October 1925; Daniel Gorman of Boise to Burke, 21 October 1925; John T. Nicholson to Burke, 16 November 1925; all in ACUA, NCWC General Secretary files.

35. Burke to Muldoon, 9 November 1925, copy, ACUA, NCWC General Secretary files.

36. *New York Times,* 5 September 1925.

37. Coolidge, "State of the Union Message," 3 December 1924 and 8 December 1925, *State of the Union Messages,* 3:2655–57, 2670–72.

38. Minutes of the Departmental Meeting, 27 November 1925, ACUA, NCWC Department of Education files.

39. J. J. Cochran to Frederick Kenkel, 30 November 1925, AUND, Kenkel Papers II, box 8.

40. U.S. Congress, Senate, S. 291, 69th Cong., 1st sess., 1925; U.S. Congress, House, H.R. 5000, 69th Cong., 1st sess., 1925.

41. J. H. Ryan, "Educational Legislation Affecting Private Schools," *Catholic Educational Review* 20 (January 1926): 5–6. For Coolidge's views, see William J. Cochran, "The Beginning of the End of Paternalism," *NCWC Bulletin* 7 (July 1925): 6–7, and 31; Fuess, *The Man from Vermont,* 370–71; Nash, "'Great Enigma,'" 149.

42. Howard to Curley, 10 and 17 December 1925, AABa, H1414 and H1418.

43. Curley to Howard, 14 and 19 December 1925, copy, AABa, H1415 and H1417.

44. Burke to Muldoon, 2 January 1926, copy, ACUA, NCWC General Secretary files.

45. Ibid.; Curley to Howard, 4 January 1926, copy, AABa, H1419.

46. Parsons to Curley, 24 December 1925, AABa, P235; Curley to Parsons, 26 December 1925, copy, AABa, P236.

47. The evidence here is cryptic but clear. John Cochran, secretary of Representative Hawes, made it plain to Kenkel that an organization had apparently changed its view toward federal education legislation. Kenkel's later correspondence makes it evident that the organization at issue was the NCWC. See J. J. Cochran to

Kenkel, 15 and 23 January 1926, AUND, Kenkel Papers II, box 8; Kenkel, circular to Central Verein, 3 February 1926, copy, AUND, Kenkel Papers II, box 8.

48. Richard Haberlin to Burke, 28 December 1925, copy, AASF, NCWC files.

49. Burke to W. H. O'Connell, 30 December 1925, copy, AASF, NCWC files.

50. "An Alarm and a Warning," *America* 34 (2 January 1926): 271; Burke to Muldoon, 2 January 1926, copy, ACUA, NCWC General Secretary files.

51. Burke to Wilfrid Parsons, 1 January 1926, copy, ACUA, NCWC General Secretary files.

52. Parsons to Burke, 4 January 1926; Burke to Parsons, 5 January 1926, copy; Parsons to Burke, 7 January 1926; Burke to Lawrence Kelly, 18 January 1926, copy; Parsons to Burke, 8 February 1926; all in ACUA, NCWC General Secretary files

53. Burke to Muldoon, 2 January 1926, copy, ACUA, NCWC General Secretary files.

54. Burke to Dowling, 2 January 1926, copy, ACUA, NCWC General Secretary files; Dowling to Burke, 6 January 1926, ACUA, NCWC General Secretary files.

55. Burke to Muldoon, 7 January 1926, copy, ACUA, NCWC General Secretary files; Curley to Howard, 8 January 1926, copy, AABa, H1420.

56. Curley to Howard, 8 January 1926, copy, AABa, H1420.

57. Howard to Curley, 8 February 1926, AABa, H1423.

58. Department of Education of the Ancient and Accepted Scottish Rite of Freemasonry, Orient of Wyoming, *Education Bill Not Abandoned* (Nebraska: Nebraska Education League, 1926), enclosure with Nicholson to Burke, 15 January 1926; William Hafey of Raleigh to Burke, 7 January 1926; Kirwin to Burke, 8 January 1926; all in ACUA, NCWC General Secretary files.

59. William Montavon, memorandum, 14 January 1926, ACUA, NCWC Legal Department files.

60. Burke telephone interview with Peter Gerry, 22 January 1926, ACUA, NCWC General Secretary files; Andrew Haley, Report number 11, 4 February 1926, ACUA, NCWC Legal Department files. The information confirming Gerry's assessment of Curtis's attitude comes from the Haley report. As will be seen, Haley, a Catholic, was a correspondent for a non-Catholic paper. The NCWC engaged him to gather intelligence on congressional attitudes.

61. Burke to J. H. Ryan, telegram, 26 January 1926, copy, ACUA, NCWC General Secretary files; Burke to E. Gibbons, 27 January 1926, ADA, NCWC files.

62. Minutes of the Executive Committee of the Department of Education, 27 January 1926, ACUA, NCWC George Johnson Papers.

63. Burke to Hanna, telegram, 28 January 1926, AASF, Diocesan files; Hanna to Burke, handwritten on back of same.

64. Burke to Lawrence Phipps, 12 February 1926, copy, AUSCC, General Secretary files; Burke to hierarchy, [27 February 1926], AASF, NCWC files.

65. Andrew Haley, Report Number 15, 12 February 1926, ACUA, NCWC Legal Department files.

66. Ibid.

67. Howard to Curley, 15 February 1926, AABa, H1425; Curley to Parsons, 17 February 1926, copy, AABa, P237; Parsons to Curley, 19 February 1926, AABa, P238; Curley to Howard, 20 February 1926, copy, AABa, H1426; Burke to Muldoon, 26 February 1926, copy, ACUA, NCWC General Secretary files.

68. Circular from Agnes Regan, 15 February 1926, copy, ACUA, Vice-Rector files; Circular from J. H. Ryan, 12 February 1926, ACUA, NCWC Department of Education files; Burke to hierarchy, [27 February 1926], AASF, NCWC files.

69. Burke to Muldoon, 26 February 1926, copy, ACUA, NCWC General Secretary files.

70. J. H. Ryan, *The Curtis-Reed Bill—A Criticism* (Washington, D.C.: NCWC, 1926); Distribution sheet for *The Curtis-Reed Bill—A Criticism,* 18 February 1926, ACUA, NCWC Department of Education files.

71. Williams, "Report of Legislative Commission, NEA," *Addresses and Proceedings of NEA, 1926,* 64 (1926): 1135–36; Burke to E. Gibbons, 27 January 1926, ADA, NCWC files; Montavon to Mark Lally, 13 February 1926, copy, ACUA, NCWC Legal Department files; Montavon interview with Williams, 20 February 1926, ACUA, NCWC Legal Department files.

72. Burke to Muldoon, 26 February 1926, copy, ACUA, NCWC General Secretary files.

73. *Washington Post,* 22 February 1926.

74. U.S. Congress, Senate, Committee on Education and Labor, *Proposed Department of Education: Joint Hearings on S. 291, H.R. 5000, and S. 2841,* 69th Cong., 1st sess., 1926, 8-47, passim. Hereafter cited as *Joint Hearings, 1926.*

75. Burke to Muldoon, 25 February 1926, copy, ACUA, NCWC General Secretary files.

76. *Joint Hearings, 1926,* 95–172, 237–94, passim.

77. Ibid., 161–64.

78. Ibid., 164–66.

79. Ibid., 280.

80. Ibid., 576.

81. Slawson, *NCWC,* 93–94, 206–7.

82. Ibid., 202–4, 228–31.

83. *Congressional Record,* 69th Cong., 1st sess., 1926, 67, pt. 4, 4590-91.

84. *Washington Post,* 26 February 1926; "The Washington Resolutions," *Journal of the NEA* 15 (April 1926): 121.

85. Washington *Star,* 6 March 1926.

86. J. H. Ryan, *Editorial Opinion and the Curtis-Reed Bill* (Washington, D.C.: National Capital Press, 1926).

87. Mark Lally to Montavon, 15, 21, and 28 January 1926, ACUA, NCWC Legal Department files; Montavon to Lally, 23 January 1926, copy, ACUA, NCWC Legal Department files; Lally to Montavon, 25 February 1926, ACUA, NCWC Legal Department files; Montavon to Lally, 4 March 1926, copy, ACUA, NCWC Legal Department files; Lally to Montavon, 8 March 1926, ACUA, NCWC Legal Department files.

88. Francis Crowley to J. H. Ryan, 1 March 1926, ACUA, Department of Education files.

89. *New York Times,* 10 March 1926; John Braeman, "The American Polity in the Age of Normalcy," in *Coolidge and Coolidge Era,* 29. After his 1924 state of the union address, Coolidge never again called for passage of the Reorganization bill and reiterated his call for a department of education and relief only in 1927 (*State of Union Messages,* 3:2669–2725).

90. *New York Times,* 10 March 1926.

91. Ibid., 13 March 1926.

92. Montavon to Charles Tobin, 13 March 1926, copy, ACUA, NCWC Legal Department files. The draft of the bill is in ACUA, NCWC Legal Department files.

93. Burke interview with Loring Black, 6 March 1926, ACUA, NCWC General Secretary files; Montavon, memorandum, 8 March 1926, ACUA, NCWC Legal Department files; Minutes of the Administrative Committee, 13 April 1926, ACUA, NCWC General Secretary files; U.S. Congress, Senate, S. 3533, 69th Cong., 1st sess., 1926.

94. NCWC news service, 15 March 1926, extra.

95. Circular from Phipps, 16 March 1926, ACUA, NCWC Legal Department files; Montavon to Marie Dougherty, undated, copy, ACUA, NCWC Legal Department files.

96. *Indiana Catholic and Record,* 19 March 1926; Saint Paul *Catholic Bulletin,* 12 April 1926; Cincinnati *Catholic Telegraph,* 18 March 1926; Cleveland *Catholic Bulletin,* 9 April 1926.

97. *Pittsburgh Catholic,* 15 April 1926.

98. Wiltbye [Blakely], "From Smith-Towner to Phipps," *America* 34 (3 April 1926): 595–96; John J. Delaney, *Dictionary of American Catholic Biography* (Garden City, N.Y.: Doubleday and Company, 1984), 48. See also W. C. Murphy, "The Phipps Bill," *America* 35 (26 June 1926): 256–68.

99. Quoted in Williams, "Report of Legislative Division," 1138–39.

100. U.S. Congress, Senate, Committee on Education and Labor, *Extension of Purpose and Duties of the Bureau of Education: Report to Accompany S. 3533,* 69th Cong., 1st sess., 1926, S. Rept. 776.

101. Williams, "Report of Legislative Division," 1135–40.

102. Strayer, "Report of the Legislative Commission, NEA," *Addresses and Proceedings of NEA, 1926* 64 (1926): 221–23.

103. McCracken, "Report of the Committee on Federal Legislation," *Educational Record* 7 (July 1926): 169–71.

104. Lally to Montavon, 9 November 1926, ACUA, NCWC Legal Department files; Circular from John Leach, 28 October 1926, ACUA, NCWC Legal Department files.

105. Patrick Henry Callahan to Kenkel, 5 March 1925, AUND, Kenkel Papers, part 1, box 25.

106. "By What Authority?" *Fortnightly Review* 33 (15 May 1926): 215–16.

107. Preuss to Kenkel, 24 May 1926, AUND, Kenkel Papers, part 1, box 26.

108. James R. Ryan to Preuss, 21 May 1926, ACBCCU, Preuss Papers, reel 6.

109. Burke to Curley, 5 December 1925, AABa, B2072.

110. J. R. Ryan to Preuss, 21 May 1926, ACBCCU, Preuss Papers, reel 6.

111. "More Light on the Activities of the N.C.W.C.," *Fortnightly Review* 33 (15 June 1926): 259–60.

112. J. R. Ryan to Preuss, 21 June 1926, ACBCCU, Preuss Papers, reel 6; J. R. Ryan to Preuss, 25 June 1926, ACBCCU, Preuss Papers, reel 6.

113. J. R. Ryan to Preuss, 25 June 1926; J. R. Ryan to Preuss, 30 July 1926; J. R. Ryan to Preuss, 30 August 1926; all in ACBCCU, Preuss Papers, reel 6. The quotation is from the last letter.

114. [J. R. Ryan and Grattan Kerans], "The N.C.W.C. and Diocesan Autonomy," *Fortnightly Review* 33 (15 August 1926): 355–56. The other two articles were "More

N.C.W.C. Meddling and Muddling," ibid. (15 July 1926): 307–8, and a letter to the editor signed *Ignotus* (Unknown), ibid. (1 September 1926): 393–94.

115. Curley to Pietro Fumasoni-Biondi, 19 March 1926, copy, AABa, Roman Documents; Douglas J. Slawson, "The National Catholic Welfare Conference and the Church-State Conflict in Mexico, 1925–1929," *The Americas* 47 (July 1990): 56–60; Thomas Spalding, *The Premier See: A History of the Archdiocese of Baltimore, 1789–1989* (Baltimore: Johns Hopkins University Press, 1989), 349–50; Parsons to Curley, 14 February 1927, AABa, P246; Slawson, *NCWC,* 130–31.

116. Burke interview with W. H. O'Connell, 12 September 1926, ACUA, NCWC General Secretary files.

117. Goldberg, "Hooded Empire in Colorado," 118 and 131; Jackson, *Klan in the City,* 226–28; Chalmers, *Hooded Americanism,* 127–29.

118. Burke interview with W. H. O'Connell, 12 September 1926, ACUA, NCWC General Secretary files.

119. Ibid.

120. Ibid.

121. Minutes of the Administrative Committee, 13 September 1926, ACUA, NCWC General Secretary files.

122. *Minutes of the Hierarchy, 1926,* 6–7, ACUA, NCWC General Secretary files.

123. [J. R. Ryan and Kerans], "Senator Phipps and the Phipps Bill," *Fortnightly Review* 33 (1 October 1926): 433; J. R. Ryan to Preuss, 23 September 1926, ACBCCU, Preuss Papers, reel 6. By late October NCWC personnel were beginning to suspect that there was an inside informant behind the *Fortnightly* articles, and Kerans was the prime suspect because of his friendship with Ryan. To deflect suspicion, Preuss published another article in which Kerans collaborated on a thinly veiled attack against himself. Within days of its publication, Kerans had become the object of "sympathy and condolence" rather than suspicion ("Notes and Gleanings," *Fortnightly Review* 33 [15 November 1926]: 514; J. R. Ryan to Preuss, 4 November 1926, ACBCCU, Preuss Papers, reel 6; Kerans to Preuss, 4 November 1926, ACBCCU, Preuss Papers, reel 6; J. R. Ryan to Preuss, 18 November 1926, ACBCCU, Preuss Papers, reel 6).

124. Davenport *Catholic Messenger,* 11 November 1926.

125. "The Phipps Federal Education Bill," *America* 36 (30 October 1926): 54.

126. Burke to Hanna, 13 December 1926, copy, ACUA, NCWC General Secretary files.

127. Minutes of the Department of Education, 13 January 1927, ACUA, NCWC Johnson Papers.

128. Parsons to Curley, 14 February 1927, with enclosure, AABa, P246; Curley to Parsons, 17 March 1927, copy, AABa, P250.

129. Curley to Dowling, 25 March 1927, copy, AABa, D1283.

130. Burke to Hanna, 13 December 1926, copy, ACUA, NCWC General Secretary files; Montavon to J. H. Ryan, 8 February 1927, copy, ACUA, NCWC Legal Department files.

131. Minutes of the Administrative Committee, 27 April 1927, ACUA, NCWC General Secretary files.

132. [J. R. Ryan and Kerans], " 'Nationalism' in Catholic America," *Fortnightly Review* 34 (15 February 1927): 79. On *Action Française,* see Adrien Dansette, *Religious*

History of Modern France, 2 vols., trans. John Dingle (New York: Herder and Herder, 1961), 2:280–84; Aubert, *Church in Secularized Society,* 549–51.

133. "The Action Francaise [*sic*] and Nationalism," *Fortnightly Review* 34 (1 March 1927): 95.

6. The Religious Issue to the Fore

1. Williams, "Policy of the National Education Association Towards Federal Legislation," *Addresses and Proceedings of NEA, 1927,* 65 (1927): 152–56.

2. Circular from Williams, 16 September 1927, ACUA, NCWC Department of Education files.

3. Broadside from the Boston *Advertiser,* 21 August 1927, in ACUA, NCWC Department of Education. The same editorial can be found in the *Los Angeles Examiner,* 21 August 1927.

4. Bureau of Census, *Historical Statistics,* 214.

5. Circular from Williams, 1 October 1927, ACUA, NCWC Department of Education files.

6. NCCW "News Sheet to Organizations," August 1927, ACUA, Vice-Rector files.

7. "Dangers of Federalization Pointed out by Rt. Rev. Msgr. James H. Ryan, D.D.," *NCWC Bulletin* 9 (November 1927): 10.

8. *New York Times,* 16 April 1927.

9. Herbert Hoover, *The Memoirs of Herbert Hoover: The Cabinet and the President, 1920–1933* (New York: Macmillan, 1952), 316.

10. *New York Times,* 26 November 1927; *State of the Union Addresses,* 3: 2724–25.

11. U.S. Congress, House, H.R. 7, 70th Cong., 1st sess., 1927; U.S. Congress, Senate, S. 1584, 70 Cong., 1st sess., 1927.

12. *Congressional Record,* 70th Cong., 1st sess., 1927, 69, pt. 1, 144–46.

13. William Davidson, "Report of Legislative Commission," *Addresses and Proceedings of NEA, 1928,* 66 (1928): 192.

14. [James S. Vance], *Proof of Rome's Political Meddling in America* (Washington, D.C.: Fellowship Forum, 1927), 8–23.

15. Ibid., 47–48, 117–21.

16. "Education Committee," *New Age* 36 (March 1928): 138–39. Four of the five Catholics on the committee were Democrats: Loring Black of New York, René DeRouen of Louisiana, John Douglass of Massachusetts, and Vincent Palmisano of Maryland. The fifth was a Republican, Louis Monast of Rhode Island, who was reputed to be a member of the Knights of Pythias, a secret society condemned by Rome in 1894 (NCWC Legal Department, Confidential Record of Congressional Hearing and Other Items, for the Information Only of Heads of Departments and Bureaus, 28 April 1928, vol. 2, no. 40, ACUA, NCWC Department of Education files).

17. "School at Auction," *New Age* 36 (April 1928): 203.

18. Blakely is quoted in "The Issue Stated," *New Age* 30 (May 1922): 281–82; Dolan, *American Catholic Experience,* 275–79; *Official Catholic Directory, 1920.*

19. Montavon to NCWC Department heads, 23 February 1928, ACUA, NCWC Legal Department files; Montavon to J. H. Ryan, 11 April 1928, copy, ACUA, NCWC Legal Department files.

20. Lally to Montavon, telegram, 27 April 1928; Montavon to Lally, telegram, undated, copy; Lally to Montavon, 4 May 1928; all in ACUA, NCWC Legal Department files.

21. U.S. Congress, House, *Hearing Before the Committee on Education, House of Representatives, Seventieth Congress, First Session, on H.R. 7, A Bill to Create a Department of Education and for Other Purposes, April 25–28 and May 2, 1928* (Washington, D.C.: U.S. Government Printing Office, 1928), 20–25. Hereafter cited as *House Hearing, 1928*. For Strayer's view on federal aid, see *Joint Hearing, 1926*, 12–13.

22. Ibid., 39–42, 135–37.

23. Ibid., 129, 166.

24. Ibid., 28–29.

25. Ibid., 30–32.

26. Ibid., 52–55.

27. NCWC Legal Department, Confidential Record, 28 April 1928, vol. 2, no. 40, ACUA, NCWC Department of Education files.

28. *House Hearing, 1928*, 367–73.

29. Ibid., 377.

30. Ibid., 377–81.

31. NCWC Legal Department, Confidential Record, 28 April 1928, vol. 2, no. 40, ACUA, NCWC Department of Education files.

32. *House Hearing, 1928*, 441–44; Thomas Slater, S.J., *A Manual of Moral Theology for English-Speaking Countries*, 2 vols. (New York: Benziger Brothers, 1909), 1:275–76. Robsion actually read two full paragraphs from the book. The above quote is the essential section.

33. *House Hearing, 1928*, 444–45.

34. Ibid., 445.

35. Ibid. 546–50; NCWC Legal Department, Confidential Record, 28 April 1928, vol. 2, no. 40, ACUA, NCWC Department of Education files.

36. "Congressional Appropriation for the Study of Secondary Education," *School Review* 37 (February 1929): 81–82.

37. [McCracken], "Pending Educations Bills," *Educational Record* 9 (January 1928): 48.

38. McCracken, "Federal Legislation," *Educational Record* 9 (October 1928): 220–21. Even though this report was not published until October, internal evidence indicates that it was written after the hearing and prior to the national party conventions.

39. Montavon to Burke, 9 June 1928, ACUA, NCWC General Secretary files.

40. Burke to Montavon, 9 June 1928, copy, ACUA, NCWC General Secretary files; Burke to Bishop Thomas Lillis of Kansas City, 9 June 1928, copy, ACUA, NCWC General Secretary files; Burke to Colonel Donavon, 11 June 1928, copy, ACUA, NCWC General Secretary files; Minutes of the Executive Committee of the NCCW, 7–10 June 1928, ACUA, NCCW files.

41. Montavon, Memorandum Report on the Republican National Convention Held at Kansas City, 12–15 June 1928, ACUA, NCWC General Secretary files. Hereafter cited as Memorandum Report.

42. Gilbert Fite, *George N. Peek and the Fight for Farm Parity* (Norman: University of Oklahoma Press, 1954), 3–202; Burner, *Politics of Provincialism,* 161, 165–78; Farrell, *Presidency of Coolidge,* 88–94; Sobel, *Coolidge,* 274–76, 326–27, 330–34.

43. Montavon, Memorandum Report.

44. Ibid.; *NCWC News Sheet,* 18 June 1928.

45. Montavon, Memorandum Report.

46 Ibid.; Montavon to Burke, 13 June 1928, ACUA, NCWC General Secretary files; *New York Times,* 15 June 1928. After six terms in the House, Vare was elected to the Senate in 1926 in a contest suspected of fraud. His seat remained vacant for three years while a committee investigated the charges. In November 1929 Governor John Fisher of Pennsylvania appointed Joseph Grundy to the vacancy. A month later the Senate Committee on Elections finally reported that Vare had made excessive campaign expenditures, and after two days of debate, the upper chamber refused to seat him. Within a week Grundy filled the vacancy (*Congressional Record,* 71st Cong., 2d sess., 1929, 72, pt. 1, 197, 539–40).

47. Mayer, *Republican Party,* 403–4; Burner, *Politics of Provincialism,* 195 n. 8.

48. Burner, *Politics of Provincialism,* 194; Hicks, *Republican Ascendancy,* 210; Mayer, *Republican Party,* 411–12; Schriftgiesser, *This Was Normalcy,* 248–49.

49. Burke to George Walsh, 23 June 1928, copy, ACUA, NCWC General Secretary files; Dumenil, *Modern Temper,* 226–35; Hicks, *Republican Ascendancy,* 203.

50. "Governor Smith and the Progressives," *New Republic* 53 (1 February 1928): 284–85; Burner, *Politics of Provincialism,* 182–90.

51. Montavon to Burke, Memorandum (Confidential) on the Democratic National Convention Held at Houston, Texas, 26–30 June 1928, ACUA, NCWC General Secretary files (hereafter cited as Memorandum [Confidential]). The most recent study of the 1928 campaign corroborates Montavon's assessment of evangelical strategy (Allan J. Lichtman, *Prejudice and the Old Politics: The Presidential Election of 1928* [Chapel Hill: University of North Carolina Press, 1979]).

52. Montavon, Memorandum (Confidential). For questions about Smith's position on public education, see Charles C. Marshall, "An Open Letter to the Honorable Alfred E. Smith," *Atlantic Monthly* 139 (April 1927): 543–46; Edmund A. Moore, *A Catholic Runs for President: The Campaign of 1928* (New York: Ronald Press Company, 1956), 25, 61, 69.

53. Montavon, Memorandum (Confidential); Democratic Party, *The Official Report of the Proceedings of the Democratic National Convention Held at Houston, Texas, 26–29 June 1928* (Indianapolis: Bookwalter-Ball-Greathouse, 1928), 198–99.

54. Democratic Party, *Official Report,* 197; Montavon, Memorandum (Confidential). The first quotation is from the former; the second from the latter.

55. Moore, *Catholic Runs for President;* Lichtman, *Prejudice and Old Politics;* Leuchtenburg, *Perils of Prosperity,* 229–40; Hicks, *Republican Ascendancy,* 201–14; Paul F. Boeller, Jr., *Presidential Campaigns* (Oxford and New York: Oxford University Press, 1985), 223–27; Kirschner, *City and Country,* 41–53; Burner, *Politics of Provincialism,* 194–97, 202–16.

56. Davidson, "Report of Legislative Commission," *Addresses and Proceedings of NEA, 1928,* 66 (1928): 190–93.

57. Mayer, *Republican Party*, 405; Farrell, *Presidency of Coolidge*, 92; Hicks, *Republican Ascendancy*, 201.

58. Davidson, "Report of Legislative Commission," *Addresses and Proceedings of NEA, 1928*, 66 (1928): 200–201.

59. Minutes of the Department of Education, 7 February 1929, ACUA, NCWC Johnson Papers.

60. U.S. Congress, House, H.R. 16165, 70th Cong., 2d sess., 1929.

61. *Congressional Record*, 70th Cong., 2d sess., 1929, 70, pt. 5, 4424–27.

62. Ibid., 4427–28. Robsion makes the first mention of congressional polls that this author has seen. Efforts to discover further information about them proved fruitless.

63. Ibid., 4428. Robsion erred regarding the number of students in school. The figure was actually 25 million (Bureau of Census, *Historical Statistics*, 207).

64. *Congressional Record*, 70th Cong., 2d sess., 1929, 70, pt. 5, 4431–32. Black's figures on local and state funding and the number of children in school are correct (Bureau of Census, *Historical Statistics*, 207–8).

7. The Triumph of Home Rule

1. Ray Lyman Wilbur and Arthur Mastick Hyde, *The Hoover Policies* (New York: Scribner's, 1937), 74-79; NCWC news service, 11 March 1929, "Washington Letter."

2. Moore, *Catholic Runs for President*; Lichtman, *Prejudice and Old Politics*; Leuchtenburg, *Perils of Prosperity*, 229–40; Hicks, *Republican Ascendancy*, 201–14; Boeller, *Presidential Campaigns*, 223–27.

3. San Francisco *Monitor*, 9 February 1929; Saint Paul *Catholic Bulletin*, 23 February 1929; Eugene Lyons, *Herbert Hoover: A Biography* (Garden City, N.Y.: Doubleday, 1964), 27, 179.

4. Davis Newton Lott, ed., *The Presidents Speak: The Inaugural Addresses of the American Presidents from Washington to Kennedy* (New York: Holt, Rinehart, and Winston, 1962), 226.

5. Minutes of the Meeting of NCWC Department Heads, 8 March 1929, ACUA, NCWC Department of Education files.

6. NCWC news service, 11 March 1929, "Washington Letter"; NCCW, "Monthly Message to Affiliated Organizations," March 1929, ACUA, NCWC NCCW files.

7. Richard Washburn Child, "Hoover—Or Some Other?" *Saturday Evening Post*, 16 March 1929, 25, 149–54.

8. NCWC news service, 1 April 1929, "Editorial Service."

9. Davenport *Catholic Messenger*, 4 April 1929; Detroit *Michigan Catholic*, 4 April 1929; Saint Paul *Catholic Bulletin*, 6 April 1929; New York *Catholic News*, 13 April 1929; Los Angeles *Tidings*, 3 May 1929.

10. U.S. Congress, House, H.R. 10, 71st Cong., 1st sess., 1929.

11. U.S. Congress, House, H.R. 2570, 71st Cong., 1st sess., 1929.

12. *Congressional Record*, 71st Cong., 1st sess., vol. 71, pt. 5, 5861; Buis, "Role of the Federal Government in Financial Support of Education," 114–17.

13. Ray Lyman Wilbur, "Local Self-Government in Education," *Educational Record* 10 (July 1929): 182–83. Because the *Record* was published quarterly, Wilbur's speech, delivered in May, was printed in the July issue.

14. Saint Paul *Catholic Bulletin*, 11 May 1929; Boston *Pilot*, 25 May 1929; Omaha *True Voice*, 10 May 1929; Cleveland *Catholic Universe Bulletin*, 17 May 1929. See also *Baltimore Catholic Review*, 10 May 1929; Davenport *Catholic Messenger*, 9 May 1929; San Antonio-Dallas *Southern Messenger*, 16 May 1929.

15. Quoted in *Editorial Opinion on Secretary Wilbur's Address "Local Self-Government in Education"* (Baltimore: Belvedere Press, 1929), 10.

16. Ibid., 10–13, and 19. Some of the other major dailies that welcomed Wilbur's message included the *Washington Post*, the Washington *Evening Star*, the Baltimore *Sun*, the Philadelphia *Public Ledger*, the Brooklyn *Eagle*, the *Hartford Daily Courant*, the *Boston Sunday Post*, the *Indianapolis Star*, the *Detroit Free Press*, and the Saint Paul *Pioneer Press*.

17. "Convention News," 33rd Annual Convention, National Congress of Parents and Teachers, Washington, 8 May 1929, copy in ACUA, NCWC Department of Education files. Crabtree's allegation about secretaries of the interior was misleading. Secretary Franklin K. Lane (1917–21) was nationalist with respect to education (Chapter 1 above; "Education a National Concern, Says Secretary Lane," *School Life* 2 [1 January 1919]: 3–4; "Sees Washington as Nation's School Center: 'Line Can be Drawn Between School Control and School Guidance,' Says Secretary Lane—Urges National University," *School Life* 4 [1 March 1920: 1–2].

18. Burke interview with Herbert Hoover, 22 May 1929, ACUA, NCWC General Secretary files; Burke to Bishop Hugh Boyle of Pittsburgh, 25 May 1929, copy, ACUA, NCWC General Secretary files.

19. Burke to Boyle, 25 May 1929, copy, ACUA, NCWC General Secretary files; Johnson to Howard, 28 May 1929, ANCEA, Howard Papers; Wilbur to Pace, 27 May 1929, copy, AAB, P62. By 1927 the Catholic Educational Association had achieved such stature that the Executive Board added the word "National" to the title.

20. Quoted in Circular from Burke to NCWC Administrative Committee, 22 June 1929, AASF, NCWC files.

21. Questions submitted to the NACE, 7 June 1929, ACUA, Vice-Rector files; "The Relations of the National Government to Education," *School and Society* 29 (15 June 1929: 781–83.

22. "Relations of the National Government to Education," 781–83.

23. Davidson, "Report of the Legislative Commission—Abstract," *Addresses and Proceedings of NEA, 1929* 67 (1929): 190–91.

24. "Report of the Committee on Resolutions to the Representative Assembly of the National Education Association, Atlanta, Georgia, 3 July 1929," ACUA, NCWC Department of Education files.

25. In the Matter of the Federal Education Measure as Considered at the National Education Association Convention: Confidential Report of Mrs. Bacon, 8 July 1929, ACUA, Department of Education files.

26. *Congressional Record*, 71st Cong., 1st sess., vol. 71, pt. 5, 5861.

27. Crabtree, "The Challenge of the Rural Schools: Secretary Crabtree's Reunion Address," *Journal of the NEA* 18 (November 1929): 273–75.

28. Martin Carmody to Burke, 24 September 1929, with enclosure: circular from Vance, August 1929, ACUA, Department of Education files. Carmody was a Chicago Knight of Columbus, who received a copy of the handbill as it was being distributed to passengers on a platform of the city's elevated railway.

29. U.S. Congress, Senate, S. 1586, 71st Cong., 1st sess., 1929; Murray, *Harding Era,* 128–29.

30. *Los Angeles Examiner,* 7 September 1929. The Hearst chain had long supported a department of education and functioned as a unit in this regard. The *Examiner* is taken as representative.

31. Ibid., 8 September 1929.

32. Ibid., 10 September 1929.

33. Ibid., 13 September 1929.

34. "Senator Capper to the Rescue," *America* 41 (21 September 1929): 557.

35. San Francisco *Monitor,* 12 October 1929.

36. Indianapolis *Indiana Catholic and Record,* 13 September 1929.

37. NCWC news service, 28 October 1929.

38. *Congressional Record,* 71st Cong., 2d sess., 1929, 72, pt. 1, 483–86, 508.

39. Saint Paul *Catholic Bulletin,* 2 November 1929; San Antonio-Dallas *Southern Messenger,* 21 November 1929; circular letter from Arthur Capper and John Robsion, undated, printed in *Fellowship Forum* (Washington), 14 December 1929, also in San Francisco *Monitor,* 29 March 1930.

40. *Baltimore Catholic Review,* 3 January 1930. Capper's papers included the *Topeka Daily Capital, Capper's Weekly, Farmer's Mail and Breeze, Household Magazine, Capper's Farmer, Missouri Ruralist, Nebraska Farm Journal,* and *Oklahoma Farmer.*

41. Chicago *New World,* 3 January 1930; *Pittsburgh Catholic,* 26 December 1929; San Antonio-Dallas *Southern Messenger,* 13 February 1930; Saint Paul *Catholic Bulletin,* 15 February 1930; Francis Crowley, "Education Bills in the 71st Congress," *Catholic Educational Review* 27 (December 1929): 581–82.

42. Johnson to Howard, 11 December 1929, ANCEA, Howard Papers.

43. Draft of an editorial by Johnson, 12 December 1929, for release January 1930, ACUA, NCWC Johnson Papers. The editorial appeared in at least the Los Angeles *Tidings* (10 January 1930). The *Forum* clippings to which Johnson referred bore the following titles: "Foreign-minded Group Continues Fight on Education Bill," "Where We Stand—A Declaration of Principles," and "Rome Successfully Blocks Public School Bill" (Crowley to Margaret Lynch, 7 May 1929, copy, ACUA, NCWC Department of Education files).

44. Minutes of Department Heads, 3 May 1929, ACUA, NCWC Department of Education files.

45. Francis M. Crowley, "Education Bills in the 71st Congress," *Catholic Educational Review* 27 (December 1929): 577–89. In the 1928 election, the twelve largest cities in the nation voted Democratic for the first time in decades, marking a major shift in American politics. This was due in part, no doubt, to an exceptionally heavy turnout of voters, many of them probably Catholic immigrants who voted for the first time. The urban-rural split in the 1928 election became evident in the movement of major metropolitan areas into the Democratic column, paving the way for Democratic victories

in the 1930s and 1940s (Leuchtenburg, *Perils of Prosperity*, 234; Hicks, *Republican Ascendancy*, 212).

46. "Dictatorial Education," *Commonweal* 11 (25 December 1929): 209–10.

47. Milwaukee *Catholic Citizen*, 15 and 22 March 1930, and 3 May 1930.

48. *Congressional Record*, 71st Cong., 2d sess., 1929, 72, pt. 2, 524, 544. To determine the positions of committeemen, the author used the *Congressional Record* and the NCWC's report on the position of members of the 1928 committee—NCWC Legal Department, Confidential Record, 28 April 1928, vol. 2, no. 40, ACUA, NCWC Department of Education files.

49. See previous chapters; also Montavon to Burke, memorandum, 3 February 1930, copy, ACUA, NCWC Legal Department files; *Congressional Record*, 71st Cong., 1st sess., 1929, pt. 1, 246; ibid., 2d sess., 1930, 72, pt. 2, 1420; ibid., pt. 3, 2492, 3136.

50. Montavon to Burke, memorandum, 3 February 1930, copy, ACUA, NCWC Legal Department files.

51. A.M.P., Memorandum, 1 March 1930, ACUA, Legal Department files.

52. Arthur Capper, "Put Education in the President's Cabinet," printed in *Congressional Record*, 71st Cong., 2d sess., 1930, 72, pt. 3, 3136–37; Chalmers, *Hooded Americanism*, 302.

53. Wilburn Cartwright, "The Capper Robsion Bill," printed in *Congressional Record*, 71st Cong., 2d sess., 1930, 72, pt. 6, 6197–98.

54. A.M.P., Memoranda, 6 and 8 March 1930, ACUA, NCWC Legal Department files. The quotation is from the second.

55. Agnes Regan to Montavon, 18 March 1930, with attachment, ACUA, NCWC Legal Department files. The wife of a Scottish Rite Mason had informed a member of the National Council of Catholic Women of what was afoot, and the latter informed Regan, executive secretary of the NCCW.

56. "Mr. Hearst on the American School," *America* 43 (19 April 1930): 30–31.

57. *Los Angeles Examiner*, 13 April 1930. Eastern papers carried this editorial the previous Sunday.

58. *Congressional Record*, 71st Cong., 2d sess., 1930, 72, pt. 8, 8077–82. Montavon later remarked that because Reed, chairman of the Education Committee, failed to push the Capper-Robsion bill, friends of the measure began to accuse him of "having fallen under the charm of the Roman Pope" (Montavon to Lally, 30 September 1930, copy, ACUA, NCWC Legal Department files).

59. *Congressional Record*, 71st Cong., 2d sess., 1930, 72, pt. 9, 9807–10.

60. Thomas Finegan, "Report of the Legislative Commission," *Addresses and Proceedings of NEA, 1931*, 69:253.

61. Known supporters included Capper of Kansas, Cartwright of Oklahoma, Patterson of Alabama, Robsion of Kentucky, Tarver of Georgia, Trammell of Florida (all the foregoing mentioned in the text above), Robert Blackburn of Kentucky, Thomas Heflin of Alabama, Elva Kendall of Kentucky, Katherine Langley of Kentucky, Joe Manlove of Missouri, J. Lincoln Newhall of Kentucky, J. Will Taylor of Tennessee, and Louis Walker of Kentucky (Davidson, "Report of Legislative Commission," 192).

62. Mann and Crabtree, Conference of the National Advisory Committee, 14 and 15 October 1929, ACUA, Vice-Rector files.

63. These were the Southern Jurisdiction of the Scottish Rite, the Conference on Negro Education, the Department of Agriculture, the Federal Board of Vocational Education, the American Vocational Association, the Association of State Directors of Vocational Education, the Conference on Indian Education, the Association of Land Grant Colleges, the Executive Committee of the NEA Department of Superintendence, the American Federation of Labor, the National Association of State Universities, the American Home Economic Association, and the National Catholic Educational Association.

64. NACE, Digest of Conference with American Vocational Association and Association of State Directors of Vocational Education, 17 January 1930, ACUA, Vice-Rector files.

65. NACE, Digest of Conference with the official representative of the Supreme Council 33 Degree, Ancient and Accepted Scottish Rite of Freemasonry, Southern Jurisdiction, 10 January 1930, enclosure with Mann and Crabtree to members of NACE, ACUA, Vice-Rector files.

66. NACE, Digest of Conference with the Cooperating Committee of the Department of Superintendence, N.E.A., 1 February 1930, enclosure with Mann and Crabtree to member of NACE, 25 February 1930, ACUA, Vice-Rector files.

67. NACE, Digest of Conference with Representatives of the National Association of State Universities and the Conference of Separate State Universities, 3 March 1930, ACUA, Vice-Rector files.

68. NACE, Digest of Conference with Representatives of the American Federation of Labor, 1 March 1930, ACUA, Vice-Rector files.

69. NACE, Digest of Conference with Representatives of the National Catholic Educational Association, 28 February 1930, NA, Department of the Interior, Central Classified file, 1907–1936, box 1534.

70. F. M. Connell to Johnson, 3 April 1930, ACUA, NCWC Johnson Papers.

71. William McNally to Johnson, 9 April 1930, ACUA, NCWC Johnson Papers.

72. Joseph McClancy to Johnson, 10 April 1930, ACUA, NCWC Johnson Papers.

73. NACE, *Federal Relations to Education: A Memorandum of Progress* (Washington, D.C.: NACE, 1930), 20–37.

74. Ibid., 12–13, 38–49.

75. Ibid., 14–15; *Historical Statistics*, 10.

76. NACE, *Federal Relations to Education*, 15–18. Criticism came from outside the committee. Like Johnson, the conservative J. Gresham Machen of Westminster Theological Seminary in Philadelphia objected to any kind of federal headquarters because it would unavoidably exercise control. Moreover, the Constitution contained no provision that allowed the federal government to enter the field of education. Machen was puzzled at how the NACE could speak so forthrightly about the local control of education and then propose federal machinery. He hoped that the steering committee would stand by its principles and come out squarely for local control of education (J. Gresham Machen to the NACE, 11 August 1930, copy, ACUA, NCWC Johnson Papers). Johnson commended Machen for his letter: "My personal opinion is that the 'Memorandum of Progress' . . . was born out of due time," wrote the priest. "For some reason or other there seems to be a feeling in the Steering Committee that something definite should be prepared as soon as possible." He also thought that the

committee was naive in its approach to the problem. "There is a tendency to think in what they call social trends and a rather superior attitude toward the Constitution. Dr. Suzzallo said in open meeting, 'What concerns me most is not what is law, even constitutional law, but what is wise.'" (Johnson to Machen, 19 August 1930, copy, ACUA, NCWC Johnson Papers.)

77. NACE, *Memorandum of Progress,* 19.

78. Mann to the members of the NACE, 4 November 1930, ACUA, Vice-Rector files.

79. Report on the Committee on Education, 26 November 1930, ACUA, NCWC Johnson Papers; John McNicholas to Thomas Welch, 24 October 1930, copy, AAC, McNicholas Papers.

80. *Minutes of the Twelfth Annual Meeting of the Hierarchy, November 1930,* 13–14, ACUA, NCWC General Secretary files.

81. Johnson, Notes on the Meeting of the Steering Committee of National Advisory Committee on Education, Washington, D.C., 10–11 December 1930, copy, ACUA, NCWC Johnson Papers.

82. Ibid.

83. "Notes on Political and Economic Aspects of Education by the States and by the Nation," NACE, February 1931, ACUA, Vice-Rector files. The quotation appears on 39–40.

84. "Confidential: Opinions of Economists Concerning Federal Aid to the States in Support of Education," NACE, April 1931, ACUA, Vice-Rector files. The quotations appear on 18 and 20.

85. NACE, *Federal Relations to Education,* Vol. I: *Committee Findings and Recommendations* (Washington, D.C.: National Capital Press, 1931), 9–89. For the authorship and critiquing committee, see Johnson, Notes on Meeting of Steering Committee of National Advisory Committee on Education, Washington, D.C., 10–11 December 1930, copy, ACUA, NCWC Johnson Papers.

86. NACE, *Federal Relations to Education,* 1:93–99.

87. Mann to the members of the NACE, 26 June 1931, ACUA, Vice-Rector files. The names of those voting negatively have been deduced from their votes on the various parts of the final report in general session (NACE, *Federal Relations to Education,* 1:89 and 99).

88. Burke to Karl Alter, 20 July 1931, copy, ACUA, NCWC General Secretary files. Burke's correspondence with McNicholas cannot be located in either ACUA or in the AAC.

89. Alter to Burke, 16 July 1931, ACUA, NCWC General Secretary files.

90. Burke to Alter, 20 July 1931, copy, ACUA, NCWC General Secretary files.

91. Alter to Burke, 27 July 1931, ACUA, NCWC General Secretary files.

92. Pace to Mann, with corrections by John T. McNicholas, 11 September 1931, copy, ACUA, Vice-Rector files; Johnson, Memorandum on the Department of Education, 3 October 1931, copy, ACUA, NCWC Johnson Papers; NACE, *Committee Findings and Recommendations,* 1:89, 99, 100.

93. Johnson, Memorandum on Department of Education, 3 October 1931, ACUA, NCWC George Johnson Papers.

94. NACE, *Committee Findings and Recommendations,* 1:103–5.

95. *Baltimore Catholic Review,* 20 November 1931.

96. *Indiana Catholic and Record,* 27 November 1931.

97. Omaha *True Voice,* 4 December 1931; *Washington Post,* 24 November 1931. See also Cleveland *Catholic Universe Bulletin,* 4 December 1931; New York *Catholic News,* 2 November 1931; Chicago *New World,* 27 November 1931; Saint Paul *Catholic Bulletin,* 28 November 1931; "The Report on Education," *Commonweal* 15 (December 1931): 117–18; *Pittsburgh Catholic,* 3 December 1931.

98. NCWC, *Editorial Opinion on Federal Relations to Education: Newspaper Comment on the Report of the National Advisory Committee on Education* (Baltimore: Belvedere Press, undated). The quotation is on page 10.

99. Burke, Memorandum, 10 November 1931, ACUA, NCWC General Secretary files.

100. Ibid.

101. *Minutes of the Thirteenth Annual Meeting of the American Hierarchy, November 1931,* ACUA, NCWC General Secretary files.

102. Burke interview with Hanna, 14 November 1931, ACUA, NCWC General Secretary files; Burke interview with Michael Ready, 18 November 1931, ACUA, NCWC General Secretary files.

103. U.S. Congress, House, H.R. 4757, 72nd Cong., 1st sess., 1931.

104. Montavon to Burke, 6 February 1932, copy, ACUA, NCWC Legal Department files.

105. Minutes of the Administrative Committee, 5 April 1932, ACUA, NCWC General Secretary files.

106. Ibid.

107. Burke Walsh to Frank Hall, 6 July 1932, copy, ACUA, NCWC General Secretary files; Hall to John Burke, memorandum, 27 June 1932, ACUA, NCWC General Secretary files. The first quotation is from the former, the second from the latter.

108. B. Walsh to Hall, 6 July 1932, copy; Burke, memorandum to himself, 28 June 1932; Hall to John Burke, 29 June 1932—all in ACUA, NCWC General Secretary files.

109. *Minutes of the Fourteenth Annual Meeting of the American Hierarchy, November 1932,* ACUA, NCWC General Secretary files.

110. See Buis, "Role of Federal Government in Support of Education," 151–654; Clegg, "Federal Aid to Education," 88–245; Munger and Fenno, *National Politics and Federal Aid,* 59–165; Gilbert E. Smith, *The Limits of Reform: Politics and Federal Aid to Education 1937–1950* (New York: Garland, 1982), 1–212.

Select Bibliography

The select bibliography is divided into two parts: primary and secondary sources. The first lists the archival collections cited throughout the book. It does not identify the many magazines, newspapers, pamphlets, anthologies of speeches, and reports of proceedings consulted and/or quoted. The second lists all historical works cited or consulted as well as some articles and books from the period under study that were treated as secondary sources.

Manuscript Collections

Archives of the Archdiocese of Baltimore
- James Gibbons Papers
- Michael Curley Papers
- Roman Documents

Archives of the Archdiocese of Saint Louis
- U.S.A. Hierarchy

Archives of the Archdiocese of San Francisco
- Diocesan files
- NCWC files

Archives of The Catholic University of America
- Annual Meeting of the Hierarchy file
- Edward A. Pace Papers
- NCWC Chairman files
- NCWC Committee on Special War Activities, Executive Secretary files
- NCWC Department of Education files

NCWC General Secretary files
NCWC George Johnson Papers
NCWC James H. Ryan Papers
NCWC Legal Department files
NCWC NCCM files
NCWC NCCW files
Nolan-Dixon Collection
Philosophy Department files
Rector files
Vice-Rector files
Archives of the Central Bureau of the Catholic Central Union
Arthur Preuss Papers
Archives of the Diocese of Charleston
William Russell Papers
Archives of the Diocese of Cleveland
Joseph Schrembs Papers
Archives of the Diocese of Portland, Maine
Louis Walsh Papers
NCWC files
Archives of Georgetown University
America Collection
Archives of the Knights of Columbus
Education file
Archives of the National Catholic Educational Association
Francis Howard Papers
Archives of the University of Notre Dame
Frederick Kenkel Papers
Manuscript Collection of the Library of Congress
Thomas J. Walsh Papers
National Archives
Interior Department files
Sulpician Archives, Baltimore
Record Group 13

Secondary Sources

Abrams, Albertina Adelheit. "The Policy of the National Education Association toward Federal Aid to Education (1857–1953)." Ph.D. diss., University of Michigan, 1954.

Abrams, Richard M. *The Burdens of Progress, 1900–1929*. Glenview, Ill.: Scott, Foresman, 1978.

Ahern, Patrick. *The Catholic University of America, 1887–1896: The Rectorship of John J. Keane*. Washington, D.C.: The Catholic University of America Press, 1948.

———. *The Life of John Keane 1839–1913*. Milwaukee: Bruce Publishing Company, 1955.

Ahlstrom, Sydney E. *A Religious History of the American People.* New Haven: Yale University Press, 1972.

Alexander, Charles C. *The Ku Klux Klan in the Southwest.* Lexington: University of Kentucky Press, 1966.

Allen, Frederick Lewis. *Only Yesterday: An Informal History of the Nineteen-Twenties.* New York: Harper and Row, 1964.

Anbinder, Tyler. *Nativism and Slavery: The Northern Know-Nothings & the Politics of the 1850s.* New York and Oxford: Oxford University Press, 1992.

Appleby, R. Scott. *"Church and Age Unite!": The Modernist Impulse in American Catholicism.* Notre Dame, Ind.: University of Notre Dame Press, 1992.

Asbury, Herbert. *The Great Illusion: An Informal History of Prohibition.* Garden City, N.Y.: Greenwood Press, 1950.

Aubert, Roger, et al. *The Church in the Industrial Age.* Vol. 9 of *History of the Church,* edited by Hubert Jedin. New York: Seabury Press, 1980–81.

———. *The Church in a Secularized Society.* Vol. 5 of *The Christian Centuries,* edited by Louis J. Rogier et al. London: Darton, Longman, and Todd, 1964–78.

Axt, Richard. *The Federal Government and the Financing of Higher Education.* New York: Columbia University Press, 1952.

Bagby, Wesley M. *The Road to Normalcy: The Presidential Campaign and Election of 1920.* Baltimore: Johns Hopkins University Press, 1962.

Bailey, Beth L. *From Front Porch to Back Seat: Courtship in Twentieth-Century America.* Baltimore: Johns Hopkins University Press, 1988.

Barry, Colman J., O.S.B. *The Catholic Church and German Americans.* Milwaukee: Bruce Publishing Company, 1953.

Beale, Howard K. *Are American Teachers Free? An Analysis of Restraints upon the Freedom of Teaching in American Schools.* New York: Octagon Books, 1972, originally published in 1936.

Behr, Edward. *Prohibition: Thirteen Years that Changed America.* New York: Arcade Publications, 1996.

Billington, Ray Allen. *The Protestant Crusade, 1800–1860: A Study of the Origins of American Nativism.* New York: Rinehart and Company, 1952.

Blumenfeld, Samuel. *NEA: Trojan Horse in American Education.* Boise, Idaho: Paradigm Company, 1984.

Boeller, Paul F., Jr. *Presidential Campaigns.* Oxford and New York: Oxford University Press, 1985.

Boyer, Richard O., and Herbert M. Morais. *Labor's Untold Story.* New York: United Electrical, Radio, and Machine Workers of America, 1965.

Braeman, John. "The American Polity in the Age of Normalcy." In *Calvin Coolidge and the Coolidge Era: Essays on the History of the 1920s,* edited by John Earl Haynes. Washington, D.C.: Library of Congress, 1998.

Buetow, Harold A. *Of Singular Benefit: The Story of Catholic Education in the United States.* London: Macmillan, 1970.

Buis, Ann Gibson. "An Historical Study of the Role of the Federal Government in the Financial Support of Education, with Special Reference to Legislation Proposals and Action." Ph.D. diss., Ohio State University, 1953.

Bureau of the Census. *Historical Statistics of the United States, Colonial Times to 1957.* Washington, D.C.: U.S. Government Printing Office, 1961.

Burner, David. *The Politics of Provincialism: The Democratic Party in Transition, 1918–1932.* New York: Alfred A. Knopf, 1968.

Butts, Freeman R., and Lawrence Cremin. *A History of Education in American Culture.* New York: Holt, Rinehart, and Winston, 1953.

Callahan, Raymond E. *Education and the Cult of Efficiency: A Study of the Social Forces That Have Shaped the Administration of the Public Schools.* Chicago: University of Chicago Press, 1964.

Campbell, Debra. "A Catholic Salvation Army: David Goldstein, Pioneer Lay Evangelist." *Church History* 52 (September 1983): 322–32.

———. "David Goldstein and the Rise of the Catholic Campaigners for Christ." *Catholic Historical Review* 72 (January 1986): 33–50.

———. " 'I Can't Imagine Our Lady on an Outdoor Platform': Women in the Catholic Street Propaganda Movement." *U.S. Catholic Historian* 3 (Spring– Summer 1983): 103–14.

———. "Part-Time Female Evangelists of the Thirties and Forties: The Rosary College Catholic Evidence Guild." *U.S. Catholic Historian* 5 (Summer–Fall 1986): 371–83.

Capper, Arthur. *The Agricultural Bloc.* New York: Harcourt, Brace and Company, 1922.

Carrión, Arturo Morales. *Puerto Rico: A Political and Cultural History.* New York: W. W. Norton & Company, 1983.

Cashman, Sean Dennis. *Prohibition: The Lie of the Land.* New York: Free Press, 1981.

Cassidy, Francis P. "Catholic Education in the Third Plenary Council of Baltimore." *Catholic Historical Review* 34 (October 1948): 257–305.

Chalmers, David M. *Hooded Americanism: The History of the Ku Klux Klan.* New York: Franklin Watts, 1976.

Chandler, Alfred D., Jr. "The Origins of Progressive Leadership." In *Letters of Theodore Roosevelt,* edited by Elting Morison et al. Cambridge, Mass.: Harvard University Press, 1954.

Church, Robert L., and Michael Sedlak. *Education in the United States: An Interpretive History.* New York: Free Press, 1976.

Clark, Truman R. *Puerto Rico and the United States, 1917–1933.* Pittsburgh: University of Pittsburgh Press, 1975.

Clegg, Ambrose A., Jr. "Federal Aid to Education: A Study of the Interest of Church and Labor Groups in Proposals for Federal Aid to Elementary and Secondary Schools, 1890–1945." Ph.D. diss., University of North Carolina, 1963.

Cochran, Thomas, and William Miller. *The Age of Enterprise: A Social History of Industrial America.* 2d ed., rev. New York: Harper Torchbooks, 1961.

Commager, Henry Steele, ed. *Documents of American History.* 2 vols. New York: Appleton-Century-Crofts, 1963.

Cremin, Lawrence A. *The Transformation of the School: Progressivism in American Education, 1876–1957.* New York: Alfred A. Knopf, 1961.

Croly, Herbert. *The Promise of American Life.* New York: Macmillan, 1909.

Cubberley, Ellwood. *Public Education in the United States: A Study and Interpretation of American Educational History.* Boston: Houghton Mifflin Company, 1934.

Cuddy, Edward. "Pro-Germanism and American Catholicism, 1914–1917." *Catholic Historical Review* 54 (October 1968): 427–54.

Curran, R. Emmett. *Michael Augustine Corrigan and the Shaping of Conservative Catholicism in America, 1878–1902.* New York: Arno Press, 1978.

Daley, Gabriel, O.S.A. *Immanence and Transcendance: A Study of Catholic Modernism and Integralism.* Oxford: Oxford University Press, 1980.

Dansette, Adrien. *Religious History of Modern France.* 2 vols. Translated by John Dingle. New York: Herder and Herder, 1961.

Degler, Carl. *The Age of Economic Revolution, 1876–1900.* Glenview, Ill.: Scott, Foresman and Company, 1967.

Dixon, Blase, T.O.R. "The Catholic University of America, 1909–1928: The Rectorship of Thomas Joseph Shahan." Ph.D. diss., The Catholic University of America, 1972.

Dolan, Jay P. *The American Catholic Experience: A History from Colonial Times to the Present.* Garden City, N.Y.: Doubleday and Company, 1985.

———. *The Immigrant Church: New York's Irish and German Catholics, 1815–1865.* Notre Dame, Ind.: University of Notre Dame Press, 1983.

Dumenil, Lynn. *Freemasonry and American Culture, 1880–1930.* Princeton, N.J.: Princeton University Press, 1984.

———. " 'The Insatiable Maw of Bureaucracy': Antistatism and Education Reform in the 1920s." *Journal of American History* 77 (September 1990): 499–524.

———. *The Modern Temper: American Culture and Society in the 1920.* New York: Hill and Wang, 1995.

———. "The Tribal Twenties: 'Assimilated' Catholics' Response to Anti-Catholicism in the 1920s." *Journal of American Ethnic History* 11 (Fall 1991): 21–49.

Ede, Alfred J. *The Lay Crusade for a Christian America: A Study of the American Federation of Catholic Societies, 1900–1919.* New York: Garland, 1988.

Ellis, John Tracy, ed. *Documents of American Catholic History.* Milwaukee: Bruce Publishing Company, 1962.

———. *The Life and Times of James Cardinal Gibbons, Archbishop of Baltimore, 1834–1921.* 2 vols. Milwaukee: Bruce Publishing Company, 1952.

Ellis, William E. "Catholicism and the Southern Ethos: The Role of Patrick Henry Callahan." *Catholic Historical Review* 69 (January 1983): 41–50.

Esslinger, Dean R. "American German and Irish Attitudes Toward Neutrality, 1914–1917: A Study of Catholic Minorities." *Catholic Historical Review* 53 (July 1967): 194–216.

Evans, John Whitney. "Catholics and the Blair Education Bill." *Catholic Historical Review* 46 (October 1960): 273–98.

Farrell, John T. "Archbishop Ireland and Manifest Destiny," *Catholic Historical Review* 33 (October 1947): 269–301.

Feldberg, Michael. *The Philadelphia Riots of 1844: A Study of Ethnic Conflict.* Westport, Conn.: Greenwood Press, 1975.

———. *The Turbulent Era: Riot and Disorder in Jacksonian America.* New York and Oxford: Oxford University Press, 1980.

Ferrell, Robert H. *The Presidency of Calvin Coolidge.* Lawrence: University Press of Kansas, 1998.

Fite, Gilbert. *George N. Peek and the Fight for Farm Parity.* Norman: University of Oklahoma Press, 1954.

Fuess, Claude M. *Calvin Coolidge: The Man from Vermont.* Hamden, Conn.: Archon Books, 1965.

Fogarty, Gerald P., S.J. *The Vatican and the Americanist Crisis: Denis J. O'Connell, American Agent in Rome, 1885–1903.* Rome: Universita Gregoriana, 1974.

———. *The Vatican and the American Hierarchy from 1870 to 1965.* Wilmington, Del.: Michael Glazier, 1985.

Foner, Eric. *Reconstruction: America's Unfinished Revolution, 1863–1877.* New York: Harper and Row, 1988.

Fuess, Claude M. *Calvin Coolidge: The Man From Vermont.* Hamden, Conn.: Archon Books, 1965.

Furniss, Norman. *The Fundamentalist Controversy, 1918–1931.* Hamden, Conn.: Archon Books, 1963.

Gannon, Michael V. "Before and After Modernism: The Intellectual Isolation of the American Priest." In *The Catholic Priest in the United States: Historical Investigations,* edited by John Tracy Ellis. Collegeville, Minn.: Saint John's University Press, 1971.

Gerstle, Gary. "Liberty, Coercion, and the Making of Americans." *Journal of American History* 84 (September 1997): 524–58.

Gleason, Philip. *The Conservative Reformers: German-American Catholics and the Social Order.* Notre Dame, Ind.: University of Notre Dame Press, 1968.

———. "In Search of Catholic Unity: American Catholic Thought 1920–1960." *Catholic Historical Review* 65 (April 1979): 185–205.

Goldberg, Robert Alan. "Hooded Empire: The Ku Klux Klan in Colorado, 1921–1932." Ph.D. diss., University of Wisconsin, 1977.

Goldfield, David. *Black, White, and Southern: Race Relations and Southern Culture, 1940 to the Present.* Baton Rouge: Louisiana State University Press, 1990.

Good, H. G. *A History of American Education.* 2d ed. New York: Macmillan, 1964.

Gorman, Mary Adele Frances. "Federation of Catholic Societies in the United States, 1870–1920." Ph.D. diss., University of Notre Dame, 1962.

Graham, Otis, Jr. *The Great Campaigns: Reform and War in America, 1900–1928.* Englewood Cliffs, N.J.: Prentice-Hall, 1971.

Grantham, Dewey W. Jr. *Hoke Smith and the Politics of the New South.* Baton Rouge: Louisiana State University Press, 1958.

———. *The South in Modern America: A Region at Odds.* New York: Harper Perennial, 1995.

———. *Southern Progressivism: The Reconciliation of Progress and Tradition.* Knoxville: University of Tennessee Press, 1983.

Graper, Elmer D. "The American Farmer Enters Politics." *Current History* 19 (February 1924): 817–26.

Green, Harvey. *The Uncertainty of Everyday Life, 1915–1945.* New York: Harper Perennial, 1993.

Gruber, Carol. *Mars and Minerva: World War I and the Uses of Higher Education.* Baton Rouge: Louisiana State University Press, 1975.

Guilday, Peter. *The National Pastorals of the American Hierarchy.* Washington, D.C.: NCWC, 1923.

Halsey, William M. *The Survival of American Innocence: Catholicism in an Era of Disillusionment, 1920–1940.* Notre Dame, Ind.: University of Notre Dame Press, 1980.

Hays, Samuel P. "The Politics of Reform in Municipal Government in the Progressive Era." *Pacific Northwest Quarterly* 55 (October 1964): 157–69.

Heckscher, August. *Woodrow Wilson.* New York: Charles Scribner and Sons, 1992.

Hennesey, James, S.J. *American Catholics: A History of the Roman Catholic Community in the United States.* Oxford and New York: Oxford University Press, 1983.

Hicks, John D. *The Republican Ascendancy, 1921–1933.* New York: Harper and Row, 1960.

Higham, John. "Hanging Together: Divergent Unities in American History." *Journal of American History* 61 (June 1974): 5–28.

———. *Strangers in the Land: Patterns of American Nativism, 1860–1925.* New York: Atheneum, 1975.

Hofstadter, Richard. *The Age of Reform: From Bryan to F.D.R.* New York: Vintage Books, 1955.

Hogan, Peter E., S.S.J. *The Catholic University of America, 1896–1903: The Rectorship of Thomas J. Conaty.* Washington, D.C.: The Catholic University of America Press, 1949.

Holsinger, Paul M. "The Oregon School Bill Controversy, 1922–1925." *Pacific Historical Review* 37 (August 1968): 327–42.

Hoover, Herbert. *The Memoirs of Herbert Hoover: The Cabinet and the President, 1920–1933.* New York: Macmillan, 1952.

Hurley, Mark. *Church-State Relationships in Education in California.* Washington, D.C.: The Catholic University of America Press, 1948.

Huthmacher, J. Joseph. "Urban Liberalism in the Age of Reform." *Mississippi Valley Historical Review* 44 (September 1962): 321–41.

Israel, Fred, ed. *The State of the Union Messages of the Presidents.* New York: Chelsea House–Robert Hector Publishers, 1966.

Jackson, Kenneth T. *The Ku Klux Klan in the City, 1915–1930.* New York: Oxford University Press, 1967.

Jedin, Hubert, ed. *History of the Church.* 10 vols. New York: Seabury Press, 1980–1981.

Johnson, Paul. "Calvin Coolidge and the Last Arcadia." In *Calvin Coolidge and the Coolidge Era: Essays on the History of the 1920s,* edited by John Earl Haynes. Washington, D.C.: Library of Congress, 1998.

Jorgenson, Lloyd P. "The Oregon School Law of 1922: Passage and Sequel." *Catholic Historical Review* 54 (October 1968): 455–66.

———. *The State and the Non-Public School, 1825–1925.* Columbia: University of Missouri Press, 1987.

Kauffman, Christopher J. *Faith and Fraternalism: The History of the Knights of Columbus, 1882–1982.* New York: Harper and Row, 1982.

———. *Mission to Rural America: The Story of W. Howard Bishop, Founder of Glenmary.* New York: Paulist Press, 1991.

Keller, Morton. *Regulating a New Society: Public Policy and Social Change in America, 1900–1933*. Cambridge, Mass.: Harvard University Press, 1994.

Kenneally, James. "The Burning of the Ursuline Convent: A Different View." *Records of the American Catholic Church History Society* 90 (March-December 1979): 15–21.

Kennedy, David M. *Over Here: The First World War and American Society.* New York and Oxford: Oxford University Press, 1980.

King, Desmond, "Making Americans: Immigration Meets Race." In *E Pluribus Unum? Contemporary and Historical Perspectives on Immigrant Political Incorporation,* edited by Gary Gerstle and John Mollenkopf. New York: Russell Sage Foundation, 2001.

Kipnis, Ira. *The American Socialist Movement, 1897–1912*. New York: Columbia University Press, 1952.

Kirschner, Don S. *City and Country: Rural Responses to Urbanization in the 1920s.* Westport, Conn.: Greenwood Publishing Company, 1970.

Klinkhamer, Marie Carolyn. "The Blaine Amendment of 1875: Private Motives for Political Action." *Catholic Historical Review* 42 (April 1956): 15–49.

Knock, Thomas J. *To End All Wars: Woodrow Wilson and the Quest for a New World Order.* New York and Oxford: Oxford University Press, 1992.

Kobler, John. *Ardent Spirits: The Rise and Fall of Prohibition.* London: Putnam, 1973.

Kursh, Harry. *The United States Office of Education: A Century of Service.* Philadelphia: Chilton Books, 1965.

Lackner, Joseph H., S.M. "Bishop Ignatius Horstmann and the School Controversy of the 1890s." *Catholic Historical Review* 75 (January 1989): 73–90.

Lafeber, Walter. *The American Age: United States Foreign Policy at Home and Abroad since 1750*. New York and London: W. W. Norton & Company, 1989.

Lannie, Vincent P. "Alienation in America: The Immigrant Catholic and Public Education in Pre–Civil War America." *Review of Politics* 32 (October 1970): 503–21.

———. *Public Money and Parochial Education: Bishop Hughes, Governor Seward, and the New York School Controversy.* Cleveland: Case Western University, 1968.

Lannie, Vincent P., and Bernard Diethorn. "For the Honor and Glory of God: The Philadelphia Bible Riots of 1840 [*sic* for 1844]." *History of Education Quarterly* 8 (Spring 1968): 44–106.

Lay, Shawn, ed. *The Invisible Empire in the West: Toward a New Historical Appraisal of the Ku Klux Klan in the 1920s*. Urbana and Chicago: University of Illinois Press, 1992.

Lee, Gordon Canfield. *The Struggle for Federal Aid, First Phase: A History of the Attempts to Obtain Federal Aid for the Common Schools, 1870–1890*. New York: Teachers College, Columbia University, 1949.

Leven, Stephen A. *Go Tell It in the Streets: An Autobiography of Bishop Stephen A. Leven.* Edmond, Okla.: M. Leven, 1984.

Levine, Lawrence W. *Defender of the Faith: William Jennings Bryan: The Last Decade, 1915–1925*. London: Oxford University Press, 1965.

Leuchtenburg, William. *The Perils of Prosperity, 1914–1932*. Chicago: University of Chicago Press, 1965.

Lewis, Charles Lee. *Philander Priestly Claxton: Crusader for Public Education.* Knoxville: University of Tennessee Press, 1948.

Lichtman, Allan J. *Prejudice and the Old Politics: The Presidential Election of 1928.* Chapel Hill: University of North Carolina Press, 1979.

Link, Arthur S. "What Happened to the Progressives in the 1920s?" *American Historical Review* 64 (July 1959): 833–51.

———. *Woodrow Wilson and the Progressive Era, 1910–1917.* New York: Harper and Row, 1954.

Link, Arthur S., and Richard L. McCormick. *Progressivism.* Arlington Heights, Ill.: Harlan Davidson, 1983.

Link, William A. *The Paradox of Southern Progressivism, 1880–1930.* Chapel Hill: University of North Carolina Press, 1992.

Liptak, Dolores, R.S.M. *Immigrants and Their Church.* New York: Macmillan, 1989.

Luebke, Frederick C. *Bonds of Loyalty: German Americans and World War I.* De Kalb: Northern Illinois University Press, 1974.

Lyons, Eugene. *Herbert Hoover: A Biography.* Garden City, N.Y.: Doubleday and Company, 1964.

McAvoy, Thomas. *The Great Crisis in American Catholic History.* Chicago: H. Regnery Co., 1957.

———. "Public Schools vs. Catholic Schools and James McMaster." *Review of Politics* 28 (January 1966): 19–46.

McCadden, Joseph J. "Bishop Hughes Versus the Public School Society of New York." *Catholic Historical Review* 50 (July 1964): 188–207.

McKenna, Patrick, C.M. "The Catholic Motor Missions in Missouri." *Vincentian Heritage* 7, no. 2 (1986): 97–134.

McKeown, Elizabeth. "Apologia for an American Catholicism: The Petition and Report of the National Catholic Welfare Council to Pius XI, April 25, 1922." *Church History* 43 (December 1974): 514–28.

———. "Catholic Identity in America." In *America in Theological Perspective,* edited by Thomas McFadden. New York: Seabury Press, 1976.

———. *War and Welfare: American Catholics and World War I.* New York: Garland, 1988.

McShane, Joseph M., S.J. "Mirrors and Teachers: A Study of Catholic Periodical Fiction Between 1930 and 1950." *U.S. Catholic Historian* 6 (1987): 181–98.

Martin, Frank T. "The Michigan School Controversy." M.A. thesis, The Catholic University of America, 1949.

Marty, Martin E. *Pilgrims in Their Own Land: 500 Years of Religion in America.* New York: Penguin Books, 1988.

May, Henry F. *The End of American Innocence: A Study of the First Years of Our Own Time, 1912–1917.* New York: Alfred A. Knopf, 1959.

Mayer, George H. *The Republican Party, 1854–1966.* New York: Oxford University Press, 1967.

Mecklin, John Moffat. *The Ku Klux Klan: A Study of the American Mind.* Reprint 1924 ed., New York: Russell and Russell, 1963.

Meiring, Bernard. *Educational Aspects of the Legislation of the Councils of Baltimore, 1829–1884.* New York: Arno Press, 1978.

Merz, Charles. *The Dry Decade.* Garden City, N.Y.: Doubleday and Company, 1931.

Miller, Robert Moats. "The Ku Klux Klan." In *Change and Continuity in Twentieth-Century America: The 1920s.* Columbus: Ohio State University Press, 1968.

Moore, Edmund A. *A Catholic Runs for President: The Campaign of 1928.* New York: Ronald Press Company, 1956.

Moore, Leonard. *Citizen Klansmen: The Ku Klux Klan in Indiana, 1921–1928.* Chapel Hill: University of North Carolina Press, 1991.

Morawska, Ewa. "Immigrants, Transnationalism, and Ethnicization: A Comparison of This Great Wave and the Last." In *E Pluribus Unum? Contemporary and Historical Perspectives on Immigrant Political Incorporation,* edited by Gary Gerstle and John Mollenkopf. New York: Russell Sage Foundation, 2001.

Mowry, George E. "The California Progressive and His Rationale: A Study in Middle Class Politics." *Mississippi Valley Historical Review* 34 (September 1949): 239–50.

———. *The Era of Theodore Roosevelt and the Birth of Modern America, 1900–1912.* New York: Harper and Row, 1958.

Moynihan, James H. *The Life of Archbishop John Ireland.* New York: Harper and Brothers, 1953.

Munger, Frank J., and Richard F. Fenno, Jr. *National Politics and Federal Aid to Education.* Syracuse, N.Y.: University of Syracuse Press, 1962.

Murphy, Paul. *The Constitution in Crisis Times, 1918-1968.* New York: Harper and Row, 1972.

Murray, Robert K. *The Harding Era: Warren G. Harding and His Administration.* Minneapolis: University of Minnesota Press, 1969.

———. *The 103rd Ballot: Democrats and the Disaster in Madison Square Garden.* New York: Harper and Row, 1976.

———. *The Politics of Normalcy: Governmental Theory and Practice in the Harding-Coolidge Era.* New York: W. W. Norton and Company, 1973.

Myers, Gustavus. *History of Bigotry in the United States.* Edited and revised by Henry M. Christman. New York: Capricorn Books, 1960.

Nash, George B. "The 'Great Enigma' and the 'Great Engineer.' " In *Calvin Coolidge and the Coolidge Era: Essays on the History of the 1920s,* edited by John Earl Haynes. Washington, D.C.: Library of Congress, 1998.

Nelson, E. Clifford. *The Lutherans in North America.* Philadelphia: Fortress Press, 1975.

Nuesse, C. Joseph. "Thomas Joseph Bouquillon (1840–1902), Moral Theologian and Precursor of the Social Sciences in The Catholic University of America." *Catholic Historical Review* 72 (October 1986): 601–19.

O'Brien, David J. *Public Catholicism.* New York: Macmillan, 1988.

O'Connell, Marvin R. *John Ireland and the American Catholic Church.* St. Paul: Minnesota Historical Society Press, 1988.

Oregon School Cases: Complete Record. Baltimore: Belvedere, 1925.

Ostrander, Gilman M. "The Revolution in Morals." In *Change and Continuity in Twentieth-Century America: The 1920s,* edited by John Braeman, Robert H. Bremner, and David Brody. Columbus: Ohio State University Press, 1968.

Pies, Timothy Mark. "The Parochial School Campaigns in Michigan, 1920-1924: The Lutheran and Catholic Involvement." *Catholic Historical Review* 72 (April 1986): 222–38.

Reardon, Bernard M. G. *Roman Catholic Modernism.* Stanford: Stanford University Press, 1970.

Reher, Margaret Mary. *Catholic Intellectual Life in America: A Historical Study of Persons and Movements.* New York: Macmillan, 1989.

———. "The Church and the Kingdom of God in America: The Ecclesiology of the Americanists." Ph.D. diss., Fordham University, 1972.

Reilly, Daniel F. *The School Controversy, 1891–1893.* Washington, D.C.: The Catholic University of America Press, 1943.

Reilly, Mary Lonan, O.S.F. *A History of the American Catholic Press Association.* Metuchen, N.J.: Scarecrow Press, 1971.

Rippley, La Vern J. "Archbishop Ireland and the School Language Controversy." *U.S. Catholic Historian* 1 (Fall 1988): 1–16.

Rogier, Louis J., et al., eds. *The Christian Centuries.* 5 vols. London: Darton, Longman, and Todd, 1964–1978.

Rosenberg, Emily. *Spreading the American Dream: American Economic and Cultural Expansionism, 1890–1945.* New York: Hill and Wang, 1982.

Russell, Francis. *The Shadow of Blooming Grove: Warren G. Harding in His Times.* New York: McGraw-Hill, 1968.

Saloutos, Theodore, and John D. Hicks. *Agricultural Discontent in the Middle West, 1900–1939.* Madison: University of Wisconsin Press, 1951.

Salvatera, David L. *American Catholicism and the Intellectual Life.* New York: Garland, 1988.

Schmiedler, Edgar, O.S.B. "Churches-on-Wheels." *Homiletic and Pastoral Review* 40 (January 1940): 62, 366–72.

———. "Little Motor Mission Rationing." *Homiletic and Pastoral Review* 43 (March 1943): 502–16.

———. "Motor Missions." *Homiletic and Pastoral Review* 38 (March 1938): 585–88.

———. "Motor Missions, 1940." *Homiletic and Pastoral Review* 41 (January 1941): 388–400.

———. "Trailing Trailer-Chapels." *Homiletic and Pastoral Review* 42 (December 1941): 237–47.

Schriftgiesser, Karl. *This Was Normalcy: An Account of Party Politics During the Twelve Republican Years, 1920–1932.* Boston: Little, Brown and Company, 1948.

Shannon, David A. *Between the Wars: America, 1919–1941.* Boston: Houghton Mifflin Company, 1965.

———. *The Socialist Party in America: A History.* New York: Macmillan, 1955.

Shaw, J. G. *Edwin Vincent O'Hara: American Prelate.* New York: Farrar, Straus and Company, 1956.

Sheed, Frank. *The Church and I.* Garden City, N.Y.: Doubeday & Company, 1974.

Sheerin, John B. *Never Look Back: The Career and Concerns of John J. Burke.* New York: Paulist Press, 1975.

Shelley, Thomas J. "The Oregon School Case and the National Catholic Welfare Conference." *Catholic Historical Review* 75 (July 1989): 439–57.

Sinclair, Andrew. *Era of Excess: A Social History of the Prohibition Movement.* New York: Harper and Row, 1964.

Sinclair, Upton. *The Goslings: A Study of the American Schools.* Pasadena: Upton Sinclair, 1924.

Slawson, Douglas J. " 'The Boston Tragedy and Comedy': The Near-Repudiation of Cardinal William O'Connell." *Catholic Historical Review* 77 (October 1991): 616–43.

———. *The Foundation and First Decade of the National Catholic Welfare Council.* Washington, D.C.: The Catholic University of America Press, 1992.

———. "John J. Burke, C.S.P.: The Vision and Character of a Public Churchman." *Journal of Paulist Studies* 4 (1995–96): 47–93.

———. "The National Catholic Welfare Conference and the Church-State Conflict in Mexico, 1925–1929." *The Americas* 47 (July 1990): 55–93.

———. "Thirty Years of Street Preaching: Vincentian Motor Missions, 1934–1965," *Church History* 62 (March 1993): 60–81.

———. "Wine for the Gods: Negotiations for the Sacramental Use of Alcohol during Prohibition." In *Building the Church in America: Studies in Honor of Monsignor Robert F. Trisco on the Occasion of His Seventieth Birthday,* edited by Joseph C. Linck, C.O., and Raymond J. Kupke. Washington, D.C.: The Catholic University of America Press, 1999.

Smith, Daniel M. *The Great Departure: The United States and World War I, 1914–1920.* New York: John Wiley and Sons, 1965.

Smith, Darrell H. *The Bureau of Education: Its History, Activities, and Organization.* Baltimore: Johns Hopkins University Press, 1923.

Smith, Gilbert E. *The Limits of Reform: Politics and Federal Aid to Education 1937–1950.* New York: Garland, 1982.

Smith, Timothy L. "Protestant Schooling and American Nationality." *Journal of American History* 53 (March 1967): 679–95.

Sobel, Robert. *Coolidge: An American Enigma.* Washington, D.C.: Regnery Publishing, 1998.

Sparr, Arnold J. "From Self-Congratulation to Self-Criticism: Main Currents in American Catholic Fiction, 1900–1960." *U.S. Catholic Historian* 6 (1987): 213–22.

Spaulding, Thomas. *The Premier See: A History of the Archdiocese of Baltimore, 1789–1989.* Baltimore: Johns Hopkins University Press, 1989.

Spring, Joel H. "Psychologists and the War: The Meaning of Intelligence in the Alpha and Beta Tests." *History of Education Quarterly* 12 (Spring 1972): 3–15.

Stephenson, Isaac. *Recollections of a Long Life, 1829–1915.* Chicago: Privately Printed, 1915.

Sterne, Evelyn Savidge. "Beyond the Boss: Immigration and American Political Culture from 1880 to 1940." In *E Pluribus Unum? Contemporary and Historical Perspectives on Immigrant Political Incorporation,* edited by Gary Gerstle and John Mollenkopf. New York: Russell Sage Foundation, 2001.

Storch, Neil T. "John Ireland's Americanism *After* 1899: The Argument from History." *Church History* 51 (December 1982): 434–44.

Tindall, George B. *The Emergence of the New South, 1913–1945.* Baton Rouge: Louisiana State University Press, 1967.

Todd, Lewis P. *Wartime Relations of the Federal Government and the Public Schools, 1917–1918*. New York: Teachers College, Columbia University, 1945.

Trani, Eugene P., and David L. Wilson. *The Presidency of Warren G. Harding.* Lawrence: Regents Press of Kansas, 1977.

Trisco, Robert. "The Church's History in the University's History." *Catholic Historical Review* 75 (October 1989): 658–67.

Tyack, David B. "The Kingdom of God and the Common School: Protestant Ministers and the Educational Awakening in the West." *Harvard Educational Review* 36 (Fall 1966): 447–69.

———. *The One Best System: A History of American Urban Education.* Cambridge, Mass.: Harvard University Press, 1974.

———. "The Perils of Pluralism: The Background of the Pierce Case." *American Historical Review* 74 (October 1968): 74–98.

———. "School for Citizens: The Politics of Civic Education from 1790 to 1990." In *E Pluribus Unum? Contemporary and Historical Perspectives on Immigrant Political Incorporation,* edited by Gary Gerstle and John Mollenkopf. New York: Russell Sage Foundation, 2001.

Tyack, David B., and Elisabeth Hansot. *Managers of Virtue: Public School Leadership in America, 1820–1980*. New York: Basic Books, 1982.

Ueda, Reed. "Historical Patterns of Immigrant Status and Incorporation in the United States." In *E Pluribus Unum? Contemporary and Historical Perspectives on Immigrant Political Incorporation,* edited by Gary Gerstle and John Mollenkopf. New York: Russell Sage Foundation, 2001.

[Vance, James S.] *Proof of Rome's Political Meddling in America.* Washington, D.C.: Fellowship Forum, 1927.

Veverka, Fayette Breaux. *"For God and Country": Catholic Schooling in the 1920s.* New York: Garland, 1988.

Walker, Daniel Howe. "American Victorianism as Culture." *American Quarterly* 27 (December 1975): 507–32.

Wangler, Thomas E. "The Americanism of J. St. Clair Etheridge." *Records of the American Catholic Historical Society of Philadelphia* 85 (March–June 1974): 88–105.

———. "The Birth of Americanism: 'Westward the Apocalyptic Candlestick.'" *Harvard Theological Review* 65 (July 1972): 415–36.

———. "Emergence of John J. Keane as a Liberal Catholic and Americanist (1878–1887)." *American Ecclesiastical Review* 166 (September 1972): 457–78.

———. "John Ireland and the Origins of Liberal Catholicism in the United States." *Catholic Historical Review* 56 (January 1971): 617–29.

———. "John Ireland's Emergence as a Liberal Catholic and Americanist: 1875–1887." *Records of the American Catholic Historical Society of Philadelphia* 81 (June 1970): 62–87.

Ward, Frank. *The Church and I.* New York: Sheed and Ward, 1974.

Ward, Maisie. *Unfinished Business.* New York: Sheed and Ward, 1964.

Warren, Donald. "The Federal Interest: Politics and Policy Study" in *Historical Inquiry in Education: A Research Agenda,* edited by John Harden Best. Washington, D.C.: American Educational Research Association, 1983.

Wesley, Edgar B. *NEA: The First Hundred Years, The Building of the Teaching Profession.* New York: Harper and Brothers, 1957.

White, William Allen. *A Puritan in Babylon: The Story of Calvin Coolidge.* New York: Macmillan, 1938.

Wilbur, Ray Lyman, and Arthur Mastick Hyde. *The Hoover Policies.* New York: Scribner's, 1937.

Williams, Michael. *American Catholics in the War: The National Catholic War Council, 1917–1921.* New York: Macmillan, 1921.

Williams, William Appleman. *The Tragedy of American Diplomacy.* Cleveland and New York: The World Publishing Company, 1959.

Woodward, C. Vann. *Origins of the New South, 1877–1913.* Baton Rouge: Louisiana State University Press, 1951.

———. *Tom Watson: Agrarian Rebel.* New York: Oxford University Press, 1963.

Zabel, Orville B. *God and Caesar in Nebraska: A Study of the Legal Relationship of Church and State, 1854–1954.* Lincoln: University of Nebraska Press, 1955.

Zwierlein, Frederick. *The Life and Letters of Bishop McQuaid.* 3 vols. Rochester, N.Y.: Art Print Shop, 1925–1927.

Index

DOUGLAS J. SLAWSON is vice president for student services at National University in San Diego, California.

www.ingramcontent.com/pod-product-compliance
Lightning Source LLC
Chambersburg PA
CBHW030921150426
42812CB00046B/445

* 9 7 8 0 2 6 8 0 4 1 1 0 6 *